T0304033

NUMERICAL METHODS AND OPTIMIZATION

AN INTRODUCTION

CHAPMAN & HALL/CRC
Numerical Analysis and Scientific Computing

Aims and scope:

Scientific computing and numerical analysis provide invaluable tools for the sciences and engineering. This series aims to capture new developments and summarize state-of-the-art methods over the whole spectrum of these fields. It will include a broad range of textbooks, monographs, and handbooks. Volumes in theory, including discretisation techniques, numerical algorithms, multiscale techniques, parallel and distributed algorithms, as well as applications of these methods in multi-disciplinary fields, are welcome. The inclusion of concrete real-world examples is highly encouraged. This series is meant to appeal to students and researchers in mathematics, engineering, and computational science.

Proposals for the series should be submitted to one of the series editors above or directly to:
CRC Press, Taylor & Francis Group
3 Park Square, Milton Park
Abingdon, Oxfordshire OX14 4RN
UK

Published Titles

Classical and Modern Numerical Analysis: Theory, Methods and Practice
Azmy S. Ackleh, Edward James Allen, Ralph Baker Kearfott, and Padmanabhan Seshaiyer

Cloud Computing: Data-Intensive Computing and Scheduling
Frédéric Magoulès, Jie Pan, and Fei Teng

Computational Fluid Dynamics
Frédéric Magoulès

A Concise Introduction to Image Processing using C++
Meiqing Wang and Choi-Hong Lai

Coupled Systems: Theory, Models, and Applications in Engineering
Juergen Geiser

Decomposition Methods for Differential Equations: Theory and Applications
Juergen Geiser

Designing Scientific Applications on GPUs
Raphaël Couturier

Desktop Grid Computing
Christophe Cérin and Gilles Fedak

Discrete Dynamical Systems and Chaotic Machines: Theory and Applications
Jacques M. Bahi and Christophe Guyeux

Discrete Variational Derivative Method: A Structure-Preserving Numerical Method for Partial Differential Equations
Daisuke Furihata and Takayasu Matsuo

Grid Resource Management: Toward Virtual and Services Compliant Grid Computing
Frédéric Magoulès, Thi-Mai-Huong Nguyen, and Lei Yu

Fundamentals of Grid Computing: Theory, Algorithms and Technologies
Frédéric Magoulès

Handbook of Sinc Numerical Methods
Frank Stenger

Introduction to Grid Computing
Frédéric Magoulès, Jie Pan, Kiat-An Tan, and Abhinit Kumar

Iterative Splitting Methods for Differential Equations
Juergen Geiser

Mathematical Objects in C++: Computational Tools in a Unified Object-Oriented Approach
Yair Shapira

Numerical Linear Approximation in C
Nabih N. Abdelmalek and William A. Malek

Numerical Methods and Optimization: An Introduction
Sergiy Butenko and Panos M. Pardalos

Numerical Techniques for Direct and Large-Eddy Simulations
Xi Jiang and Choi-Hong Lai

Parallel Algorithms
Henri Casanova, Arnaud Legrand, and Yves Robert

Parallel Iterative Algorithms: From Sequential to Grid Computing
Jacques M. Bahi, Sylvain Contassot-Vivier, and Raphaël Couturier

Particle Swarm Optimisation: Classical and Quantum Perspectives
Jun Sun, Choi-Hong Lai, and Xiao-Jun Wu

XML in Scientific Computing
C. Pozrikidis

NUMERICAL METHODS AND OPTIMIZATION
AN INTRODUCTION

SERGIY BUTENKO
TEXAS A&M UNIVERSITY
COLLEGE STATION, USA

PANOS M. PARDALOS
UNIVERSITY OF FLORIDA
GAINESVILLE, USA

CRC Press
Taylor & Francis Group
Boca Raton London New York

CRC Press is an imprint of the
Taylor & Francis Group, an **informa** business

A CHAPMAN & HALL BOOK

CRC Press
Taylor & Francis Group
6000 Broken Sound Parkway NW, Suite 300
Boca Raton, FL 33487-2742

© 2014 by Taylor & Francis Group, LLC
CRC Press is an imprint of Taylor & Francis Group, an Informa business

No claim to original U.S. Government works

International Standard Book Number-13: 978-1-4665-7777-0 (Hardback)

Dedicated to the memory of our grandmothers,
Uliana and Sophia,
who taught us how to count.

Preface

This text provides a basic introduction to numerical methods and optimization for undergraduate and beginning graduate students in engineering and operations research. It is based on the materials used by the authors during many years of teaching undergraduate and introductory graduate courses in industrial and systems engineering at Texas A&M University and the University of Florida. The book is intended for use as a text or supplement for introductory one-semester courses on numerical methods, optimization, and deterministic operations research.

Combining the topics from entry-level numerical methods and optimization courses into a single text aims to serve a dual purpose. On the one hand, this allows us to enrich a standard numerical methods syllabus with additional chapters on optimization, and on the other hand, students taking an introductory optimization or operations research course may appreciate having numerical methods basics (typically assumed as a background) handy. In particular, the fact that students in engineering and operations research represent diverse educational backgrounds, some with no previous coursework related to numerical methods, served as a motivation for this work.

In presenting the material, we assumed minimum to no previous experience of a reader with the subjects of discussion. Some mathematical proofs are included as samples of rigorous analysis; however, in many cases the presentation of facts and concepts is restricted to examples illustrating them. While the content of the text is not tied to any particular software, the book is accompanied by MATLAB® notes and codes available for download from the publisher's website, which also contains other supporting materials.

We would like to thank the numerous students who have taken our courses throughout the years for their valuable feedback on preliminary versions of parts of this text. This work would not have been possible without the love and support of our families. Finally, we would like to thank our publisher, Sunil Nair, for his patient assistance and encouragement.

Sergiy Butenko
Panos Pardalos

Contents

Part I

Basics

Part I

Basics

Chapter 1

Preliminaries

1.1 Sets and Functions

A *set* is a collection of objects of a certain kind, which are called *elements* of the set. We typically denote a set by a capital Greek or Latin letter, i.e., $A, B, \Sigma, \ldots$, and an element of a set by a small letter, such as $a, b, c, \ldots$. A set may be given by the list of its elements, for example, $A = \{a, b, c\}$ or $B = \{1, 2\}$, or by some clearly defined rules, which can be used to determine whether a given object is an element of the given set. For example, the set of all real numbers between 0 and 1, inclusive, can be equivalently represented in several different ways as follows:

$$X = \{x : x^2 - x \leq 0\} = \{x : 0 \leq x \leq 1\} = \{x : x \in [0, 1]\} = [0, 1].$$

We write $a \in A$ if a is an element of set A, and $A \subseteq B$ if A is a *subset* of B (i.e., every element of A is also an element of B). If A is a *proper subset* of B, meaning that A is a subset of B and $A \neq B$, then we write $A \subset B$. For example, for the set $X = [0, 1]$ above, it is correct to write $0 \in X$ or $\{0\} \subset X$, but writing $0 \subset X$ or $\{0\} \in X$ is incorrect. Some other basic set-theoretic notations, terms, and definitions are summarized in Table 1.1.

> **Definition 1.1 (Mapping)** *Let X, Y be two sets. A rule associating a single element $y = f(x)$ of Y with each element $x \in X$ is called a mapping of X into Y, denoted by $f : X \to Y$, and f is said to map X into Y (and, if $f(x) = y$, x into y).*

> **Definition 1.2 (Real Function)** *If $X \subseteq \mathbb{R}$, then a mapping $f : X \to \mathbb{R}$ is called a real function on X, where X is called the domain of f and $Y = \{y = f(x) : x \in X$ is called the range of f.*

The following terminology applies to mappings in general and to real functions in particular. Consider a mapping $f : X \to Y$. If $f(x) = y$, then we call y the image of x under the mapping f. For a set $X' \subseteq X$, we call the set

TABLE 1.1: Common set-theoretic notations, terminology, and the corresponding definitions.

Notation	Terminology	Definition
$a \in A$	a belongs to A	a is an element of A
$A \ni a$	A contains a	A has a as its element
$\forall a \in A$	for all a in A	for every element a of A
$\exists a \in A$	there exists a in A	there exists an element a of A
$a \notin A$	a does not belong to A	a is not an element of A
$A = B$	A and B are equal	A and B consist of the same elements
$A \neq B$	A and B are not equal	there is an element that belongs to exactly one of A and B
$A \subseteq B$	A is a subset of B	every element of A is an element of B
$A \subset B$	A is a proper subset of B	$A \subseteq B$ and $A \neq B$
$A \supseteq B$	A is a superset of B	every element of B is an element of A
$A \supset B$	A is a proper superset of B	$A \supseteq B$ and $A \neq B$
$\emptyset$	empty set	set with no elements
$A \cup B$	union of A and B	$\{x : x \in A \text{ or } x \in B\}$
$A \cap B$	intersection of A and B	$\{x : x \in A \text{ and } x \in B\}$
$A \cap B = \emptyset$	A and B are disjoint	A and B have no elements in common
$A \setminus B$	difference of A and B	$\{x : x \in A \text{ and } x \notin B\}$
$A \Delta B$	symmetric difference of A and B	$(A \setminus B) \cup (B \setminus A)$
$\displaystyle\bigcup_{\alpha \in \mathcal{F}} A_\alpha$	union of family of sets $\{A_\alpha : \alpha \in \mathcal{F}\}$	$\{x : \exists \alpha \in \mathcal{F} \text{ such that } x \in A_\alpha\}$
$\displaystyle\bigcap_{\alpha \in \mathcal{F}} A_\alpha$	intersection of family $\{A_\alpha : \alpha \in \mathcal{F}\}$	$\{x : x \in A_\alpha \ \forall \alpha \in \mathcal{F}\}$
$A \times B$	Cartesian product of A and B	$\{(a, b) : a \in A, b \in B\}$
$\mathbb{R}$	reals	set of all real numbers
$\mathbb{Z}$	integers	set of all integer numbers
$\mathbb{Z}_+$	positive integers	set of all positive integer numbers

$\{y = f(x) : x \in X'\}$ the image of X' under the mapping f. For a given $y \in Y$, any $x \in X$ such that $f(x) = y$ is called a preimage of y. If y has a unique preimage x, we denote it by $f^{-1}(y)$. Given $Y' \subseteq Y$, the set of all $x \in X$ such that $f(x) \in Y'$ is called the preimage of Y' and is denoted by $f^{-1}(B)$. We say that f maps X *into* Y if $f(X) \subseteq Y$, and we say that f maps X *onto* Y if $f(X) = Y$. If $f(X) = Y$ and every element $y \in Y$ has a unique preimage $f^{-1}(y)$, then f defines a *one-to-one correspondence* (*bijection*) between X and Y. In this case, $f^{-1} : Y \to X$ is called the inverse mapping of f.

Two sets X and Y are called *equivalent* (*equinumerable*; *equipotent*), denoted by $X \sim Y$, if there exists a one-to-one correspondence between them. Obviously, two finite sets are equivalent if and only if they have the same number of elements. The equivalence between the sets is used to compare infinite sets. In particular, we say that a set X is *countable* if it is equivalent to the set $\mathbb{Z}_+$ of all positive integers. An infinite set that is not countable is called *uncountable*.

Example 1.1 *The set $\mathbb{Z}$ of all integers is countable. This can be shown using the following one-to-one correspondence:*

$$f(n) = \begin{cases} (n-1)/2 & \text{if } n \text{ is odd} \\ -n/2 & \text{if } n \text{ is even} \end{cases} \quad \forall n \in \mathbb{Z}_+.$$

In the following example, we use the method known as *Cantor's diagonal method* to show that set $[0,1]$ is uncountable.

Example 1.2 *To prove that the set $[0,1]$ of all reals between 0 and 1 is uncountable, assume that there is a function $f : \mathbb{Z}_+ \to [0,1]$ defining a one-to-one correspondence between $\mathbb{Z}_+$ and $[0,1]$. Let $f(n) = \alpha_n = 0.a_{n1}a_{n_2}\ldots a_{nn}\ldots$ be the decimal representation of the image of $n \in \mathbb{Z}_+$, where $a_{nk} = 0$ if α_n requires less than k digits to be represented exactly. We construct number $\beta = 0.b_1b_2\ldots b_n\ldots$ by assigning $b_n = 2$ if $a_{nn} = 1$ and $b_n = 1$, otherwise. Then for any n we have $f(n) \neq \beta$, thus β cannot be counted using f (i.e., $\beta \in [0,1]$ does not have a preimage under f). This contradicts the assumption that a one-to-one correspondence exists.*

Given a set of n integers $N = \{1,\ldots,n\}$, a *permutation* $p : N \to N$ is a bijection of the elements of N.

Example 1.3 *If $N = \{1,2,3\}$, then a permutation given by $p(1) = 3, p(2) = 2$ and $p(3) = 1$ can be written as*

$$p = \begin{pmatrix} 1 & 2 & 3 \\ 3 & 2 & 1 \end{pmatrix}.$$

A *binary (dyadic) relation* on a set A is a collection of elements of $A \times A$ (ordered pairs of elements of A). A binary relation $\sim$ on a set A is called an *equivalence relation* if it satisfies the following properties: (1) $a \sim a \ \forall a \in A$

(reflexivity); (2) $\forall a, b \in A$, if $a \sim b$, then $b \sim a$ (symmetry); (3) $\forall a, b, c \in A$, if $a \sim b$ and $b \sim c$, then $a \sim c$ (transitivity). The *equivalence class* $[a]$ of an element a is defined as the set $[a] = \{e \in A : e \sim x\}$. Each element of A belongs to exactly one equivalence class induced by any given equivalence relation on A. The set of all equivalence classes in A induced by an equivalence relation $\sim$ is called the *quotient set* of A by $\sim$, denoted by $A/\sim$.

1.2   Fundamental Theorem of Algebra

Polynomial functions, or simply *polynomials*, are at the core of many numerical methods and analytical derivations.

Definition 1.3 (Polynomial) *A function $p(x)$ of a single variable x is a polynomial of degree n on its domain if it can be written in the form*

$$p(x) = a_n x^n + a_{n-1} x^{n-1} + \cdots + a_1 x + a_0,$$

where $a_n, a_{n-1}, \ldots, a_1, a_0$ are constant coefficients.

A polynomial of degree $n = 1$ is called a *linear function*, and a polynomial of degree $n = 2$ is called a *quadratic function*. A root of a polynomial $p(x)$ is a number $\bar{x}$ such that $p(\bar{x}) = 0$. A root $\bar{x}$ of $p(x)$ is said to be of *multiplicity* k if there is a polynomial $q(x)$ such that $q(\bar{x}) \neq 0$ and $p(x) = (x - \bar{x})^k q(x)$. A *simple* root is a root of multiplicity $k = 1$.

Theorem 1.1 (Fundamental Theorem of Algebra) *Given a polynomial $p(x) = x^n + a_{n-1} x^{n-1} + \ldots + a_1 x + a_0$ (where the coefficients $a_0, \ldots, a_{n-1}$ can be real or complex numbers), there exist n (not necessarily distinct) complex numbers $x_1, \ldots, x_n$ such that $p(x) = (x - x_1)(x - x_2) \cdot \ldots \cdot (x - x_n)$. In other words, every complex polynomial of degree n has exactly n roots, counted with multiplicity.*

Given a polynomial $p(x) = x^n + a_{n-1} x^{n-1} + \ldots + a_2 x^2 + a_1 x + a_0$ with *integer* coefficients a_i, $i = 0, \ldots, n-1$, it is easy to see that if $p(x)$ has an integer root r, then r divides a_0.

Example 1.4 *We find the roots of the polynomial $p(x) = x^3 - 6x^2 + 10x - 4$. First, we look for integer roots of $p(x)$. If $p(x)$ has an integer root x_1, then this root has to be equal to one of the integers that divide -4, i.e., $\pm 1, \pm 2, \pm 4$. It is easy to check that $p(2) = 0$, so $x_1 = 2$. Next, we can divide $p(x)$ by $x - 2$,* *obtaining $p(x) = (x-2)(x^2 - 4x + 2)$. To find the two other roots, we need to solve the quadratic equation $x^2 - 4x + 2 = 0$, yielding $x_{2,3} = 2 \pm \sqrt{2}$.*

1.3 Vectors and Linear (Vector) Spaces

A *vector space* or a *linear space* is an important mathematical abstraction that can be used to study many objects of interest. We give a formal definition of vector spaces and mention some fundamental properties describing their structure. We will consider *real* vector spaces only, i.e., by a *scalar* we will always mean a *real* number.

Let a set V be given, and let the operations of addition and scalar multiplication be defined on V, such that for any $x, y \in V$ and any scalar α we have $x + y \in V$ and $\alpha x \in V$.

Definition 1.4 (Linear (Vector) Space) *A set V together with operations of addition and scalar multiplication is a linear space if the following axioms are satisfied for any $x, y, z \in V$ and any scalars α, β.*

1. *$x + y = y + x$.*

2. *$(x + y) + z = x + (y + z)$.*

3. *There exists an element $0 \in V$ such that $x + 0 = x$.*

4. *There exists an element $-x \in V$ such that $x + (-x) = 0$.*

5. *$\alpha(x + y) = \alpha x + \alpha y$.*

6. *$(\alpha + \beta)x = \alpha x + \beta x$.*

7. *$(\alpha\beta)x = \alpha(\beta x)$.*

8. *$1 \cdot x = x$.*

The elements of V are called vectors.

In this text we deal primarily with n-dimensional real vectors, defined next.

Definition 1.5 (Real Vector) *A real n-dimensional real vector is an ordered set of n real numbers $\{x_1, x_2, \ldots, x_n\}$ and is usually written in the form*

$$x = \begin{bmatrix} x_1 \\ x_2 \\ \vdots \\ x_n \end{bmatrix}$$

(column vector) or

$$x = [x_1, x_2, \ldots, x_n]$$

(row vector). We will also write

$$x = [x_i]_{i=1}^n$$

The numbers $x_1, x_2, \ldots, x_n$ are called the components of x.

For the sake of clarity, unless otherwise specified, by a vector we will mean a column vector. For a vector x, the corresponding row vector will be denoted by x^T, which represents the transpose of x (see Definition 1.25 at page 14).

Definition 1.6 (Real Coordinate Space) *The set of all n-dimensional vectors is called an n-dimensional real coordinate space and is denoted by $\mathbb{R}^n$.*

To consider the real coordinate space as a linear space, we introduce the operations of addition and scalar multiplication in $\mathbb{R}^n$ as follows.

Definition 1.7 (Sum of Vectors; Scalar Multiplication) *Given vectors*

$$x = [x_i]_{i=1}^n, \; y = [y_i]_{i=1}^n \in \mathbb{R}^n,$$

their sum is defined as

$$x + y = [x_i + y_i]_{i=1}^n.$$

For a vector $x \in \mathbb{R}^n$ and a scalar α, the scalar multiplication αx is defined as

$$\alpha x = [\alpha x_i]_{i=1}^n.$$

Example 1.5 *For $x = [1, 2, 3]^T, y = [4, 5, 6]^T$ and $\alpha = 2$ we have*

$$x + y = [1 + 4, 2 + 5, 3 + 6]^T = [5, 7, 9]^T; \quad \alpha x = [2 \cdot 1, 2 \cdot 2, 2 \cdot 3]^T = [2, 4, 6]^T.$$

Definition 1.8 (Subspace) *Let V be a linear space. If $S \subseteq V$ such that $x + y \in S$ for any $x, y \in S$ and $\alpha x \in S$ for any $x \in S$ and any scalar α, then S is called a subspace of V.*

Definition 1.9 (Linear Combination) *Given vectors $v^{(1)}, \ldots, v^{(k)}$ in V, a sum in the form $c_1 v^{(1)} + \ldots + c_k v^{(k)}$, where $c_1, \ldots, c_k$ are scalars, is called a linear combination of $v^{(1)}, \ldots, v^{(k)}$.*

Definition 1.10 (Span) *The set of all linear combinations of $v^{(1)}, \ldots, v^{(k)}$ is called the span of $v^{(1)}, \ldots, v^{(k)}$ and is denoted by $Span(v^{(1)}, \ldots, v^{(k)})$.*

Definition 1.11 (Linear Independence) *Vectors $v^{(1)}, \ldots, v^{(k)} \in V$ are called linearly independent if*

$$c_1 v^{(1)} + \ldots + c_k v^{(k)} = 0$$

implies that $c_1 = \ldots = c_k = 0$. Otherwise, the vectors are linearly dependent.

Example 1.6 *Consider the following vectors $v^{(1)}, v^{(2)} \in \mathbb{R}^2$:*

$$v^{(1)} = \begin{bmatrix} 1 \\ 2 \end{bmatrix} \text{ and } v^{(2)} = \begin{bmatrix} 1 \\ 3 \end{bmatrix}.$$

These vectors are linearly independent, since

$$c_1 v^{(1)} + c_2 v^{(2)} = 0 \quad \Leftrightarrow \quad \begin{array}{ccccc} c_1 & + & c_2 & = & 0 \\ 2c_1 & + & 3c_2 & = & 0 \end{array} \quad \Leftrightarrow \quad c_1 = c_2 = 0.$$

The vectors

$$v^{(1)} = \begin{bmatrix} 0 \\ 2 \end{bmatrix} \text{ and } v^{(2)} = \begin{bmatrix} 0 \\ 3 \end{bmatrix}$$

are linearly dependent, because for $c_1 = -3$ and $c_2 = 2$ we have $c_1 v^{(1)} + c_2 v^{(2)} = 0$.

Theorem 1.2 *Given $v^{(1)}, \ldots, v^{(k)} \in V$, a vector $v \in Span(v^{(1)}, \ldots, v^{(k)})$ has a unique representation as a linear combination of $v^{(1)}, \ldots, v^{(k)}$ if and only if $v^{(1)}, \ldots, v^{(k)}$ are linearly independent.*

Definition 1.12 (Basis) *We say that vectors $v^{(1)}, \ldots, v^{(k)} \in V$ form a basis of the space V if and only if $v^{(1)}, \ldots, v^{(k)}$ are linearly independent and $Span(v^{(1)}, \ldots, v^{(k)}) = V$.*

Theorem 1.3 *The following statements are valid.*

1. *If $V = Span(v^{(1)}, \ldots, v^{(k)})$ then any set of $m > k$ vectors in V is linearly dependent.*

2. *Any two bases $\{v^{(1)}, \ldots, v^{(k)}\}$ and $\{u^{(1)}, \ldots, u^{(m)}\}$ of V contain equal number of vectors, $m = k$. This number is called the dimension of V. The subspace $\{0\}$ is said to have dimension 0.*

3. *If V is a linear space of dimension $k > 0$, then no set of less than k vectors can span V and the following statements are equivalent:*

   (a) $v^{(1)}, \ldots, v^{(k)}$ are linearly independent;

   (b) $Span(v^{(1)}, \ldots, v^{(k)}) = V$;

   (c) $v^{(1)}, \ldots, v^{(k)}$ form a basis of V.

1.3.1   Vector norms

Definition 1.13 (Norm) *A function $p : V \to \mathbb{R}$, denoted by $p(x) = \|x\|$ for $x \in V$, is called a norm in V if for any $x, y \in V$ and any scalar α the following properties hold:*

1. *$\|x\| \geq 0$ with equality if and only if $x = 0$;*

2. *$\|\alpha x\| = |\alpha| \|x\|$;*

3. *$\|x + y\| \leq \|x\| + \|y\|$.*

Definition 1.14 (Normed Linear Spaces) *A vector space equipped with a norm $p(x) = \|x\|$ is called a normed linear space.*

Definition 1.15 (p-norm) *A p-norm of a vector $x = [x_i]_{i=1}^n$ is defined as*

$$\|x\|_p = \left(\sum_{i=1}^n |x_i|^p \right)^{1/p}$$

for any real $p \geq 1$.

The most commonly used values for p are $1, 2$, and ∞, and the corresponding norms are

- 1-norm: $\|x\|_1 = \sum\limits_{i=1}^n |x_i|$;

- 2-norm: $\|x\|_2 = \sqrt{\sum\limits_{i=1}^n |x_i|^2}$;

- ∞-norm: $\|x\|_\infty = \max\limits_{1 \leq i \leq n} |x_i|$.

Definition 1.16 (Inner Product) *The inner product of two vectors $x, y \in \mathbb{R}^n$ is defined as*

$$x^T y = \sum_{i=1}^n x_i y_i.$$

Definition 1.17 (Euclidean Space) *The linear space $\mathbb{R}^n$ equipped with the inner product is called the* Euclidean n-space.

Definition 1.18 (Orthogonal) *A set of vectors $\{v^{(1)}, \ldots, v^{(k)}\}$ is called orthogonal if*

$$v^{(i)T} v^{(j)} = 0 \text{ for } i \neq j, \quad i, j = 1, \ldots, k.$$

Definition 1.19 (Orthonormal) *A set of orthogonal vectors $\{v^{(1)}, \ldots, v^{(k)}\}$ is said to be orthonormal if*

$$v^{(i)T} v^{(j)} = \begin{cases} 0 & \text{for } i \neq j. \\ 1 & \text{for } i = j. \end{cases}$$

For an arbitrary set of linearly independent vectors $p^{(0)}, \ldots, p^{(k-1)} \in \mathbb{R}^n$, where $k \leq n$, the *Gram-Schmidt orthogonalization* procedure generates the set of vectors $d^{(0)}, \ldots, d^{(k-1)} \in \mathbb{R}^n$ as follows:

$$
\begin{aligned}
d^{(0)} &= p^{(0)}; \\
d^{(s)} &= p^{(s)} - \sum_{i=0}^{s-1} \frac{p^{(s)T} d^{(i)}}{d^{(i)T} d^{(i)}} d^{(i)}, \quad s = 1, \ldots, k-1.
\end{aligned}
\tag{1.1}
$$

Then $\{d^{(0)}, \ldots, d^{(k-1)}\}$ is an orthogonal set of vectors (Exercise 1.5).

1.4 Matrices and Their Properties

> **Definition 1.20 (Matrix)** *A real matrix is a rectangular array of real numbers composed of rows and columns. We write*
>
> $$
> A = [a_{ij}]_{m \times n} = \begin{bmatrix} a_{11} & a_{12} & \cdots & a_{1n} \\ a_{21} & a_{22} & \cdots & a_{2n} \\ \vdots & \vdots & \ddots & \vdots \\ a_{m1} & a_{m2} & \cdots & a_{mn} \end{bmatrix}
> \tag{1.2}
> $$
>
> *for a matrix of m rows and n columns, and we say that the matrix A is of order m × n.*

Note that if $m = 1$ or $n = 1$ in the above definition, then A becomes a vector (row vector and column vector, respectively).

We will deal with real matrices (i.e., $A \in \mathbb{R}^{m \times n}$) and $m \times n$ will always denote the rows×columns. We say that two matrices $A = [a_{ij}]_{m \times n}$ and $B = [b_{ij}]_{m \times n}$ are *equal*, $A = B$ when $a_{ij} = b_{ij}, \forall i, j$.

1.4.1 Matrix addition and scalar multiplication

> **Definition 1.21 (Matrix Addition)** *Given two matrices, $A = [a_{ij}]_{m \times n}$ and $B = [b_{ij}]_{m \times n} \in \mathbb{R}^{m \times n}$, their sum is defined as*
>
> $$
> A + B = \begin{bmatrix} a_{11} + b_{11} & a_{12} + b_{12} & \cdots & a_{1n} + b_{1n} \\ a_{21} + b_{21} & a_{22} + b_{22} & \cdots & a_{2n} + b_{2n} \\ \vdots & \vdots & \ddots & \vdots \\ a_{m1} + b_{m1} & a_{m2} + b_{m2} & \cdots & a_{mn} + b_{mn} \end{bmatrix}.
> $$

The addition of matrices is not defined for matrices of different order.

Definition 1.22 (Scalar Multiplication) *Given a matrix* $A = [a_{ij}]_{m \times n}$ *and some scalar* $\alpha \in \mathbb{R}$, *scalar multiplication* αA *is defined as*

$$\alpha A = \begin{bmatrix} \alpha a_{11} & \alpha a_{12} & \cdots & \alpha a_{1n} \\ \alpha a_{21} & \alpha a_{22} & \cdots & \alpha a_{2n} \\ \vdots & \vdots & \ddots & \vdots \\ \alpha a_{m1} & \alpha a_{m2} & \cdots & \alpha a_{mn} \end{bmatrix}.$$

Next we list the algebraic rules of matrix addition and and scalar multiplication.

Theorem 1.4 *Suppose that* A, B, C *are* $m \times n$ *matrices,* O *is the* $m \times n$ *zero matrix, and* α, β *are scalars. The following rules of matrix addition and scalar multiplication are valid:*

(i)	$A + B = B + A$	*commutativity*
(ii)	$A + O = A$	*additive identity*
(iii)	$A + (-A) = O$	*additive inverse*
(iv)	$(A + B) + C = A + (B + C)$	*associativity*
(v)	$\alpha(A + B) = \alpha A + \alpha B$	*distributive property for matrices*
(vi)	$(\alpha + \beta)A = \alpha A + \beta A$	*distributive property for scalars*
(vii)	$(\alpha\beta)A = \alpha(\beta A)$	*associativity for scalars*

1.4.2 Matrix multiplication

Definition 1.23 (Matrix Product) *Given* $A \in \mathbb{R}^{m \times n}, B \in \mathbb{R}^{p \times q}$ *where* $n = p$ *(i.e., the number of columns in* A *is equal to the number of rows in* B*), the matrix product* AB *is defined as:*

$$AB = C = [c_{ij}]_{m \times q},$$

where

$$c_{ij} = \sum_{k=1}^{n} a_{ik} b_{kj}, \ i = 1, 2, \ldots, m, \ j = 1, 2, \ldots, q.$$

Matrix multiplication is not defined when $n \neq p$.

Example 1.7 *Let*

$$A = \begin{bmatrix} 2 & 1 \\ 4 & 3 \end{bmatrix}, \ x = \begin{bmatrix} 5 \\ 1 \end{bmatrix}.$$

Since the number of columns in A *and the number of rows in* x *are both equal* *to 2, the matrix multiplication Ax is defined and the resulting matrix will be of order (2×1):*

$$Ax = \begin{bmatrix} 2 & 1 \\ 4 & 3 \end{bmatrix} \cdot \begin{bmatrix} 5 \\ 1 \end{bmatrix} = \begin{bmatrix} 2 \cdot 5 + 1 \cdot 1 \\ 4 \cdot 5 + 3 \cdot 1 \end{bmatrix} = \begin{bmatrix} 11 \\ 23 \end{bmatrix}.$$

Definition 1.24 (Identity Matrix) *The identity matrix I_n of order n is defined by $I_n = [\delta_{ij}]_{n \times n}$, where*

$$\delta_{ij} = \begin{cases} 1, & i = j, \\ 0, & i \neq j. \end{cases}$$

In other words, the identity matrix is a square matrix which has ones on the main diagonal and zeros elsewhere.

Note that $I_n A = A I_n = A$ for $A \in \mathbb{R}^{n \times n}$.

Example 1.8

$$I_3 = \begin{bmatrix} 1 & 0 & 0 \\ 0 & 1 & 0 \\ 0 & 0 & 1 \end{bmatrix}; \quad \begin{bmatrix} 1 & 2 & 3 \\ 4 & 5 & 6 \\ 7 & 8 & 9 \end{bmatrix} \begin{bmatrix} 1 & 0 & 0 \\ 0 & 1 & 0 \\ 0 & 0 & 1 \end{bmatrix} = \begin{bmatrix} 1 & 2 & 3 \\ 4 & 5 & 6 \\ 7 & 8 & 9 \end{bmatrix}.$$

Theorem 1.5 *Let α be a scalar, and assume that A, B, C and the identity matrix I are matrices such that the following sums and products are defined. Then the following properties hold.*

(i)	$(AB)C = A(BC)$	*associativity*
(ii)	$IA = AI = A$	*identity matrix*
(iii)	$A(B+C) = AB + AC$	*left distributive property*
(iv)	$(A+B)C = AC + BC$	*right distributive property*
(v)	$\alpha(AB) = (\alpha A)B = A(\alpha B)$	*scalar associative property*

Note that matrix multiplication is not commutative, i.e., $AB \neq BA$ in general.

1.4.3   The transpose of a matrix

Definition 1.25 (Matrix Transpose) *Given a matrix $A \in \mathbb{R}^{m \times n}$, the transpose of A is an $n \times m$ matrix whose rows are the columns of A. The transpose matrix of A is denoted by A^T.*

Preliminaries 15

> **Definition 1.26 (Symmetric Matrix)** *A square matrix $A \in \mathbb{R}^{n \times n}$, is said to be symmetric if $A^T = A$.*

Example 1.9 *For*

$$A = \begin{bmatrix} 2 & 4 & 8 \\ 1 & 3 & 7 \end{bmatrix}$$

the transpose is

$$A^T = \begin{bmatrix} 2 & 1 \\ 4 & 3 \\ 8 & 7 \end{bmatrix}.$$

For

$$B = \begin{bmatrix} 1 & 5 & 9 \\ 5 & 2 & 6 \\ 9 & 6 & 3 \end{bmatrix}$$

we have $B^T = B$, therefore B is a symmetric matrix.

The following are algebraic rules for transposes.

> **Theorem 1.6** *Given matrices A and B and scalar $\alpha \in \mathbb{R}$, we have*
> *(i)* $(A^T)^T = A$
> *(ii)* $(\alpha A)^T = \alpha A^T$
> *(iii)* $(A + B)^T = A^T + B^T$
> *(iv)* $(AB)^T = B^T A^T$

1.4.4 Triangular and diagonal matrices

A matrix $A = [a_{ij}]_{n \times n}$ is called

- *upper triangular* if $a_{ij} = 0$ for all $i > j$.

- *lower triangular* if $a_{ij} = 0$ for all $i < j$.

In both cases the matrix A is called *triangular*. Matrix A is called *diagonal* if $a_{ij} = 0$ when $i \neq j$.

Example 1.10 *The matrices*

$$U = \begin{bmatrix} 1 & 2 & 3 \\ 0 & 4 & 5 \\ 0 & 0 & 2 \end{bmatrix}, \quad L = \begin{bmatrix} 1 & 0 & 0 \\ 2 & 4 & 0 \\ 3 & 6 & 2 \end{bmatrix}, \quad and \quad D = \begin{bmatrix} 1 & 0 & 0 \\ 0 & 4 & 0 \\ 0 & 0 & 8 \end{bmatrix}$$

are upper triangular, lower triangular, and diagonal, respectively.

1.4.5   Determinants

For a square matrix $A \in \mathbb{R}^{n \times n}$ the *determinant* of A is a real number, which is denoted by $\det(A)$. If

$$
A = [a_{ij}]_{n \times n} = \begin{bmatrix} a_{11} & a_{12} & \cdots & a_{1n} \\ a_{21} & a_{22} & \cdots & a_{2n} \\ \vdots & \vdots & \ddots & \vdots \\ a_{n1} & a_{n2} & \cdots & a_{nn} \end{bmatrix},
$$

then the determinant of A is usually written as

$$
\det(A) = \begin{vmatrix} a_{11} & a_{12} & \cdots & a_{1n} \\ a_{21} & a_{22} & \cdots & a_{2n} \\ \vdots & \vdots & \ddots & \vdots \\ a_{n1} & a_{n2} & \cdots & a_{nn} \end{vmatrix}
$$

and is defined as follows.

- If $n = 1$ and $A = a_{11}$, then $\det(A) = a_{11}$.

- If $n \geq 2$, then

$$
\det(A) = \sum_{j=1}^{n} (-1)^{i+j} a_{ij} \cdot \det(A_{ij}), \tag{1.3}
$$

where A_{ij} is the matrix obtained from A by removing its ith row and j^{th} column. Note, that i in the above formula is chosen arbitrarily, and (1.3) is called the ith row expansion. We can similarly write the jth column expansion as

$$
\det(A) = \sum_{i=1}^{n} (-1)^{i+j} a_{ij} \cdot \det(A_{ij}). \tag{1.4}
$$

So, for the 2×2 matrix

$$
A = \begin{bmatrix} a_{11} & a_{12} \\ a_{21} & a_{22} \end{bmatrix}
$$

we have

$$
\det(A) = a_{11}a_{22} - a_{12}a_{21}.
$$

If $n \geq 3$, then calculation of the determinant of A is reduced to calculation of several 2×2 determinants, as in the following example.

Example 1.11 *Let* $n = 3$ *and*

$$
A = \begin{bmatrix} 4 & 1 & 0 \\ 1 & 3 & -1 \\ 0 & 1 & 2 \end{bmatrix}.
$$

Using the first row expansion, we have

$$\det(A) = 4 \cdot \begin{vmatrix} 3 & -1 \\ 1 & 2 \end{vmatrix} - 1 \cdot \begin{vmatrix} 1 & -1 \\ 0 & 2 \end{vmatrix} + 0 = 4(6+1) - 1(2-0) = 26.$$

Given a triangular matrix $A = [a_{ij}]_{n \times n}$, its determinant is equal to the product of its diagonal elements:

$$\det(A) = \prod_{i=1}^{n} a_{ii}.$$

A *minor* of a matrix A is the determinant of a square matrix obtained from A by removing one or more of its rows or columns. For an $m \times n$ matrix A and k-element subsets $I \subset \{1, \ldots, m\}$ and $J \subset \{1, \ldots, n\}$ of indices, we denote by $[A]_{I,J}$ the minor of A corresponding to the $k \times k$ matrix obtained from A by removing the rows with index not in I and the columns with index not in J. Then $[A]_{I,J}$ is called a *principal minor* if $I = J$, and a *leading principal minor* if $I = J = \{1, \ldots, k\}$.

1.4.6 Trace of a matrix

Definition 1.27 (Matrix Trace) *The trace* $\mathrm{tr}(A)$ *of a matrix* $A = [a_{ij}]_{n \times n}$ *is the sum of its diagonal elements, i.e.,*

$$\mathrm{tr}(A) = a_{11} + \ldots + a_{nn} = \sum_{i=1}^{n} a_{ii}.$$

Some basic properties of the trace of a matrix are given next.

Theorem 1.7 *Let* $A = [a_{ij}]_{n \times n}$ *and* $B = [b_{ij}]_{n \times n}$ *be* $n \times n$ *matrices and let* α *be a scalar. Then*

(i) $\mathrm{tr}(A + B) = \mathrm{tr}(A) + \mathrm{tr}(B)$

(ii) $\mathrm{tr}(\alpha A) = \alpha \mathrm{tr}(A)$

(iii) $\mathrm{tr}(A^T) = \mathrm{tr}(A)$

(iv) $\mathrm{tr}(AB) = \mathrm{tr}(BA)$

(v) $\mathrm{tr}(A^T B) = \mathrm{tr}(AB^T) = \mathrm{tr}(B^T A) = \mathrm{tr}(BA^T) = \sum_{i,j=1}^{n} a_{ij} b_{ij}$

1.4.7   Rank of a matrix

Definition 1.28 (Row (Column) Space) *Given an $m \times n$ matrix A, its row (column) space is the subspace of $\mathbb{R}^n$ ($\mathbb{R}^m$) spanned by the row (column) vectors of A.*

The column space of A, which is given by $\{y \in \mathbb{R}^m : y = Ax, x \in \mathbb{R}^n\}$, is also called the *range space* of A. The set $\{x \in \mathbb{R}^n : Ax = 0\}$ is called the *kernel* or *null space* of A and is denoted by $\text{Ker}(A)$ or $\text{Null}(A)$.

Theorem 1.8 (Rank of a matrix) *The dimension of the row space of any matrix A is equal to the dimension of its column space and is called the rank of A (rank(A)).*

Example 1.12 *The rank of*

$$A = \begin{bmatrix} 1 & 2 & 3 \\ 4 & 5 & 6 \\ 5 & 7 & 9 \end{bmatrix}$$

is 2, since the first two rows are linearly independent, implying that rank$(A) \geq 2$, and the third row is the sum of the first two rows, implying that rank$(A) \neq 3$.

An $m \times n$ matrix A is said to have *full rank* if rank$(A) = \min\{m, n\}$. Clearly, A has full rank if and only if either rows or columns of A are linearly independent.

1.4.8   The inverse of a nonsingular matrix

Definition 1.29 (Nonsingular Matrix) *Matrix $A = [a_{ij}]_{n \times n}$ is said to be nonsingular if there exists some matrix A^{-1} such that $AA^{-1} = A^{-1}A = I_n$. The matrix A^{-1} is called the multiplicative inverse of A. If A^{-1} does not exist, then A is said to be singular.*

Note that A^{-1} is unique for any nonsingular $A \in \mathbb{R}^{n \times n}$, since if we assume that B and C are both inverses of A, then we have

$$B = BI_n = B(AC) = (BA)C = I_nC = C,$$

that is, B is equal to C.

The following properties of nonsingular matrices are valid.

Theorem 1.9 *Let A and B be nonsingular $n \times n$ matrices. Then*

(i) $(A^{-1})^{-1} = A$

(ii) $(AB)^{-1} = B^{-1}A^{-1}$

1.4.9 Eigenvalues and eigenvectors

Definition 1.30 (Eigenvalue; Eigenvector) *Given a matrix $A \in \mathbb{R}^{n \times n}$, a scalar λ is called an eigenvalue of A if there exists a nonzero vector v, called an eigenvector of A corresponding to λ, such that*

$$Av = \lambda v.$$

An eigenvalue λ of A together with a corresponding eigenvector v are referred to as an eigenpair of A.

Hence, the eigenvalues of a matrix $A \in \mathbb{R}^{n \times n}$ are the roots of the characteristic polynomial $p(\lambda) = \det(A - \lambda I_n)$.

Theorem 1.10 *The following properties hold:*

(a) Each eigenvalue has at least one corresponding eigenvector.

(b) If an eigenvalue λ is a root of $p(\lambda)$ of multiplicity r, then there are at most r linearly independent eigenvectors corresponding to λ.

(c) A set of eigenvectors, each of which corresponds to a different eigenvalue, is a set of linearly independent vectors.

Example 1.13 *Find the eigenvalues and the corresponding eigenvectors of the matrix*

$$A = \begin{bmatrix} 2 & 2 & 1 \\ 1 & 3 & 1 \\ 1 & 2 & 2 \end{bmatrix}.$$

The characteristic polynomial of A is

$$\varphi(\lambda) = \begin{vmatrix} 2-\lambda & 2 & 1 \\ 1 & 3-\lambda & 1 \\ 1 & 2 & 2-\lambda \end{vmatrix} = -\lambda^3 + 7\lambda^2 - 11\lambda + 5 = -(\lambda-1)^2(\lambda-5).$$

Solving the characteristic equation $\varphi(\lambda) = 0$, we obtain the eigenvalues of A:

$$\lambda_1 = \lambda_2 = 1 \text{ and } \lambda_3 = 5.$$

In order to find an eigenvector v corresponding to eigenvalue λ_i, we solve the system $(A - \lambda_i I)v = 0$, $i = 1, 2, 3$.

For $\lambda_1 = \lambda_2 = 1$ we have:

$$\begin{bmatrix} 1 & 2 & 1 \\ 1 & 2 & 1 \\ 1 & 2 & 1 \end{bmatrix} \begin{bmatrix} v_1 \\ v_2 \\ v_3 \end{bmatrix} = \begin{bmatrix} 0 \\ 0 \\ 0 \end{bmatrix}.$$

This system consists of three equivalent equations $v_1 + 2v_2 + v_3 = 0$. Thus, $v_1 = -2v_2 - v_3$. Setting $[v_2, v_3] = [1, 0]$ and $[v_2, v_3] = [0, 1]$ we obtain two linearly independent eigenvectors corresponding to $\lambda_1 = \lambda_2 = 1$. These vectors are $v^{(1)} = [-2, 1, 0]^T$ and $v^{(2)} = [-1, 0, 1]^T$.

Similarly, for $\lambda_3 = 5$ we consider the system $(A - 5I)v = 0$:

$$\begin{bmatrix} -3 & 2 & 1 \\ 1 & -2 & 1 \\ 1 & 2 & -3 \end{bmatrix} \begin{bmatrix} v_1 \\ v_2 \\ v_3 \end{bmatrix} = \begin{bmatrix} 0 \\ 0 \\ 0 \end{bmatrix}.$$

Solving this system we obtain an infinite set of solutions $v_1 = v_2 = v_3$. We can set $v_1 = 1$ to obtain an eigenvector $v^{(3)} = [1, 1, 1]^T$ corresponding to $\lambda_3 = 5$.

Theorem 1.11 *The following statements are valid:*

(a) The determinant of any matrix is equal to the product of its eigenvalues.

(b) The trace of any matrix is equal to the sum of its eigenvalues.

(c) The eigenvalues of a nonsingular matrix A are nonzero, and the eigenvalues of A^{-1} are reciprocals of the eigenvalues of A.

Example 1.14 *The eigenvalues of*

$$A = \begin{bmatrix} 2 & 2 & 1 \\ 1 & 3 & 1 \\ 1 & 2 & 2 \end{bmatrix}$$

were computed in the previous example, and they are $\lambda_1 = \lambda_2 = 1, \lambda_3 = 5$. We can conclude, therefore, that

$$\det(A) = 1 \cdot 1 \cdot 5 = 5,$$

$$\mathrm{tr}(A) = 1 + 1 + 5 = 7,$$

and the eigenvalues of A^{-1} are $\mu_1 = \mu_2 = 1, \mu_3 = 1/5$.

Theorem 1.12 *The eigenvalues of a real symmetric matrix are all real numbers.*

Definition 1.31 (Orthogonal Matrix) *An $n \times n$ matrix A is orthogonal if $A^T A = I_n$, i.e., $A^T = A^{-1}$.*

Theorem 1.13 (Eigenvalue Decomposition) *A real symmetric matrix A can be represented as*

$$A = RDR^T,$$

where D is a diagonal matrix consisting of the eigenvalues of A; R is the matrix having an eigenvector corresponding to the ith diagonal element of D as its ith column; and R is orthogonal.

Example 1.15 *Consider the matrix*

$$A = \begin{bmatrix} 3 & 1 & 0 \\ 1 & 2 & 1 \\ 0 & 1 & 3 \end{bmatrix}.$$

The characteristic polynomial of A is

$$\varphi(\lambda) = \begin{vmatrix} 3-\lambda & 1 & 0 \\ 1 & 2-\lambda & 1 \\ 0 & 1 & 3-\lambda \end{vmatrix} = (3-\lambda)(5-5\lambda+\lambda^2) - (3-\lambda).$$

Hence, $\varphi(\lambda) = (3-\lambda)(4-5\lambda+\lambda^2)$, and the roots of $\varphi(\lambda)$ are

$$\lambda_1 = 1, \lambda_2 = 3 \text{ and } \lambda_3 = 4.$$

Next we find normalized eigenvectors corresponding to each eigenvalue.

Denote by $v^{(1)} = [v_1^{(1)}, v_2^{(1)}, v_3^{(1)}]^T$ the normalized eigenvector corresponding to $\lambda_1 = 1$. Then

$$(A-I)v^{(1)} = 0 \Leftrightarrow \begin{bmatrix} 2 & 1 & 0 \\ 1 & 1 & 1 \\ 0 & 1 & 2 \end{bmatrix} \begin{bmatrix} v_1^{(1)} \\ v_2^{(1)} \\ v_3^{(1)} \end{bmatrix} = \begin{bmatrix} 0 \\ 0 \\ 0 \end{bmatrix}.$$

This gives $v^{(1)} = c_1[1, -2, 1]^T$ for some constant c_1. Choosing c_1 such that $v^{(1)}$ is normalized, i.e., $\|v^{(1)}\|_2 = c_1\sqrt{6} = 1$, we obtain $c_1 = 1/\sqrt{6}$ and

$$v^{(1)} = [1/\sqrt{6}, -2/\sqrt{6}, 1/\sqrt{6}]^T.$$

Denote by $v^{(2)} = [v_1^{(2)}, v_2^{(2)}, v_3^{(2)}]^T$ the normalized eigenvector correspond-ing to $\lambda_2 = 3$. We have

$$(A - 3I)v^{(2)} = 0 \Leftrightarrow \begin{bmatrix} 0 & 1 & 0 \\ 1 & -1 & 1 \\ 0 & 1 & 0 \end{bmatrix} \begin{bmatrix} v_1^{(2)} \\ v_2^{(2)} \\ v_3^{(2)} \end{bmatrix} = \begin{bmatrix} 0 \\ 0 \\ 0 \end{bmatrix},$$

yielding the solution $v^{(2)} = c_2[1, 0, -1]^T$ for some c_2. Selecting c_2 satisfying $\|v^{(2)}\|_2 = c_2\sqrt{2} = 1$, we get $c_2 = 1/\sqrt{2}$ and the vector

$$v^{(2)} = [1/\sqrt{2}, 0, -1/\sqrt{2}]^T.$$

Similarly, the normalized eigenvector $v^{(3)} = [v_1^{(3)}, v_2^{(3)}, v_3^{(3)}]^T$ corresponding to $\lambda_3 = 4$ satisfies

$$(A - 4I)v^{(3)} = 0 \Leftrightarrow \begin{bmatrix} -1 & 1 & 0 \\ 1 & -2 & 1 \\ 0 & 1 & -1 \end{bmatrix} \begin{bmatrix} v_1^{(3)} \\ v_2^{(3)} \\ v_3^{(3)} \end{bmatrix} = \begin{bmatrix} 0 \\ 0 \\ 0 \end{bmatrix}.$$

The solution of this system is $v^{(3)} = c_3[1, 1, 1]^T$, where c_3 is found from the equation $\|v^{(3)}\|_2 = c_3\sqrt{3} = 1$, i.e., $c_3 = 1/\sqrt{3}$ and

$$v^{(3)} = [1/\sqrt{3}, 1/\sqrt{3}, 1/\sqrt{3}]^T.$$

Now we are ready to write matrices D and R from the eigenvalue decompo-sition theorem. The matrix D is a diagonal matrix containing eigenvalues of A,

$$D = \begin{bmatrix} 1 & 0 & 0 \\ 0 & 3 & 0 \\ 0 & 0 & 4 \end{bmatrix},$$

and R has corresponding normalized eigenvectors as its columns,

$$R = \begin{bmatrix} 1/\sqrt{6} & 1/\sqrt{2} & 1/\sqrt{3} \\ -2/\sqrt{6} & 0 & 1/\sqrt{3} \\ 1/\sqrt{6} & -1/\sqrt{2} & 1/\sqrt{3} \end{bmatrix}.$$

It is easy to check that $A = RDR^T$.

1.4.10   Quadratic forms

Definition 1.32 (Quadratic Form) *A quadratic form is*

$$Q(x) = \sum_{i=1}^{n} \sum_{j=1}^{n} a_{ij} x_i x_j,$$

where a_{ij}, $i, j = 1, \ldots, n$ are the coefficients of the quadratic form, $x = [x_1, \ldots, x_n]^T$ is the vector of variables.

Let $A = [a_{i,j}]_{n \times n}$ be the matrix of coefficients of $Q(x)$. Then $Q(x) = x^T A x = x^T Q x$, where $Q = (A + A^T)/2$ is a symmetric matrix. Therefore, any quadratic form is associated with the symmetric matrix of its coefficients.

Definition 1.33 (Definite (Semidefinite) Matrix) *A symmetric matrix Q and the corresponding quadratic form $q(x) = x^T Q x$ is called*

- *positive definite if $q(x) = x^T Q x > 0$ for all nonzero $x \in \mathbb{R}^n$;*

- *positive semidefinite if $q(x) = x^T Q x \geq 0$ for all $x \in \mathbb{R}^n$;*

- *negative definite if $q(x) = x^T Q x < 0$ for all nonzero $x \in \mathbb{R}^n$;*

- *negative semidefinite if $q(x) = x^T Q x \leq 0$ for all $x \in \mathbb{R}^n$.*

If a matrix (quadratic form) is neither positive nor negative semidefinite, it is called indefinite.

Theorem 1.14 (Sylvester's Criterion) *A real symmetric matrix A is positive definite if and only if all its leading principal minors are positive.*

Consider a quadratic form $Q(x) = x^T Q x$. By the eigenvalue decomposition theorem, $Q = R^T D R$, therefore, denoting by $y = Rx$ we obtain

$$Q(x) = x^T Q x = x^T R^T D R x = y^T D y = \sum_{i=1}^{n} \lambda_i y_i^2.$$

This leads to the following statement.

Theorem 1.15 *A symmetric matrix Q is*

- *positive (negative) definite if and only if all the eigenvalues of Q are positive (negative);*

- *positive (negative) semidefinite if and only if all the eigenvalues of Q are nonnegative (nonpositive).*

Example 1.16 *The eigenvalues of matrix*

$$A = \begin{bmatrix} 3 & 1 & 0 \\ 1 & 2 & 1 \\ 0 & 1 & 3 \end{bmatrix}$$

were computed in Example 1.15:

$$\lambda_1 = 1, \lambda_2 = 3 \text{ and } \lambda_3 = 4.$$

They are all positive, hence matrix A is positive definite.

Theorem 1.16 (Rayleigh's Inequality) *For a positive semidefinite* $n \times n$ *matrix Q, for any* $x \in \mathbb{R}^n$:

$$\lambda_{\min}(Q)\|x\|^2 \le x^T Q x \le \lambda_{\max}(Q)\|x\|^2, \tag{1.5}$$

where $\lambda_{\min}(Q)$ *and* $\lambda_{\max}(Q)$ *are the smallest and the largest eigenvalues of Q, respectively.*

1.4.11   Matrix norms

Given a vector p-norm, the corresponding *induced norm* or *operator norm* on the space $\mathbb{R}^{m \times n}$ of $m \times n$ matrices is defined by

$$\|A\|_p = \max_{x \neq 0} \frac{\|Ax\|_p}{\|x\|_p}. \tag{1.6}$$

The most commonly used values of p are $p = 1, 2$, and ∞. For $p = 1$ and $p = \infty$, (1.6) simplifies to 1-norm,

$$\|A\|_1 = \max_{1 \le j \le n} \sum_{i=1}^{m} |a_{ij}|$$

and ∞-norm,

$$\|A\|_\infty = \max_{1 \le i \le m} \sum_{j=1}^{n} |a_{ij}|,$$

respectively. For $p = 2$ and $m = n$, the induced p-norm is called the *spectral norm* and is given by

$$\|A\|_2 = \sqrt{\lambda_{\max}(A^T A)},$$

i.e., the square root of the largest eigenvalue of $A^T A$.

The *Frobenius norm* of an $m \times n$ matrix A is

$$\|A\|_F = \sqrt{\sum_{i=1}^{m} \sum_{j=1}^{n} a_{ij}^2}.$$

Example 1.17 *For*

$$A = \begin{bmatrix} 1 & -6 \\ 3 & 2 \end{bmatrix},$$

we have

$$\|A\|_1 = \max\{1+3,\ 6+2\} = 8;$$

$$\|A\|_2 = \sqrt{\lambda_{\max}\left(\begin{bmatrix} 10 & 0 \\ 0 & 40 \end{bmatrix}\right)} = \sqrt{40} = 2\sqrt{10};$$

$$\|A\|_\infty = \max\{1+6,\ 3+2\} = 7;$$

$$\|A\|_F = \sqrt{1+36+9+4} = \sqrt{50}.$$

1.5 Preliminaries from Real and Functional Analysis

For a scalar set $S \subset \mathbb{R}$, the number m is called a *lower bound* on S if $\forall s \in S : s \geq m$. Similarly, the number M is an *upper bound* on S if $\forall s \in S : s \leq M$. A set that has a lower bound is called *bounded from below*; a set that has an upper bound is *bounded from above*. A set that is bounded from both below and above is called *bounded*. If a set S is bounded from below, its greatest lower bound $glb(S)$, i.e., the lower bound that has the property that for any $\epsilon > 0$ there exists an element $s' \in S$ such that $glb(S) + \epsilon > s'$, is called the *infimum* of S and is denoted by $\inf(S) = \inf_{s \in S} s$. Similarly, for a bounded from above set S, the least upper bound $lub(S)$, which is the upper bound such that for any $\epsilon > 0$ there exists an element $s' \in S$ satisfying the inequality $lub(S) - \epsilon < s'$, is called the *supremum* of S and is denoted by $\sup(S) = \sup_{s \in S} s$. If S has no lower bound, we set $\inf(S) = -\infty$; if there is no upper bound, we set $\sup(S) = \infty$. If $s^* = \inf_{s \in S} s$ and $s^* \in S$, we use the term *minimum* and write $s^* = \min_{s \in S} s$. Similarly, we use the term *maximum* and write $s^* = \max_{s \in S} s$ if $s^* = \sup_{s \in S} s$ and $s^* \in S$.

1.5.1   Closed and open sets

We consider the Euclidean n-space $\mathbb{R}^n$ with the norm given by $\|x\| = \sqrt{x^T x} = \sqrt{\sum_{i=1}^{n} x_i^2}$. Let $X \subset \mathbb{R}^n$.

Definition 1.34 (Interior Point; Open Set) *We call $\bar{x} \in X$ an interior point of X if there exists $\varepsilon > 0$ such that $\bar{x}$ is included in X together with a set of points around it, $\{x : \|x - \bar{x}\| < \varepsilon\} \subset X$ for some $\varepsilon > 0$. A set X is called* open *if all of its points are interior.*

Example 1.18 *For $\bar{x} \in \mathbb{R}^n$ and $\varepsilon > 0$, the open ε-neighborhood of $\bar{x}$ or the open ε-ball centered at $\bar{x}$ defined by*

$$B(\bar{x}, \varepsilon) = \{x \in \mathbb{R}^n : \|x - \bar{x}\| < \varepsilon\},$$

is an open set.

Definition 1.35 (Limit Point; Closed Set) *We call $\bar{x}$ a* limit point *of X if every ε-neighborhood of $\bar{x}$ contains infinitely many points of X. A set X is called* closed *if it contains all of its limit points.*

Example 1.19 *For $\bar{x} \in \mathbb{R}^n$ and $\varepsilon > 0$, the closed ε-neighborhood of $\bar{x}$ or the closed ε-ball centered at $\bar{x}$ given by*

$$\bar{B}(\bar{x}, \varepsilon) = \{x \in \mathbb{R}^n : \|x - \bar{x}\| \leq \varepsilon\}$$

is a closed set.

Definition 1.36 (Bounded Set; Compact Set) *A set $X \subset \mathbb{R}^n$ is called* bounded *if there exists $C > 0$ such that $\|x\| \leq C$ for every $x \in X$. A set is called* compact *if it is closed and bounded.*

1.5.2   Sequences

A sequence in $\mathbb{R}^n$ is given by a set $\{x_k : k \in I\}$, where I is a countable set of indices. We will write $\{x_k : k \geq 1\}$ or simply $\{x_k\}$ if I is the set of positive integers. As in Definition 1.35, we call $\bar{x}$ a *limit (cluster, accumulation) point* of a sequence $\{x_k\}$ if every ε-neighborhood of $\bar{x}$ contains infinitely many points of the sequence. If a sequence $\{x_k\}$ has a unique limit point $\bar{x}$, we call $\bar{x}$ the *limit of the sequence*. In other words, a point $\bar{x} \in \mathbb{R}^n$ is said to be the limit for a sequence $\{x_k\} \subset \mathbb{R}^n$ if for every $\epsilon > 0$ there exists an integer K such that for any $k \geq K$ we have $\|x_k - \bar{x}\| < \epsilon$. In this case we write

$$x_k \to \bar{x}, k \to \infty \text{ or } \lim_{k\to\infty} x_k = \bar{x}$$

and say that the sequence *converges* to $\bar{x}$.

A *subsequence* of $\{x_k\}$ is an infinite subset of this sequence and is typically denoted by $\{x_{k_i} : i \geq 1\}$ or simply $\{x_{k_i}\}$. Then a point $\bar{x} \in \mathbb{R}^n$ is a limit point for a sequence $\{x_k\} \subset \mathbb{R}^n$ if and only if there is a subsequence of $\{x_k\}$ that converges to $\bar{x}$.

Definition 1.37 (Limit Inferior; Limit Superior) *The* limit inferior *and* limit superior *of a sequence $\{x_k : k \geq 1\} \subset \mathbb{R}$ are the infimum and supremum of the sequence's limit points, respectively:*

$$\liminf_{k\to\infty} x_k = \lim_{k\to\infty}(\inf_{m\geq k} x_m) = \sup_{k\geq 0}(\inf_{m\geq k} x_m)$$
$$\limsup_{k\to\infty} x_k = \lim_{k\to\infty}(\sup_{m\geq k} x_m) = \inf_{k\geq 0}(\sup_{m\geq k} x_m)$$

Definition 1.38 (Rate of Convergence) *Consider a sequence $\{x_k\}$ such that $x_k \to x^*, k \to \infty$. If there exist $C > 0$ and positive integers R, K, such that for any $k \geq K$:*

$$\frac{\|x_{k+1} - x^*\|}{\|x_k - x^*\|^R} \leq C,$$

then we say that $\{x_k : k \geq 0\}$ has the rate of convergence R.

- If $R = 1$, we say that the convergence is *linear*.
- If $\lim_{k\to\infty} \frac{\|x_{k+1}-x^*\|}{\|x_k-x^*\|} = 0$, the rate of convergence is *superlinear*.
- If $\lim_{k\to\infty} \frac{\|x_{k+1}-x^*\|}{\|x_k-x^*\|} = \mu$ does not hold for any $\mu < 1$, then the rate of convergence is called *sublinear*.
- If $R = 2$, the convergence is *quadratic*.

1.5.3 Continuity and differentiability

Let $X \subseteq \mathbb{R}$. Given a real function $f : X \to \mathbb{R}$ and a point $x_0 \in X$, f is called *continuous* at x_0 if $\lim_{x\to x_0} f(x) = f(x_0)$, and f is called *differentiable* at x_0 if there exists the limit $f'(x_0) = \lim\limits_{x \to x_0} \frac{f(x)-f(x_0)}{x-x_0}$, in which case $f'(x_0)$ is called the *first-order derivative*, or simply the *derivative* of f at x_0. If f is differentiable at x_0, then it is continuous at x_0. The opposite is not true in general; in fact, there exist continuous functions that are not differentiable at any point of their domain. If f is continuous (differentiable) at every point of X, then f is called continuous (differentiable) on X. The set of all continuous functions on X is denoted by $C(X)$. The kth order derivative $f^{(k)}(x_0)$ of f at x_0 can be defined recursively as the first-order derivative of $f^{(k-1)}$ at x_0 for $k \geq 1$, where $f^{(0)} \equiv f$. If f' is a continuous function at x_0, then we call f *continuously differentiable* or *smooth* at x_0. Similarly, we call f k-times continuously differentiable at x_0 if $f^{(k)}$ exists and is continuous at x_0. The set of all k times continuously differentiable functions on X is denoted by $C^{(k)}(C)$. Then $C^{(k)}(X) \subset C^{(k-1)}(X)$, $k \geq 1$ (where $C^{(0)}(X) \equiv C(X)$).

Theorem 1.1 (Intermediate Value Theorem) *Let* $f : [a,b] \to \mathbb{R}$ *be a continuous function. Then* f *takes all values between* $f(a)$ *and* $f(b)$ *on* $[a,b]$, *i.e., for any* $z \in [f(a), f(b)]$ *there exists* $c \in [a,b]$ *such that* $z = f(c)$.

Theorem 1.17 (Mean Value Theorem) *If* f *is a continuous function on* $[a,b]$ *and differentiable on* (a,b), *then there exists* $c \in (a,b)$ *such that*

$$\frac{f(b) - f(a)}{b - a} = f'(c) \tag{1.7}$$

In particular, if $f(a) = f(b)$, *then* $f'(c) = 0$ *for some* $c \in (a,b)$.

From a geometric perspective, the mean value theorem states that under the given conditions there always exists a point c between a and b such that the tangent line to f at c is parallel to the line passing through the points $(a, f(a))$ and $(b, f(b))$. This is illustrated in Figure 1.1. For the function f in this figure, there are two points $c_1, c_2 \in (a,b)$ that satisfy (1.7).

Now let $X \subseteq \mathbb{R}^n$. Then a function $f = [f_1, \ldots, f_m]^T : X \to \mathbb{R}^m$ is called continuous at point $\bar{x} \in X$ if $\lim\limits_{x \to \bar{x}} f(x) = f(\bar{x})$. The function is continuous on X, denoted by $f \in C(X)$, if it is continuous at every point of X. In this case the first-order derivative at $\bar{x} \in \mathbb{R}^n$, called *Jacobian*, is given by the following $m \times n$ matrix of partial derivatives:

$$J_f(\bar{x}) = \begin{bmatrix} \frac{\partial f_1(\bar{x})}{\partial x_1} & \cdots & \frac{\partial f_1(\bar{x})}{\partial x_n} \\ \vdots & \ddots & \vdots \\ \frac{\partial f_m(\bar{x})}{\partial x_1} & \cdots & \frac{\partial f_m(\bar{x})}{\partial x_n} \end{bmatrix}.$$

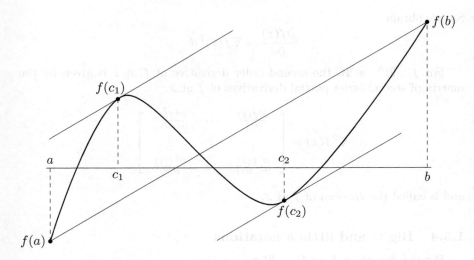

FIGURE 1.1: An illustration of the mean value theorem.

If all partial derivatives exist and are continuous on an open set containing $\bar{x}$, then f is *continuously differentiable*. If $m = 1$, i.e., we have $f : \mathbb{R}^n \to \mathbb{R}$, the Jacobian at $\bar{x} \in \mathbb{R}^n$ is given by the vector of partial derivatives

$$\nabla f(\bar{x}) = \left[\frac{\partial f(\bar{x})}{\partial x_1}, \dots, \frac{\partial f(\bar{x})}{\partial x_n} \right]^T$$

at $\bar{x}$ and is called the *gradient* of f at $\bar{x}$.

For $f : \mathbb{R}^n \to \mathbb{R}$ and $\bar{x}, d \in \mathbb{R}^n$, the directional derivative of $f(x)$ at $\bar{x}$ in the direction d is defined as

$$\frac{\partial f(\bar{x})}{\partial d} = \lim_{\alpha \to 0} \frac{f(\bar{x} + \alpha d) - f(\bar{x})}{\alpha}.$$

If $\|d\| = 1$, then $\frac{\partial f(\bar{x})}{\partial d}$ is also called the rate of increase of f at $\bar{x}$ in the direction d.

Assume that f is differentiable. Denoting by $\phi(\alpha) = f(\bar{x} + \alpha d)$, we have

$$\begin{aligned} \frac{\partial f(\bar{x})}{\partial d} &= \lim_{\alpha \to 0} \frac{f(\bar{x} + \alpha d) - f(\bar{x})}{\alpha} \\ &= \lim_{\alpha \to 0} \frac{\phi(\alpha) - \phi(0)}{\alpha} \\ &= \phi'(0). \end{aligned}$$

On the other hand, using the *chain rule*, according to which the derivative of the composite function $f(g(\alpha))$ is equal to the product of the derivatives of f and g, we have

$$\phi'(\alpha) = \nabla f(\bar{x} + \alpha d)^T d \quad \Rightarrow \quad \phi'(0) = \nabla f(\bar{x})^T d.$$

So, we obtain

$$\frac{\partial f(\bar{x})}{\partial d} = \nabla f(\bar{x})^T d.$$

For $f : \mathbb{R}^n \to \mathbb{R}$, the second-order derivative of f at $\bar{x}$ is given by the matrix of second-order partial derivatives of f at $\bar{x}$,

$$\nabla^2 f(\bar{x}) = \begin{bmatrix} \frac{\partial^2 f(\bar{x})}{\partial x_1^2} & \cdots & \frac{\partial^2 f(\bar{x})}{\partial x_1 \partial x_n} \\ \vdots & \ddots & \vdots \\ \frac{\partial^2 f(\bar{x})}{\partial x_n \partial x_1} & \cdots & \frac{\partial^2 f(\bar{x})}{\partial x_n^2} \end{bmatrix}$$

and is called the *Hessian* of f at $\bar{x}$.

1.5.4 Big O and little o notations

For two functions $f, g : \mathbb{R} \to \mathbb{R}$ we say that

- $f(x) = O(g(x))$ as $x \to \infty$ if there exist numbers $x_0 > 0$ and $C > 0$ such that $|f(x)| \le C|g(x)|$ for all $x \ge x_0$.

- $f(x) = o(g(x))$ as $x \to \infty$ if $\lim_{x \to \infty} \frac{|f(x)|}{|g(x)|} = 0$.

For two functions $f, g : \mathbb{R}^n \to \mathbb{R}$ and $\bar{x} \in \mathbb{R}^n$ we say that

- $f(x) = O(g(x))$ as $x \to \bar{x}$ if $|f(x)|/|g(x)|$ is asymptotically bounded near $\bar{x}$, i.e., there exist numbers $\epsilon > 0$ and $C > 0$ such that $|f(x)| \le C|g(x)|$ for all x with $\|x - \bar{x}\| \le \epsilon$.

- $f(x) = o(g(x))$ as $x \to \bar{x}$ if $\lim_{x \to \bar{x}} \frac{|f(x)|}{|g(x)|} = 0$.

It is easy to show that if $\lim_{x \to \bar{x}} \frac{|f(x)|}{|g(x)|} = c$, where $c < \infty$ is a positive number, then $f(x) = O(g(x))$ and $g(x) = O(f(x))$ as $x \to \bar{x}$.

Example 1.20 *For $f(x) = 10x^2 + 100x^3$, $g(x) = x^2$, we have $f(x) = O(g(x))$ as $x \to 0$ since*

$$\lim_{x \to 0} \frac{f(x)}{g(x)} = \lim_{x \to 0} \frac{10x^2 + 100x^3}{x^2} = 10.$$

For $f(x) = 6x^3 - 10x^2 + 15x$ and $g(x) = x^3$, we have $f(x) = O(g(x))$ as $x \to \infty$ since

$$\lim_{x \to \infty} \frac{f(x)}{g(x)} = \lim_{x \to \infty} \frac{6x^3 - 10x^2 + 15x}{x^3} = 6.$$

For $f(x) = x^2$ and $g(x) = x$, we have $f(x) = o(g(x))$ as $x \to 0$.

1.5.5 Taylor's theorem

Theorem 1.2 (Taylor's Theorem) *Let* $f \in C^{(n+1)}([\alpha, \beta])$ *and* $x_0 \in [\alpha, \beta]$. *Then, for each* $x \in (\alpha, \beta)$, *there exists* $\xi = \xi(x)$ *that lies between* x_0 *and* x *such that*

$$f(x) = P_n(x) + R_n(x), \tag{1.8}$$

where

$$P_n(x) = \sum_{k=0}^{n} \frac{f^{(k)}(x_0)}{k!}(x - x_0)^k \tag{1.9}$$

and

$$R_n(x) = \frac{f^{(n+1)}(\xi)}{(n+1)!}(x - x_0)^{n+1}. \tag{1.10}$$

The polynomial $P_n(x)$ *is called the Taylor polynomial of degree* n.

For $f : \mathbb{R}^n \to \mathbb{R}$, we will use Taylor's theorem for multivariate functions in the following forms.

If $f \in C^{(1)}(\mathbb{R}^n)$, then

$$f(x) = f(\bar{x}) + \nabla f(\tilde{x})^T(x - \bar{x}), \tag{1.11}$$

where $\tilde{x}$ is a point on the line segment connecting $\bar{x}$ and x.

If $f \in C^{(2)}(\mathbb{R}^n)$, then

$$f(x) = f(\bar{x}) + \nabla f(\bar{x})^T(x - \bar{x}) + \frac{1}{2}(x - \bar{x})^T \nabla^2 f(\tilde{x})(x - \bar{x}), \tag{1.12}$$

where $\tilde{x}$ is a point on the line segment between $\bar{x}$ and x. The error term is often represented using the big O and little o notations:

$$f(x) = f(\bar{x}) + \nabla f(\bar{x})^T(x - \bar{x}) + O(\|x - \bar{x}\|^2); \tag{1.13}$$

$$f(x) = f(\bar{x}) + \nabla f(\bar{x})^T(x - \bar{x}) + o(\|x - \bar{x}\|). \tag{1.14}$$

If $f \in C^{(3)}(\mathbb{R}^n)$, then

$$f(x) = f(\bar{x}) + \nabla f(\bar{x})^T(x - \bar{x}) + \frac{1}{2}(x - \bar{x})^T \nabla^2 f(\bar{x})(x - \bar{x}) + R_2(x), \tag{1.15}$$

where the error term $R_2(x)$ can be expressed as

$$R_2(x) = o(\|x - \bar{x}\|^2).$$

Example 1.21 *Let* $f(x) = x_1^3 + x_2^3 + 3x_1x_2 + x_1 - x_2 + 1$, $\bar{x} = [0,0]^T$. *Then*

$$\nabla f(x) = \begin{bmatrix} 3x_1^2 + 3x_2 + 1 \\ 3x_2^2 + 3x_1 - 1 \end{bmatrix}, \quad \nabla^2 f(x) = \begin{bmatrix} 6x_1 & 3 \\ 3 & 6x_2 \end{bmatrix},$$

and

$$\begin{aligned}
f(x) &= 1 + [1, -1]\begin{bmatrix} x_1 \\ x_2 \end{bmatrix} + \tfrac{1}{2}[x_1, x_2]\begin{bmatrix} 6\tilde{x}_1 & 3 \\ 3 & 6\tilde{x}_2 \end{bmatrix}\begin{bmatrix} x_1 \\ x_2 \end{bmatrix} \\
&= 1 + x_1 - x_2 + 3\tilde{x}_1 x_1^2 + 3\tilde{x}_2 x_2^2 + 3x_1x_2.
\end{aligned}$$

for some $\tilde{x}$ *between* x *and* 0, *i.e.,* $\tilde{x} = \alpha x + (1 - \alpha)0 = \alpha x$ *for some* $\alpha \in (0, 1)$. *For* $x = [1, 2]^T$, *we have* $f(x) = 15$ *and we can find* $\tilde{x}$ *by solving the following equation for* α:

$$3\alpha + 24\alpha + 6 = 15 \quad \Leftrightarrow \quad \alpha = 1/3.$$

Thus, $\tilde{x} = [1/3, 2/3]^T$ *and, according to* (1.12),

$$f(x) = 1 + x_1 - x_2 + x_1^2 + 2x_2^2 + 3x_1x_2$$

at $x = [1, 2]^T$.

The linear approximation

$$f(x) \approx f(\bar{x}) + \nabla f(\bar{x})^T(x - \bar{x}) = 1 + x_1 - x_2$$

in (1.13) *and* (1.14) *works well if* x *is very close to* $\bar{x}$. *For example, for* $x = [0.1, 0.1]$ *with* $\|x - \bar{x}\| = 0.1 \times \sqrt{2} \approx 0.1414$, *we have* $f(x) = 1.032$, *while the linear approximation gives* $f(x) \approx 1$ *with the error* $R_1(x) = 0.032$.

Using the quadratic approximation (1.15) *for* $\bar{x} = 0$, *we have*

$$f(x) \approx f(\bar{x}) + \nabla f(\bar{x})^T(x - \bar{x}) + \frac{1}{2}(x - \bar{x})^T\nabla^2 f(\bar{x})(x - \bar{x}) = 1 + x_1 - x_2 + 3x_1x_2,$$

and for $x = [0.1, 0.1]^T$ *we obtain* $f(x) \approx 1.03$ *with the error* $R_2(x) = 0.002$.

Exercises

1.1. Let $f : X \to Y$ be an arbitrary mapping and $X', X'' \subseteq X$, $Y', Y'' \subseteq Y$. Prove that

(a) $f^{-1}(Y' \cup Y'') = f^{-1}(Y') \cup f^{-1}(Y'')$;

(b) $f^{-1}(Y' \cap Y'') = f^{-1}(Y') \cap f^{-1}(Y'')$;

(c) $f(X' \cup X'') = f(X') \cup f(X'')$;

(d) $f(X' \cap X'')$ may not be equal to $f(X') \cap f(X'')$.

1.2. Prove that the following sets are countable:

    (a) the set of all odd integers;

    (b) the set of all even integers;

    (c) the set $2, 4, 8, 16, \ldots, 2^n, \ldots$ of powers of 2.

1.3. Show that

    (a) every infinite subset of a countable set is countable;

    (b) the union of a countable family of countable sets $A_1, A_2, \ldots$ is countable;

    (c) every infinite set has a countable subset.

1.4. Show that each of the following sets satisfies axioms from Definition 1.4 and thus is a linear space.

    (a) $\mathbb{R}^n$ with operations of vector addition and scalar multiplication.

    (b) $\mathbb{R}^{m \times n}$ with operations of matrix addition and scalar multiplication.

    (c) $C[a, b]$–the set of all continuous real-valued functions defined on the interval $[a, b]$, with addition and scalar multiplication defined as

$$(f + g)(x) = f(x) + g(x), x \in [a, b] \text{ for any } f, g \in C[a, b];$$

$$(\alpha f)(x) = \alpha f(x), x \in [a, b] \text{ for each } f \in C[a, b] \text{ and any scalar } \alpha;$$

respectively.

    (d) $\mathcal{P}_n$–the set of all polynomials of degree at most n with addition and scalar multiplication defined as

$$(p + q)(x) = p(x) + q(x), \text{ for any } p, q \in \mathcal{P}_n;$$

$$(\alpha p)(x) = \alpha p(x), \text{ for each } p \in \mathcal{P}_n; \text{ and any scalar } \alpha;$$

respectively.

1.5. (**Gram-Schmidt orthogonalization**) Let $p^{(0)}, \ldots, p^{(k-1)} \in \mathbb{R}^n$, where $k \leq n$, be an arbitrary set of linearly independent vectors. Show that the set of vectors $d^{(0)}, \ldots, d^{(k-1)} \in \mathbb{R}^n$ given by

$$
\begin{aligned}
d^{(0)} &= p^{(0)}; \\
d^{(s)} &= p^{(s)} - \sum_{i=0}^{s-1} \frac{{p^{(s)}}^T d^{(i)}}{{d^{(i)}}^T d^{(i)}} d^{(i)}, \quad s = 1, \ldots, k-1
\end{aligned}
$$

is orthogonal.

1.6. Show that for $p = 1, 2, \infty$ the norm $\| \cdot \|_p$ is *compatible*, that is for any $A \in \mathbb{R}^{m \times n}, x \in \mathbb{R}^n$:
$$\|Ax\|_p \leq \|A\|_p \|x\|_p.$$

1.7. Let
$$a = \begin{bmatrix} a_1 \\ \vdots \\ a_n \end{bmatrix} \text{ and } b = \begin{bmatrix} b_1 \\ \vdots \\ b_n \end{bmatrix}$$
be two vectors in $\mathbb{R}^n$.

(a) Compute the matrix
$$M = ab^T = \begin{bmatrix} a_1 \\ \vdots \\ a_n \end{bmatrix} [b_1, \ldots, b_n].$$

(b) How many additions and multiplications are needed to compute M?

(c) What is the rank of M?

1.8. Show that a square matrix A is orthogonal if and only if both the columns and rows of A form sets of orthonormal vectors.

1.9. A company manufactures three different products using the same machine. The sell price, production cost, and machine time required to produce a unit of each product are given in the following table.

	Product 1	Product 2	Product 3
Sell price	$30	$35	$50
Production cost	$18	$22	$30
Machine time	20 min	25 min	35 min

The table below represents a two-week plan for the number of units of each product to be produced.

	Week 1	Week 2
Product 1	120	130
Product 2	100	80
Product 3	50	60

Use a matrix product to compute, for each week, the revenue received from selling all items manufactured in a given week, the total production cost for each week, and the total machine time spent each week. Present your answers in a table.

1.10. Given matrices

$$A = \begin{bmatrix} 1 & 2 & 6 \\ -2 & 7 & -6 \\ 2 & 10 & 5 \\ 0 & 4 & 8 \end{bmatrix}, \quad B = \begin{bmatrix} 2 & -1 \\ 1 & -3 \\ 0 & 2 \end{bmatrix} \quad \text{and} \quad C = \begin{bmatrix} 1 & 0 & 0 \\ 0 & 1 & 0 \\ 0 & 0 & 1 \end{bmatrix},$$

(a) find the transpose of each matrix;

(b) consider all possible pairs $\{(A, A), (A, B), (B, A), \ldots, (C, C)\}$ and compute a product of each pair for which multiplication is a feasible operation.

1.11. For the matrices

$$A = \begin{bmatrix} -1 & 8 & 4 \\ 2 & -3 & -6 \\ 0 & 3 & 7 \end{bmatrix}; \quad B = \begin{bmatrix} -4 & 9 & 2 \\ 3 & -5 & 4 \\ 8 & 1 & -6 \end{bmatrix},$$

find (a) $A - 2B$, (b) AB, (c) BA.

1.12. Compute the p-norm for $p = 1, 2, \infty$ of the matrices A and B in Exercise 1.11.

1.13. Find the quadratic Taylor's approximation of the following functions:

(a) $f(x) = x_1^4 + x_2^4 - 4x_1x_2 + x_1^2 - 2x_2^2$ at $\bar{x} = [1, 1]^T$;

(b) $f(x) = \exp\left(x_1^2 + x_2^2\right)$ at $\bar{x} = [0, 0]^T$;

(c) $f(x) = \frac{1}{1+x_1^2+x_2^2}$ at $\bar{x} = [0, 0]^T$.

Chapter 2

Numbers and Errors

Numbers are usually represented by certain symbols, referred to as *numerals*, according to rules describing the corresponding *numeral system*. For example, the Roman numeral "X" describes the number ten, which is written as "10" in the decimal system. Clearly, our capability to perform nontrivial numerical operations efficiently depends on the numeral system chosen to represent the numbers, and *positional systems* are commonly used for this purpose. In a positional notation, each number is represented by a string of symbols, with each symbol being a multiplier for some *base* number. In general, the base number can be different for different positions, which makes counting more complicated.

Example 2.1 *Consider the common English length measures:*

$$1 \ yard = 3 \ feet$$
$$1 \ foot = 12 \ inches$$

If we take a length of 5 yards 1 foot and 7 inches, we cannot just juxtapose the three symbols and say that this length is equal to 517 inches, because the base number for the feet position is 12 inches, while the base number for the yard position is 3 feet = 36 inches. Note, that if we measure some length in metric units, for example, 5 meters, 1 decimeter, and 7 centimeters, then we can write this length simply by juxtaposing the symbols: 517 centimeters. This is because the base of the decimeter position is 10 centimeters, and the base of the meter position is 10 decimeters = 100 = 10^2 centimeters.

As the above example shows, a positional notation is simpler if different powers of the same number (called the *base* or *radix*) are used as the base number for each position. Hence, a good positional notation is characterized by a single base or radix β, a set of symbols representing all integers from 0 (null symbol) to $\beta - 1$, a radix point to separate integral and fractional portions of the number, and finally, by a plus and minus sign. It is important that every number has a unique representation in every system.

The *decimal system*, defined next, is what we use in everyday life.

Definition 2.1 (Decimal System) *In the decimal system the base $\beta = 10$, and a real number R is represented as follows:*

$$
\begin{aligned}
R &= \pm a_n a_{n-1} \ldots a_0.a_{-1} \ldots a_{-m} \\
&= \pm(a_n a_{n-1} \ldots a_0.a_{-1} \ldots a_{-m})_{10} \\
&= \pm(a_n 10^n + a_{n-1}10^{n-1} + \ldots + a_0 10^0 + a_{-1}10^{-1} + \ldots + a_{-m}10^{-m}),
\end{aligned}
$$

where all $a_i \in \{0, 1, \ldots, 9\}$. The radix point is called the decimal point in the decimal system.

Example 2.2 *Consider the number $R = 352.17$:*

$$
\begin{aligned}
352.17 &= 3 \times 100 + 5 \times 10 + 2 \times 1 + 1 \times 0.1 + 7 \times 0.01 \\
&= 3 \times 10^2 + 5 \times 10^1 + 2 \times 10^0 + 1 \times 10^{-1} + 7 \times 10^{-2}.
\end{aligned}
$$

Here $n = 2, m = 2$, and $a_2 = 3, a_1 = 5, a_0 = 2, a_{-1} = 1, a_{-2} = 7$.

However, the base 10 is not the only choice, and other bases can be used. The majority of modern computers use the base-2 or *binary* system.

Definition 2.2 (Binary System) *In the binary system the base $\beta = 2$, and a real number R is represented as follows:*

$$
\begin{aligned}
R &= \pm(a_n a_{n-1} \ldots a_0.a_{-1} \ldots a_{-m})_2 \\
&= \pm(a_n 2^n + a_{n-1}2^{n-1} + \ldots + a_0 2^0 + a_{-1}2^{-1} + \ldots + a_{-m}2^{-m}),
\end{aligned}
$$

where all $a_i \in \{0, 1\}$. In this case the radix point is called the binary point.

Example 2.3 *Consider the number*

$$
(10110.1)_2 = 1 \times 2^4 + 0 \times 2^3 + 1 \times 2^2 + 1 \times 2^1 + 0 \times 2^0 + 1 \times 2^{-1}.
$$

Here $n = 4, m = 1$, and $a_4 = 1, a_3 = 0, a_2 = 1, a_1 = 1, a_0 = 0, a_{-1} = 1$.

In general, in a system with the base β, a real R is represented as

$$
\begin{aligned}
R &= (a_n \ldots a_0.a_{-1} \ldots a_{-m})_\beta \\
&= a_n \beta^n + \ldots + a_0 \beta^0 + a_{-1}\beta^{-1} + \ldots + a_{-m}\beta^{-m}.
\end{aligned}
$$

It is easy to see that the number $n + 1$ of digits required to represent a positive integer N as a base-β number $(a_n \ldots a_0)_\beta$ is no more than $\log_\beta N + 1$. Indeed, if $n + 1 > \log_\beta N + 1$, we would have $N \geq \beta^n > \beta^{\log_\beta N} = N$, a

THERE ARE 10 TYPES

OF PEOPLE–

THOSE WHO UNDERSTAND BINARY,

AND THOSE WHO DON'T.

FIGURE 2.1: Mathematical humor playing on the equivalence between decimal 2 and binary $(10)_2$.

contradiction. Therefore, $n + 1 \leq \log_\beta N + 1$. On the other hand, since all $a_i \leq \beta - 1, i = 0, \ldots, n$, we have

$$N = a_n \beta^n + \ldots + a_0 \beta^0 \leq (\beta - 1) \sum_{i=0}^{n} \beta^i = \beta^{n+1} - 1,$$

yielding $n + 1 \geq \log_\beta (N + 1)$. Thus, we have

$$\log_\beta (N + 1) \leq n + 1 \leq \log_\beta N + 1. \tag{2.1}$$

2.1 Conversion between Different Number Systems

Converting a binary number to a decimal can be carried out in a straightforward fashion–we can just compute the decimal number resulting from the expression in Definition 2.2. For example, the binary $(10)_2$ is equivalent to the decimal 2 (see Figure 2.1):

$$(10)_2 = 1 \times 2^1 + 0 \times 2^0 = 2.$$

Next, we discuss how the conversion can be done more efficiently.

2.1.1   Conversion of integers

Note that we can use the following recursion to find the decimal representation of a base-β number $(a_n a_{n-1} \ldots a_0)_\beta$:

$$
\begin{aligned}
b_n &= a_n; \\
b_{n-1} &= a_{n-1} + b_n \beta &= a_{n-1} + a_n \beta; \\
b_{n-2} &= a_{n-2} + b_{n-1} \beta &= a_{n-2} + a_{n-1} \beta + a_n \beta^2; \\
&\ \ \vdots & \vdots \\
b_0 &= a_0 + b_1 \beta &= a_0 + a_1 \beta + a_2 \beta^2 + \ldots + a_n \beta^n.
\end{aligned}
$$

Obviously, $N = b_0$ is the decimal representation sought. This recursion is summarized in Algorithm 2.1, where the same variable N is used in place of $b_i, i = 0, \ldots, n$.

Algorithm 2.1 Converting a base-β integer to its decimal form.

1: **Input:** base β and a number $(a_n a_{n-1} \ldots a_0)_\beta$
2: **Output:** decimal representation N of $(a_n a_{n-1} \ldots a_0)_\beta$
3: $N = a_n$
4: **for** $i = n - 1, n - 2, \ldots, 0$ **do**
5: $\quad N = a_i + N\beta$
6: **end for**
7: **return** N

Algorithm 2.1 can be used to convert integers from the binary to the decimal system.

Example 2.4 *For* $(10111)_2$, $\beta = 2, n = 4, a_4 = 1$, *so Algorithm 2.1 starts with* $N = 1$ *and gives*

$$
\begin{aligned}
i = 3 : \quad & N = 0 + 1 \times 2 = 2; \\
i = 2 : \quad & N = 1 + 2 \times 2 = 5; \\
i = 1 : \quad & N = 1 + 5 \times 2 = 11; \\
i = 0 : \quad & N = 1 + 11 \times 2 = 23.
\end{aligned}
$$

To carry out the conversion, we used 4 additions and 4 multiplications. If we perform the same conversion using the formula in Definition 2.2, we have

$$
(10110)_2 = 1 \times 2^4 + 0 \times 2^3 + 1 \times 2^2 + 1 \times 2^1 + 1 \times 2^0.
$$

We need 4 additions and 5 multiplications, excluding the multiplications required to compute the different powers of 2. Obviously, we are better off if we use Algorithm 2.1.

To convert a positive integer in the decimal form to a base-β number, we use the following observation. Let $N = a_n a_{n-1} \ldots a_0$ be the given decimal number and let $N_\beta = (b_r b_{r-1} \ldots b_0)_\beta$ be its base-β representation. Then the remainder upon dividing N by β must be the same in both decimal and base-β systems. Dividing $(b_r b_{r-1} \ldots b_0)_\beta$ by β gives the remainder of b_0. Thus, b_0 is just the remainder we obtain when we divide N by β. The remainder upon dividing N by β can be expressed as

$$N - \beta \lfloor N/\beta \rfloor,$$

where $\lfloor N/\beta \rfloor$ is the integer part of N/β. If we apply the same procedure to $\lfloor N/\beta \rfloor$, we can compute b_1, and so on, as in Algorithm 2.2.

Algorithm 2.2 Converting a decimal integer N to its base-β representation.

1: **Input:** a decimal number N and base β
2: **Output:** base-β representation $(b_r b_{r-1} \ldots b_0)_\beta$ of N
3: $N_0 = N, i = 0$
4: **repeat**
5: $\quad b_i = N_0 - \beta \lfloor N_0/\beta \rfloor$
6: $\quad N_0 = \lfloor N_0/\beta \rfloor$
7: $\quad i = i + 1$
8: **until** $N_0 < \beta$
9: $r = i, b_r = N_0$
10: **return** $(b_r b_{r-1} \ldots b_0)_\beta$

Example 2.5 *We convert the decimal* 195 *to the binary form using Algorithm 2.2. We have:*

$$i = 0 : N_0 = 195, \ b_0 = 195 - 2\lfloor 195/2 \rfloor = 195 - 2 \times 97 = 1;$$

$$i = 1 : N_0 = 97, \ b_1 = 97 - 2\lfloor 97/2 \rfloor = 1;$$

$$i = 2 : N_0 = 48, \ b_2 = 48 - 2\lfloor 48/2 \rfloor = 0;$$

$$i = 3 : N_0 = 24, \ b_3 = 24 - 2\lfloor 24/2 \rfloor = 0;$$

$$i = 4 : N_0 = 12, \ b_4 = 12 - 2\lfloor 12/2 \rfloor = 0;$$

$$i = 5 : N_0 = 6, \ b_5 = 6 - 2\lfloor 6/2 \rfloor = 0;$$

$$i = 6 : N_0 = 3, \ b_6 = 3 - 2\lfloor 3/2 \rfloor = 1;$$

$$i = 7 : N_0 = 1 < \beta = 2, \ so \ b_7 = 1, \ and \ the \ algorithm \ terminates.$$

The output is $(11000011)_2$.

2.1.2   Conversion of fractions

Consider a positive real number x. We can represent x as:

$$x = x_I + x_F,$$

where $x_I = \lfloor x \rfloor$ is the integer part x, and $x_F = \{x\}$ is the fractional part of x. For example, for $x = 5.3$ we have $x_I = 5$ and $x_F = 0.3$. We can convert x_I and x_F to another base separately and then combine the results. Note that the x_F part can be written in the form

$$x_F = \sum_{k=1}^{\infty} a_{-k} 10^{-k},$$

where $a_{-k} \in \{0, 1, 2, \ldots 9\}$, for $k = 1, \ldots, \infty$. If $a_{-k} = 0$ for $k \geq N$, where N is some finite integer, then the fraction is said to *terminate*. For example,

$$\frac{1}{8} = 0.125 = 1 \times 10^{-1} + 2 \times 10^{-2} + 5 \times 10^{-3}$$

terminates, whereas

$$\frac{1}{6} = 0.16666\ldots = 1 \times 10^{-1} + 6 \times 10^{-2} + 6 \times 10^{-3} + \cdots$$

does not.

Next we discuss how $x_F < 1$ is converted from base β to a decimal. Similarly to Algorithm 2.1, we can develop a recursive procedure for this purpose, assuming that the input base-β fraction terminates. The corresponding procedure is presented in Algorithm 2.3.

Algorithm 2.3 Converting a base-β number $x_F < 1$ to its decimal form.

1: **Input:** base β and a number $(0.a_{-1} \ldots a_{-m})_\beta$
2: **Output:** decimal representation D of $(0.a_{-1} \ldots a_{-m})_\beta$
3: $D = a_{-m}/\beta$
4: **for** $i = m - 1, m - 2, \ldots, m$ **do**
5:     $D = (a_{-i} + D)/\beta$
6: **end for**
7: **return** D

Example 2.6 *We use Algorithm 2.3 to convert* $(0.110101)_2$ *to a decimal. We have* $m = 6$, $a_{-6} = 1$, *so we start with* $D = 1/2 = 0.5$.

$$i = 5: \quad D = (0 + 0.5)/2 = 0.25;$$

$$i = 4: \quad D = (1 + 0.25)/2 = 0.625;$$

$$i = 3: \quad D = (0 + 0.625)/2 = 0.3125;$$

$$i = 2: \quad D = (1 + 0.3125)/2 = 0.65625;$$

$$i = 1: \quad D = (1 + 0.65625)/2 = 0.828125.$$

Thus, $D = 0.828125$.

To represent a decimal $x_F < 1$ given by

$$x_F = \sum_{k=1}^{\infty} a_{-k} 10^{-k}$$

as a base-β number, we can equivalently represent x_F as the base-β fraction

$$x_F = (0.a_{-1}a_{-2}\cdots)_\beta = \sum_{k=1}^{\infty} a_{-k}\beta^{-k},$$

where $a_{-k} \in \{0, 1, \ldots, \beta - 1\}, k \geq 1$. Given $0 \leq x_F \leq 1$, we can find $(0.a_{-1}a_{-2}\cdots)_\beta$ as follows. Let $F_1 = x_F$, then

$$F_1 = \sum_{k=1}^{\infty} a_{-k}\beta^{-k} \Rightarrow$$

$$\beta F_1 = \sum_{k=1}^{\infty} a_{-k}\beta^{-k+1} = a_{-1} + \sum_{k=1}^{\infty} a_{-(k+1)}\beta^{-k},$$

so a_{-1} is the integer part of βF_1.

Denoting by $F_2 = \sum_{k=1}^{\infty} a_{-(k+1)}\beta^{-k}$ the fractional part of βF_1, we have

$$\beta F_2 = \sum_{k=1}^{\infty} a_{-(k+1)}\beta^{-k+1} = a_{-2} + \sum_{k=1}^{\infty} a_{-(k+2)}\beta^{-k},$$

and a_{-2} is the integer part of F_2, etc. In summary, we obtain the procedure described in Algorithm 2.4, where we assume that the given decimal fraction x_F terminates and use the same variable F for all $F_i, i \geq 1$.

Example 2.7 *We convert* $x_F = 0.625$ *to the binary form. We have* $\beta = 2$, $F = 0.625$, *and Algorithm 2.4 proceeds as follows.*

$$i = 1: \quad 2F = 2(0.625) = 1.25, a_{-1} = \lfloor 1.25 \rfloor = 1, F = 1.25 - 1 = 0.25;$$

$$i = 2: \quad 2F = 2(0.25) = 0.5, a_{-2} = \lfloor 0.5 \rfloor = 0, F = 0.5 - 0 = 0.5;$$

$$i = 3: \quad 2F = 2(0.5) = 1, a_{-3} = \lfloor 1 \rfloor = 1, F = 1 - 1 = 0.$$

Since $F = 0$, *the algorithm terminates, and we have* $0.625 = (0.101)_2$.

Unfortunately, a terminating decimal fraction may not terminate in its binary representation.

Algorithm 2.4 Converting a decimal $x_F < 1$ to its base-β representation.

1: **Input:** a decimal number $x_F < 1$ and base β, maximum allowed number R of base-β digits
2: **Output:** base-β representation $(0.b_{-1}b_{-2}\ldots b_{-r})_\beta$ of x_F, $r \le R$
3: $F = x_F, i = 1$
4: **repeat**
5: $\quad b_{-i} = \lfloor \beta F \rfloor$
6: $\quad F = \beta F - b_{-i}$
7: $\quad i = i + 1$
8: **until** $F = 0$ or $i = R + 1$
9: $r = i - 1, b_r = N_0$
10: **return** $(0.b_{-1}b_{-2}\ldots b_{-r})_\beta$

Example 2.8 *For $x_F = 0.1$ and $\beta = 2$, we have $F = 0.1$ and*

$$i = 1: \qquad 2F = 2(.1) = 0.2, a_{-1} = 0, \quad F = 0.2;$$
$$i = 2: \qquad 2F = 2(.2) = 0.4, a_{-2} = 0, \quad F = 0.4;$$
$$i = 3: \qquad 2F = 2(.4) = 0.8, a_{-3} = 0, \quad F = 0.8;$$
$$i = 4: \qquad 2F = 2(.8) = 1.6, a_{-4} = 1, \quad F = 0.6;$$
$$i = 5: \qquad 2F = 2(.6) = 1.2, a_{-5} = 1, \quad F = 0.2.$$

We see that $F = 0.2$ assumes the same value as for $i = 1$, therefore, a cycle occurs. Hence we have
$$0.1 = (0.0\overline{0011})_2,$$
i.e., $0.1 = (0.0001100110011\ldots0011\ldots)_2$.

Note that if we had to convert $(0.0\overline{0011})_2$ to a decimal, we could use a trick called shifting, *which works as follows. If we multiply $(0.0\overline{0011})_2$ by 2, we obtain $(0.\overline{0011})_2$. Because of the cycle, multiplying $(0.0\overline{0011})_2$ by $2^5 = 32$ gives a number with the same fractional part, $(11.\overline{0011})_2$. Hence, if we denote the value of $(0.0\overline{0011})_2$ by A, we have*

$$32A - 2A = (11)_2 = 3 \quad \Leftrightarrow \quad 30A = 3 \quad \Leftrightarrow \quad A = 3/30 = 0.1.$$

2.2 Floating Point Representation of Numbers

In computers, the numbers are usually represented using the *floating-point* form. A real number $x \in \mathbb{R}$ is represented as a floating-point number in base β as

$$x = m \cdot \beta^e,$$

where
$$m = \pm(0.d_1 d_2 \ldots d_n)_\beta,$$

and m is called the *mantissa* or *significand*; β is the *base*; n is the *length* or *precision*; and e is the *exponent*, where $s < e < S$ (usually $s = -S$).

Example 2.9 *The number 23.684 could be represented as $0.23684 \cdot 10^2$ in a floating-point decimal system.*

The mantissa is usually normalized when its leading digit is zero. The next example shows why.

Example 2.10 *Suppose the number $1/39 = 0.025641026...$ is stored in a floating-point system with $\beta = 10$ and $n = 5$. Thus, it would be represented as*

$$0.02564 \cdot 10^0.$$

Obviously, the inclusion of the useless zero to the right of the decimal point leads to dropping of the digit 1 in the sixth decimal place. But if we normalize the number by representing it as

$$0.25641 \cdot 10^{-1},$$

we retain an additional significant figure while storing the number.

When mantissa m is normalized, its absolute value is limited by the inequality
$$\frac{1}{\beta} \leq |m| < 1.$$

For example, in a base-10 system the mantissa would range between 0.1 and 1.

2.3 Definitions of Errors

Using numerical methods, we typically find a solution that approximates the true values with a certain error. For some computation, let x_T be the true value, and let x_C be the computed approximation. We are typically interested in either absolute or relative errors, as defined next.

Definition 2.3 (Absolute and Relative Error) *The absolute error is defined as*
$$|x_T - x_C|,$$

while the relative error *is defined as*

$$\frac{|x_T - x_C|}{|x_T|}.$$

Many of the problems solved by numerical methods involve infinite sets of values, each of which can potentially require an infinite number of digits to be represented exactly.

Example 2.11 *Consider a problem, in which one wants to find the smallest value taken by some function $y = f(x)$ defined over the interval $[0, 1]$ (Figure 2.2). The set of all values taken by $f(x)$ is given by $X = \{f(x), x \in [0, 1]\}$. Obviously, the set X is infinite, and most of its elements are irrational numbers, requiring infinitely many digits to be represented exactly.*

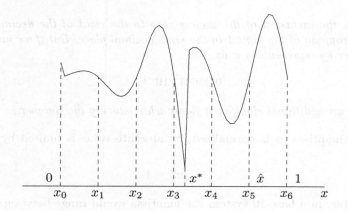

FIGURE 2.2: A continuous function defined on the interval $[0, 1]$.

In reality, any digital computer is finite by its nature. Thus, the infinite set of values should be approximated by a finite set, in which, in its turn, each element should be approximated by a numerical value representable with a finite number of digits. This leads to two types of errors in numerical computation: the *round-off* and *truncation error*.

Definition 2.4 (Truncation Error) *By the truncation error (sometimes called discretization error) we mean the error appearing as a result of the approximation of an infinite set of values by a finite set.*

Example 2.12 *In the problem from Example 2.11, we can consider the set of $n + 1$ mesh points $\{x_i = \frac{i}{n}, i = 0, 1, \ldots, n\}$, and a finite set $X_n = \{f(x_i), i =$*

$0, 1, \ldots, n\}$ *of $n + 1$ values instead of X. This way we are approximating the original problem by the problem of finding the smallest among the $n + 1$ numbers in X_n. Suppose that this smallest number in X_n equals $\hat{f} = 4/9$, and the exact minimum of X is $f^* = 1/12$ (see Figure 2.2). Then $|f^* - \hat{f}| = |1/12 - 4/9| = 13/36$ is the absolute truncation error of the considered method.*

Definition 2.5 (Round-off Error) *The error caused by the approximation of a numerical value with a value represented by fewer digits is called the round-off error.*

Example 2.13 *Assume that we want to write our answer $\breve{f}$ to the problem in the previous example as a five-digit decimal number instead of $\hat{f} = 4/9$. Then $\breve{f} = 0.4444$ and the absolute round-off error is $|\hat{f} - \breve{f}| = 4/9 - 4444/10000 = 1/22500 \approx 0.000044444$, where the last number is an approximation of the absolute round-off error.*

2.4 Round-off Errors

2.4.1 Rounding and chopping

Since any computer can only use a finite number of digits to represent a number, most of the real numbers cannot be represented exactly. There are two common ways to approximate a real number x by a floating-point number $fl(x)_{n,\beta}$–*rounding* and *chopping*.

Definition 2.6 (Rounding, Chopping) *If rounding is employed then $fl(x)_{n,\beta}$ is set equal to the nearest to x floating-point number in the system with base β and precision n. In case of a tie, round to the even digit (symmetric rounding).*
Alternatively, if chopping is employed, then $fl(x)_{n,\beta}$ is set equal to the nearest to x floating-point number between 0 and x.

Example 2.14 *Using rounding, we have*

$$fl\left(\frac{2}{3}\right)_{2,10} = fl(0.6666\ldots) = +(0.67) \times 10^0,$$

$$fl(-456789\ldots)_{2,10} = -(0.46) \times 10^6,$$

whereas using chopping we obtain the following approximations:

$$fl\left(\frac{2}{3}\right)_{2,10} = fl(0.666\ldots) = +(0.66) \times 10^0,$$

$$fl(-456789)_{2,10} = -(0.45) \times 10^6.$$

Observe that if $|x| \geq \beta^S$, then x cannot be represented. An attempt to employ such x will result in what is called an *overflow* error. If $0 \leq |x| \leq \beta^{s-n}$, then x is too small to be represented and we have *underflow*. That is, in both cases the limits of the computer are exceeded.

Relative round-off error of both rounding and chopping is defined as $\delta(x)$ such that

$$fl(x) = x(1 + \delta(x)) \Rightarrow \delta(x) = \frac{fl(x) - x}{x}.$$

Thus we can bound the error for a given x as

$$|\delta(x)| \leq \tfrac{1}{2}\beta^{1-n}, \text{ for rounding,}$$

$$|\delta(x)| \leq \beta^{1-n}, \text{ for chopping.}$$

The maximum possible $|\delta|$ is called *unit round-off*. The *machine epsilon* is defined as the smallest positive number ε such that $1 + \varepsilon \neq 1$.

Example 2.15 *A pseudocode to determine the machine epsilon for a binary computer can be written as follows:*

Algorithm 2.5 Determining machine epsilon for a binary computer.

1: $eps = 1$
2: **repeat**
3:     $eps = eps/2$
4: **until** $(1 + eps/2 = 1)$
5: **return** eps

The machine epsilon can be used in stopping or convergence criteria in implementations of numerical algorithms. This is a convenient way to ensure that your program is portable; that is, it can be executed on any computer.

2.4.2   Arithmetic operations

Consider the common arithmetic operations with floating-point numbers, such as addition, multiplication, and division by a nonzero number. Then the length of the floating-point number resulting from an operation on two floating-point numbers is not necessarily the same as that of the initial numbers.

Example 2.16 *Let*

$$x = 0.20 \cdot 10^1 = 2, \ y = 0.77 \cdot 10^{-6}, \ z = 0.30 \cdot 10^1 = 3.$$

Then

$$
\begin{aligned}
x + y &= (0.20000077) \cdot 10^1, \\
x \times y &= (0.154) \cdot 10^{-5}, \\
\frac{x}{z} &= (0.666\ldots) \cdot 10^{10}.
\end{aligned}
$$

Let $x \circ y$ denote the arithmetic operation on numbers x, y (i.e. $+, -, \times, \ldots$), and $x \otimes y$ denote the floating-point operation between x and y. Then

$$x \circ y \neq x \otimes y$$

in general. However $\otimes$ is constructed such that

$$x \otimes y = fl(x \circ y).$$

Considering the error, we have

$$x \otimes y = (x \circ y)(1 + \delta(x \circ y)),$$

where $|\delta|$ is no greater than the unit round-off.

2.4.3 Subtractive cancellation and error propagation

The following two examples illustrate the so-called *subtractive cancellation*, which is the round-off induced when subtracting two nearly equal floating-point numbers.

Example 2.17 *Suppose that we have* $x = 0.7654499, y = 0.7653001$, *then* $z = x - y = .1498 \times 10^{-3}$. *We have*

$$
\begin{aligned}
x_C = (x)_{4,10} &= (0.7654) \times 10^0, \\
y_C = (y)_{4,10} &= (0.7653) \times 10^0, \\
z_C = x_C - y_C &= (0.1000) \times 10^{-3},
\end{aligned}
$$

and the relative round-off error is $\frac{|z - z_C|}{|z|} = 0.3324 = 33.24\%$.

One of the main problems arising in numerical computations is that floating-point arithmetics gives errors that may *explode* in the computations that follow, and produce erroneous results. Once an error occurs, it "contaminates" the rest of the calculation. This is called *error propagation*.

Exercises

2.1. Find the binary representations of the numbers

    (a) $a = 32170$;

    (b) $b = 7/16$;

    (c) $c = 75/128$.

2.2. Find infinite repeating binary representations of the following numbers:

    (a) $a = 1/3$;

    (b) $b = 1/10$;

    (c) $c = 1/7$.

2.3. Find the binary representation of 10^6 and $\frac{4}{3}$.

2.4. Convert $(473)_{10}$ into a number with the base

    (a) 2;

    (b) 6;

    (c) 8.

2.5. Find the decimal representations of the following numbers:

    (a) $a = (101011010101)_2$;

    (b) $b = (16341)_8$;

    (c) $c = (4523)_6$.

2.6. Prove that any number 2^{-n}, where n is a positive integer, can be represented as a n-digit decimal number $0.a_1 a_2 \ldots a_n$.

2.7. Prove that $2 = 1.999999\ldots$.

2.8. For a decimal integer with m digits, how many bits are needed for its binary representation?

2.9. Create a hypothetical binary floating-point number set consisting of 7-bit words, in which the first three bits are used for the sign and the magnitude of the exponent, and the last four are used for the sign and magnitude of the mantissa. On the sign place, 0 indicates that the quantity is positive and 1, negative. For example, the number represented in the following figure is $-(.100)_2 \times 2^{(1 \times 2^1)} = -(1 \times 2^{-1}) \times 4 = -2$.

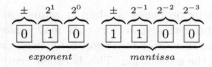

(a) How many numbers are in this set?

(b) Draw a line and show the decimal equivalents of all numbers in the set on this line. What is the distance between two consecutive numbers?

(c) Use your illustration to determine the unit round-off for this system.

2.10. Consider the quadratic equation $ax^2 + bx + c = 0$. Its roots can be found using the formula

$$x_{1,2} = \frac{-b \pm \sqrt{b^2 - 4ac}}{2a}.$$

Let $a = 1, b = 100 + 10^{-14}$, and $c = 10^{-12}$. Then the exact roots of the considered equation are $x_1 = -10^{-14}$ and $x_2 = -100$.

(a) Use the above formula to compute the first root on a computer. What is the relative round-off error?

(b) Use the following equivalent formula for the first root:

$$x_1 = \frac{-2c}{b + \sqrt{b^2 - 4ac}}.$$

What do you observe?

2.11. Use a computer to perform iterations in the form $x_{k+1} = (0.1x_k - 0.2)30, k \geq 0$ starting with $x_0 = 3$. You could use, e.g., a spreadsheet application. What do you observe after (a) 50 iterations; (b) 100 iterations? Explain.

Part II

Numerical Methods for Standard Problems

Chapter 3

Elements of Numerical Linear Algebra

In this chapter, we will consider the problems of solving a system of linear equations and computing eigenvalues of a matrix. These problems arise in many important applications, such as polynomial interpolation, solution of ordinary differential equations, numerical optimization and others.

Example 3.1 *Suppose that a company is planning to invest the total of $1 million into three projects, A, B, and C. The estimated one-year returns for each project under two different market conditions (scenarios), I and II, are summarized in Table 3.1.*

TABLE 3.1: Returns under two different scenarios.

Project	Return under scenario	
	I	II
A	1.4	0.9
B	0.8	2.0
C	1.2	1.0

The manager's goal is to distribute funds between the projects in a way that would yield the total return of 1.2 under both scenarios.

Let us denote by x_1, x_2, and x_3 the amount of money (in millions of dollars) to be invested in project A, B, and C, respectively. Then the problem can be formulated as the following system of linear equations:

$$
\begin{aligned}
1.4x_1 &+ 0.8x_2 &+ 1.2x_3 &= 1.2 \\
0.9x_1 &+ 2x_2 &+ 1.0x_3 &= 1.2 \\
x_1 &+ x_2 &+ x_3 &= 1.
\end{aligned}
$$

In this system, the first two equations express the goal of obtaining the return of 1.2 under scenarios I and II, respectively. The third equation indicates that the total amount of the investment to be made is $1 million.

An $m \times n$ *system of linear algebraic equations* has the following general form:

$$
\begin{array}{rcl}
a_{11}x_1 + a_{12}x_2 + \cdots + a_{1n}x_n &=& b_1 \\
a_{21}x_1 + a_{22}x_2 + \cdots + a_{2n}x_n &=& b_2 \\
\vdots && \vdots \\
a_{m1}x_1 + a_{m2}x_2 + \cdots + a_{mn}x_n &=& b_m.
\end{array}
\tag{3.1}
$$

In the above system, we have

- m *linear equations* $a_{i1}x_1 + a_{i2}x_2 + \cdots + a_{in}x_n = b_i$, $i = 1, \ldots, m$;

- n *unknowns* x_j, $j = 1, \ldots, n$;

- $m \times n$ *coefficients* $a_{ij} \in \mathbb{R}, i = 1, 2, \ldots, m, j = 1, \ldots, n$; and

- m *constant terms* $b_i \in \mathbb{R}, i = 1, \ldots, m$.

If all of the constant terms are zero, $b_i = 0, i = 1, \ldots, m$, then the system is called *homogeneous*.

Note that system (3.1) can be equivalently represented in the *matrix* form as follows:

$$
Ax = b,
$$

where

$$
A = \begin{bmatrix} a_{11} & a_{12} & \cdots & a_{1n} \\ a_{21} & a_{22} & \cdots & a_{2n} \\ \vdots & \vdots & \ddots & \vdots \\ a_{m1} & a_{m2} & \cdots & a_{mn} \end{bmatrix}, \quad x = \begin{bmatrix} x_1 \\ x_2 \\ \vdots \\ x_n \end{bmatrix}, \quad \text{and} \quad b = \begin{bmatrix} b_1 \\ b_2 \\ \vdots \\ b_m \end{bmatrix}.
$$

Example 3.2 *The system from Example 3.1 has the following matrix form:*

$$
\begin{bmatrix} 1.4 & 0.8 & 1.2 \\ 0.9 & 2 & 1.0 \\ 1 & 1 & 1 \end{bmatrix} \begin{bmatrix} x_1 \\ x_2 \\ x_3 \end{bmatrix} = \begin{bmatrix} 1.2 \\ 1.2 \\ 1 \end{bmatrix}.
$$

A system of linear equations is called *consistent* if there exists a set of values $x_i^*, i = 1, \ldots n$ of its unknowns that satisfies each equation in the system, in which case the vector $x^* = [x_1^*, \ldots, x_n^*]^T$ is called a *solution* of the system. If a system of linear equations has no solution, it is called *inconsistent*.

Two systems of linear equations are *equivalent* if their solution sets coincide.

Example 3.3 *The system from Example 3.1,*

$$
\begin{array}{rcrcrcl}
1.4x_1 &+& 0.8x_2 &+& 1.2x_3 &=& 1.2 \\
0.9x_1 &+& 2x_2 &+& 1.0x_3 &=& 1.2 \\
x_1 &+& x_2 &+& x_3 &=& 1
\end{array}
$$

has the solution $x_1^* = 1/2, x_2^* = 1/4, x_3^* = 1/4$, thus it is consistent.

Assume that a consistent system $Ax = b$ has two distinct solutions, x^* and $\hat{x}$. Then for any real α, $x_\alpha = \alpha x^* + (1 - \alpha)\hat{x}$ is also a solution of the system. Thus, a consistent system of linear equations has either a unique solution or infinitely many solutions.

For systems with the same number of equations as the number of unknowns, the following theorem holds.

Theorem 3.1 *For a matrix $A \in \mathbb{R}^{n \times n}$ and a vector $b \in \mathbb{R}^{n \times 1}$, the following statements are equivalent:*

  (i) *The system $Ax = b$ has a unique solution.*

 (ii) *The matrix A is nonsingular.*

(iii) $\det(A) \neq 0$.

Note that if A is nonsingular and $b = 0$ (i.e., the system is homogeneous), then the unique solution of $Ax = 0$ is $x = 0$.

The numerical methods for solving systems of linear equations are classified into *direct* and *iterative* methods. Direct methods provide the exact answer (assuming no round-off errors) in a finite number of steps. Iterative methods are not guaranteed to compute an exact solution in a finite number of steps, however they can yield high-quality approximate solutions and are easier to implement than the direct methods.

3.1 Direct Methods for Solving Systems of Linear Equations

Some direct methods proceed by first reducing the given system to a special form, which is known to be easily solvable, and then solving the resulting "easy" system. Triangular systems of linear equations, discussed next, represent one such easily solvable case.

3.1.1 Solution of triangular systems of linear equations

An $n \times n$ system $Ax = b$ is called triangular if the matrix A is triangular. If $A = [a_{ij}]_{n \times n}$ is upper triangular and nonsingular, we can solve the system

$Ax = b$ easily using *back substitution*. That is, if the system at hand is

$$\begin{bmatrix} a_{11} & a_{12} & \cdots & a_{1(n-1)} & a_{1n} \\ 0 & a_{22} & \cdots & a_{2(n-1)} & a_{2n} \\ \vdots & \vdots & \ddots & \vdots & \vdots \\ 0 & 0 & \cdots & a_{(n-1)(n-1)} & a_{(n-1)n} \\ 0 & 0 & \cdots & 0 & a_{nn} \end{bmatrix} \begin{bmatrix} x_1 \\ x_2 \\ \vdots \\ x_{n-1} \\ x_n \end{bmatrix} = \begin{bmatrix} b_1 \\ b_2 \\ \vdots \\ b_{n-1} \\ b_n \end{bmatrix},$$

then we can solve for x starting from x_n:

$$a_{nn}x_n = b_n \Rightarrow x_n = \frac{b_n}{a_{nn}},$$

$$a_{(n-1)(n-1)}x_{n-1} + a_{(n-1)n}x_n = b_n \Rightarrow x_{n-1} = \frac{b_{n-1} - a_{(n-1)n}x_n}{a_{(n-1)(n-1)}},$$

and

$$x_k = \frac{b_k - \sum_{j=k+1}^{n} a_{kj}x_j}{a_{kk}}, \quad k = n, n-1, \ldots, 1.$$

Example 3.4 *Consider the system*

$$\begin{array}{rrrrl} 3x_1 & +2x_2 & +x_3 & = & 1 \\ & x_2 & -x_3 & = & 2 \\ & & 2x_3 & = & 4. \end{array}$$

Solving the above system using back substitution we get

$$x_3 = \frac{4}{2} = 2,$$

$$x_2 = \frac{b_2 - a_{23}x_3}{a_{22}} = \frac{2+2}{1} = 4,$$

$$x_1 = \frac{b_1 - (a_{12}x_2 + a_{13}x_3)}{a_{11}} = \frac{1 - (2 \cdot 4 + 1 \cdot 2)}{3} = -3.$$

Therefore the solution to the system is $x = [-3, 4, 2]^T$.

Similarly, if we have a lower triangular system

$$\begin{bmatrix} a_{11} & 0 & \cdots & 0 & 0 \\ a_{21} & a_{22} & \cdots & 0 & 0 \\ \vdots & \vdots & \ddots & \vdots & \vdots \\ a_{(n-1)1} & a_{(n-1)2} & \cdots & a_{(n-1)(n-1)} & 0 \\ a_{n1} & a_{n2} & \cdots & a_{n(n-1)} & a_{nn} \end{bmatrix} \begin{bmatrix} x_1 \\ x_2 \\ \vdots \\ x_{n-1} \\ x_n \end{bmatrix} = \begin{bmatrix} b_1 \\ b_2 \\ \vdots \\ b_{n-1} \\ b_n \end{bmatrix},$$

we can use the *forward substitution* to solve it:

$$x_k = \frac{b_k - \sum_{j=1}^{k-1} a_{kj}x_j}{a_{kk}}, \quad k = 1, 2, \ldots, n.$$

3.1.2 Gaussian elimination

Given a nonsingular matrix $A = [a_{ij}]_{n \times n}$ and a vector $b \in \mathbb{R}^n$, Gaussian elimination transforms the system $Ax = b$ into an equivalent system $\bar{A}x = \bar{b}$, where $\bar{A}$ is a triangular matrix. Then $\bar{A}x = \bar{b}$ can be solved by back substitution to get the solution of $Ax = b$.

Note that nonsingularity of A and equivalence of the systems $Ax = b$ and $\bar{A}x = \bar{b}$ yield that $\bar{A}$ is also nonsingular, thus its diagonal elements are nonzero.

In the Gaussian elimination method, the equivalent transformation is achieved by applying the following three *elementary row operations*:

(1) changing the order of two rows;

(2) multiplying a row by a nonzero constant;

(3) replacing a row by its sum with a multiple of another row.

When any of these operations is applied to the *augmented matrix*

$$[A|b] = \left[\begin{array}{cccc|c} a_{11} & a_{12} & \cdots & a_{1n} & b_1 \\ a_{21} & a_{22} & \cdots & a_{2n} & b_2 \\ \vdots & \vdots & \ddots & \vdots & \vdots \\ a_{n1} & a_{n2} & \cdots & a_{nn} & b_n \end{array} \right],$$

which is obtained by appending the vector of constant terms to the matrix of coefficients, we obtain an equivalent system.

We introduce the following notation for the matrix $[A|b]$:

$$M^{(1)} = [A|b] = \left[\begin{array}{cccc|c} a_{11}^{(1)} & a_{12}^{(1)} & \cdots & a_{1n}^{(1)} & a_{1(n+1)}^{(1)} \\ a_{21}^{(1)} & a_{22}^{(1)} & \cdots & a_{2n}^{(1)} & a_{2(n+1)}^{(1)} \\ \vdots & \vdots & \ddots & \vdots & \vdots \\ a_{n1}^{(1)} & a_{n2}^{(1)} & \cdots & a_{nn}^{(1)} & a_{n(n+1)}^{(1)} \end{array} \right].$$

Without loss of generality we can assume that $a_{11}^{(1)} \neq 0$ (if this is not the case, we can switch rows so that $a_{11}^{(1)} \neq 0$). In the Gaussian elimination method, we first "eliminate" elements $a_{i1}^{(1)}$, $i = 2, \ldots, m$ in the first column, by turning them into 0. This can be done by replacing the ith row, $i = 2, \ldots, m$ with the row which is obtained by multiplying the first row by $-\frac{a_{i1}^{(1)}}{a_{11}^{(1)}}$ and adding the result to the ith row, thus obtaining

$$a_{ij}^{(2)} = a_{ij}^{(1)} - \frac{a_{i1}^{(1)}}{a_{11}^{(1)}} a_{1j}^{(1)}, \quad i = 2, \ldots, n, \ j = 1, 2, \ldots, n, n+1.$$

Denote the resulting matrix by $M^{(2)}$:

$$M^{(2)} = \begin{bmatrix} a_{11}^{(1)} & a_{12}^{(1)} & \cdots & a_{1n}^{(1)} & a_{1(n+1)}^{(1)} \\ 0 & a_{22}^{(2)} & \cdots & a_{2n}^{(2)} & a_{2(n+1)}^{(2)} \\ \vdots & \vdots & \ddots & \vdots & \vdots \\ 0 & a_{n2}^{(2)} & \cdots & a_{nn}^{(2)} & a_{n(n+1)}^{(2)} \end{bmatrix}.$$

Next, assuming that $a_{22}^{(2)} \neq 0$, we use the second row to obtain zeros in rows $3, \ldots, n$ of the second column of matrix $M^{(2)}$:

$$a_{ij}^{(3)} = a_{ij}^{(2)} - \frac{a_{i2}^{(2)}}{a_{22}^{(2)}} a_{2j}^{(2)}, \ i = 3, \ldots, n, \ j = 2, \ldots, n, n+1.$$

We denote the matrix obtained after this transformation by $M^{(3)}$.

Continuing this way, at the kth step, $k \leq n-1$, we have matrix $M^{(k)}$, and assuming that $a_{kk}^{(k)} \neq 0$, eliminate the elements in rows $k+1, \ldots, n$ of the kth column:

$$a_{ij}^{(k+1)} = a_{ij}^{(k)} - \frac{a_{ik}^{(k)}}{a_{kk}^{(k)}} a_{kj}^{(k)}, \ i = k+1, \ldots, n, \ j = k, \ldots, n, n+1.$$

The resulting matrix $M^{(k+1)}$ has the following form:

$$M^{(k+1)} = \begin{bmatrix} a_{11}^{(1)} & a_{12}^{(1)} & \cdots & a_{1k}^{(1)} & a_{1(k+1)}^{(1)} & \cdots & a_{1n^{(1)}} & a_{1(n+1)}^{(1)} \\ 0 & a_{22}^{(2)} & \cdots & a_{2k}^{(2)} & a_{2(k+1)}^{(2)} & \cdots & a_{2n}^{(2)} & a_{2(n+1)}^{(2)} \\ \vdots & \vdots & \ddots & \vdots & \vdots & \cdots & \vdots & \vdots \\ 0 & 0 & \cdots & a_{kk}^{(k)} & a_{k(k+1)}^{(k)} & \cdots & a_{kn}^{(k)} & a_{k(n+1)}^{(k)} \\ 0 & 0 & \cdots & 0 & a_{(k+1)(k+1)}^{(k+1)} & \cdots & a_{(k+1)n}^{(k+1)} & a_{(k+1)(n+1)}^{(k+1)} \\ \vdots & \vdots & \cdots & \vdots & \vdots & \ddots & \vdots & \vdots \\ 0 & 0 & \cdots & 0 & a_{n(k+1)}^{(k+1)} & \cdots & a_{nn}^{(k+1)} & a_{n(n+1)}^{(k+1)} \end{bmatrix}.$$

When $k = (n-1)$, we obtain a matrix $M^{(n)}$ corresponding to a triangular system $\bar{A}x = \bar{b}$:

$$M^{(n)} = \underbrace{\begin{bmatrix} a_{11}^{(1)} & a_{12}^{(1)} & \cdots & a_{1n}^{(1)} \\ 0 & a_{22}^{(2)} & \cdots & a_{2n}^{(2)} \\ \vdots & \vdots & \ddots & \vdots \\ 0 & 0 & \cdots & a_{nn}^{(n)} \end{bmatrix}}_{\bar{A}} \underbrace{\begin{bmatrix} a_{1(n+1)}^{(1)} \\ a_{2(n+1)}^{(2)} \\ \vdots \\ a_{n(n+1)}^{(n)} \end{bmatrix}}_{\bar{b}}.$$

Given two augmented matrices A_1 and A_2, the fact that they correspond to two equivalent systems can be expressed as $A_1 \sim A_2$. During the execution of the Gaussian elimination procedure, we construct a chain of matrices $M^{(1)}, M^{(2)}, \ldots, M^{(n)}$, which correspond to equivalent systems. Therefore, we can write

$$M^{(1)} \sim M^{(2)} \sim \ldots \sim M^{(n)}.$$

Example 3.5 *Consider the system of linear equations*

$$
\begin{aligned}
x_1 + 2x_2 - 4x_3 &= -4, \\
5x_1 + 11x_2 - 21x_3 &= -22, \\
3x_1 - 2x_2 + 3x_3 &= 11.
\end{aligned}
$$

Combining the coefficient matrix and the vector of constant terms into a single matrix we have

$$M = [A|b] = \begin{bmatrix} 1 & 2 & -4 & -4 \\ 5 & 11 & -21 & -22 \\ 3 & -2 & 3 & 11 \end{bmatrix}.$$

We reduce A to an upper triangular matrix, by applying the elementary row operations to the augmented matrix M. Let $M^{(1)} = [A^{(1)}|b^{(1)}] = [A|b] = M$ be the initial augmented matrix of the considered system. Let us denote by $R_i^{(k)}$ the ith row of $M^{(k)} = [A^{(k)}|b^{(k)}]$. To eliminate the first element in the second row of $M^{(1)}$, we multiply the first row by -5 and add the result to the second row to obtain the second row of $M^{(2)}$. This can be written as

$$\underbrace{\begin{bmatrix} 1 & 2 & -4 & -4 \\ 5 & 11 & -21 & -22 \\ 3 & -2 & 3 & 11 \end{bmatrix}}_{M^{(1)}} \sim^{R_2^{(2)}=-5R_1^{(1)}+R_2^{(1)}} \begin{bmatrix} 1 & 2 & -4 & -4 \\ 0 & 1 & -1 & -2 \\ 3 & -2 & 3 & 11 \end{bmatrix}.$$

We proceed by eliminating the first element of the third row, thus obtaining matrix $M^{(2)}$:

$$\begin{bmatrix} 1 & 2 & -4 & -4 \\ 0 & 1 & -1 & -2 \\ 3 & -2 & 3 & 11 \end{bmatrix} \sim^{R_3^{(2)}=-3R_1^{(1)}+R_3^{(1)}} \underbrace{\begin{bmatrix} 1 & 2 & -4 & -4 \\ 0 & 1 & -1 & -2 \\ 0 & -8 & 15 & 23 \end{bmatrix}}_{M^{(2)}}.$$

Finally,

$$\underbrace{\begin{bmatrix} 1 & 2 & -4 & -4 \\ 0 & 1 & -1 & -2 \\ 0 & -8 & 15 & 23 \end{bmatrix}}_{M^{(2)}} \sim^{R_3^{(3)}=8R_2^{(2)}+R_3^{(2)}} \underbrace{\begin{bmatrix} \overset{\bar{A}}{1} & 2 & -4 & \overset{\bar{b}}{-4} \\ 0 & 1 & -1 & -2 \\ 0 & 0 & 7 & 7 \end{bmatrix}}_{M^{(3)}}.$$

The resulting matrix $\bar{A}$ is upper-triangular. Therefore, using back substitution we can derive the solution to the system, $x = [2, -1, 1]^T$.

3.1.2.1   Pivoting strategies

The coefficient $a_{kk}^{(k)}$ in A that is used to eliminate $a_{ik}^{(k)}$, $i = k+1, k+2, \ldots, n$ in the Gaussian elimination procedure is called the *pivotal element*, and the kth row is called the *pivotal row*. The process of eliminating all the below-diagonal elements described above is called a *pivot*.

In the Gaussian elimination procedure above, at each step we assumed that the diagonal element, which was used as the pivotal element, was nonzero. However, it may be the case that at some step k the diagonal element in the kth row is equal to zero: $a_{kk}^{(k)} = 0$, therefore it cannot be used to eliminate the elements below in the kth column. In this situation we can use the following *trivial pivoting* strategy to change the pivoting element. We examine the elements below $a_{kk}^{(k)}$ in the kth column, until we find some $a_{lk}^{(k)} \neq 0$, $l > k$, and then simply switch rows k and l.

However, in some cases more sophisticated pivoting strategies are needed to minimize the round-off error of computations. The next example demonstrates the effect of the choice of a pivotal element in Gaussian elimination on the precision of the computed solution.

Example 3.6 *Consider the following system:*

$$
\begin{aligned}
0.102x_1 + 2.45x_2 &= 2.96 \\
20.2x_1 - 11.4x_2 &= 89.6
\end{aligned}
$$

which has the solution $[x_1, x_2] = [5, 1]$. *We use three-digit arithmetic to solve this system using Gaussian elimination. We have*

$$
\begin{bmatrix} 0.102 & 2.45 & | & 2.96 \\ 20.2 & -11.4 & | & 89.6 \end{bmatrix} \sim_{R_2^{(2)} = R_2^{(1)} - 198 \cdot R_1^{(1)}} \begin{bmatrix} 0.102 & 2.45 & | & 2.96 \\ 0 & -497 & | & -496 \end{bmatrix},
$$

hence the solution is $x_1 = 5.05$ *and* $x_2 = 0.998$. *However, if we switch the rows, we obtain*

$$
\begin{bmatrix} 20.2 & -11.4 & | & 89.6 \\ 0.102 & 2.45 & | & 2.96 \end{bmatrix} \sim_{R_2^{(2)} = R_2^{(1)} - 0.00505 \cdot R_1^{(1)}} \begin{bmatrix} 20.2 & -11.4 & | & 89.6 \\ 0 & 2.51 & | & 2.51 \end{bmatrix},
$$

yielding the correct solution $[x_1, x_2] = [5, 1]$.

In general, if the pivotal element, say, $a_{kk}^{(k)}$, is very small in absolute value, then $\left| \frac{a_{lk}}{a_{kk}} \right|$ may be very large for some $l > k$. When used in computations as a multiplier, the last fraction may cause undesirable round-off errors. Therefore, one should avoid choosing small numbers as pivotal elements. This motivates the following *partial pivoting* strategy.

At the kth step of Gaussian elimination, there are $n - k + 1$ candidates for the pivotal element:

$$
a_{kk}^{(k)}, a_{(k+1)k}^{(k)}, \ldots, a_{nk}^{(k)}.
$$

We choose $a_{lk}^{(k)}$ such that

$$|a_{lk}^{(k)}| = \max_{k \le p \le n} |a_{pk}^{(k)}|$$

as the pivotal element by switching the kth and l^{th} rows. Then we have

$$\left| \frac{a_{pk}^{(k)}}{a_{lk}^{(k)}} \right| \le 1, \ \forall p \ge k.$$

3.1.3 Gauss-Jordan method and matrix inversion

The Gauss-Jordan method is a modification of Gaussian elimination. It uses elementary row operations in the augmented matrix in order to transform the matrix of coefficients into the identity matrix.

Example 3.7 *Consider the system*

$$\begin{array}{rcrcrcr} x_1 & + & 2x_2 & - & 4x_3 & = & -4, \\ 5x_1 & + & 11x_2 & - & 21x_3 & = & -22, \\ 3x_1 & - & 2x_2 & + & 3x_3 & = & 11. \end{array} \quad (3.2)$$

In Example 3.5 we used Gaussian elimination to reduce the augmented matrix of this system to the equivalent form

$$\left[\begin{array}{ccc|c} 1 & 2 & -4 & -4 \\ 0 & 1 & -1 & -2 \\ 0 & 0 & 7 & 7 \end{array} \right].$$

We can further reduce the upper triangular matrix to the identity matrix by eliminating the non-diagonal elements as follows.

First, we can divide the third row by 7 in order to have unity in the diagonal in the third row:

$$\left[\begin{array}{ccc|c} 1 & 2 & -4 & -4 \\ 0 & 1 & -1 & -2 \\ 0 & 0 & 1 & 1 \end{array} \right].$$

Now the first row can be replaced by its sum with the third row multiplied by 4, and the second row can be replaced by its sum with the third row:

$$\left[\begin{array}{ccc|c} 1 & 2 & 0 & 0 \\ 0 & 1 & 0 & -1 \\ 0 & 0 & 1 & 1 \end{array} \right].$$

Finally, adding the first row to the second row multiplied by -2 and writing the result as the new first row, we obtain

$$\left[\begin{array}{ccc|c} 1 & 0 & 0 & 2 \\ 0 & 1 & 0 & -1 \\ 0 & 0 & 1 & 1 \end{array} \right].$$

Thus the solution can be extracted from the right-hand-side vector of the last matrix: $[x_1, x_2, x_3] = [2, -1, 1]$.

The Gauss-Jordan method provides a natural way of computing the inverse of a nonsingular matrix. Assume that we have a nonsingular matrix

$$A = \begin{bmatrix} a_{11} & a_{12} & \cdots & a_{1n} \\ a_{21} & a_{22} & \cdots & a_{2n} \\ \vdots & \vdots & \ddots & \vdots \\ a_{n1} & a_{n2} & \cdots & a_{nn} \end{bmatrix}.$$

Let us look for its inverse $X = A^{-1}$ in the following form:

$$X = \begin{bmatrix} x_{11} & x_{12} & \cdots & x_{1n} \\ x_{21} & x_{22} & \cdots & x_{2n} \\ \vdots & \vdots & \ddots & \vdots \\ x_{n1} & x_{n2} & \cdots & x_{nn} \end{bmatrix},$$

such that

$$AX = I_n,$$

where I_n denotes the $n \times n$ identity matrix. Then in order to find the first column of X, we can solve the system

$$\begin{bmatrix} a_{11} & a_{12} & \cdots & a_{1n} \\ a_{21} & a_{22} & \cdots & a_{2n} \\ \vdots & \vdots & \ddots & \vdots \\ a_{n1} & a_{n2} & \cdots & a_{nn} \end{bmatrix} \begin{bmatrix} x_{11} \\ x_{21} \\ \vdots \\ x_{n1} \end{bmatrix} = \begin{bmatrix} 1 \\ 0 \\ \vdots \\ 0 \end{bmatrix}.$$

In general, to find the jth column X_j of X, $j = 1, 2, \ldots, n$, we can solve the system

$$AX_j = I_n^j,$$

where I_n^j is the jth column of the identity matrix I_n. We can solve all of these systems simultaneously for all $j = 1, 2, \ldots, n$ by applying the Gauss-Jordan method to the matrix A augmented with the identity matrix $I_n = [I_n^1 \cdots I_n^n]$:

$$\left[\begin{array}{cccc|cccc} a_{11} & a_{12} & \cdots & a_{1n} & 1 & 0 & \cdots & 0 \\ a_{21} & a_{22} & \cdots & a_{2n} & 0 & 1 & \cdots & 0 \\ \vdots & \vdots & \ddots & \vdots & \vdots & \vdots & \ddots & \vdots \\ a_{n1} & a_{n2} & \cdots & a_{nn} & 0 & 0 & \cdots & 1 \end{array} \right].$$

When the coefficient matrix is reduced to the identity matrix using elementary row operations, the right-hand side of the resulting augmented matrix will contain the inverse of A. So, starting with $[A|I_n]$, we use the Gauss-Jordan method to obtain $[I_n|A^{-1}]$.

Example 3.8 *Consider the coefficient matrix of the system* (3.2):

$$A = \begin{bmatrix} 1 & 2 & -4 \\ 5 & 11 & -21 \\ 3 & -2 & 3 \end{bmatrix}.$$

Augment this matrix with the 3×3 *identity matrix:*

$$\left[\begin{array}{ccc|ccc} 1 & 2 & -4 & 1 & 0 & 0 \\ 5 & 11 & -21 & 0 & 1 & 0 \\ 3 & -2 & 3 & 0 & 0 & 1 \end{array} \right].$$

Let us eliminate the elements in rows 2 and 3 of the first column using $a_{11} = 1$ *as the pivotal element:*

$$\left[\begin{array}{ccc|ccc} 1 & 2 & -4 & 1 & 0 & 0 \\ 0 & 1 & -1 & -5 & 1 & 0 \\ 0 & -8 & 15 & -3 & 0 & 1 \end{array} \right].$$

We can use the second row to eliminate non-diagonal elements in the second column:

$$\left[\begin{array}{ccc|ccc} 1 & 0 & -2 & 11 & -2 & 0 \\ 0 & 1 & -1 & -5 & 1 & 0 \\ 0 & 0 & 7 & -43 & 8 & 1 \end{array} \right].$$

Finally, we can divide row 3 by 7 and use this row to eliminate coefficients for x_3 *in other rows:*

$$\left[\begin{array}{ccc|ccc} 1 & 0 & 0 & -9/7 & 2/7 & 2/7 \\ 0 & 1 & 0 & -78/7 & 15/7 & 1/7 \\ 0 & 0 & 1 & \underbrace{-43/7 & 8/7 & 1/7}_{A^{-1}} \end{array} \right].$$

Therefore, the inverse of A is

$$A^{-1} = \begin{bmatrix} -9/7 & 2/7 & 2/7 \\ -78/7 & 15/7 & 1/7 \\ -43/7 & 8/7 & 1/7 \end{bmatrix}.$$

Now to find the solution of (3.2) *all we need to do is to multiply* A^{-1} *by the right-hand-side vector b:*

$$x = A^{-1}b = \begin{bmatrix} -9/7 & 2/7 & 2/7 \\ -78/7 & 15/7 & 1/7 \\ -43/7 & 8/7 & 1/7 \end{bmatrix} \begin{bmatrix} -4 \\ -22 \\ 11 \end{bmatrix} = \begin{bmatrix} 2 \\ -1 \\ 1 \end{bmatrix}.$$

3.1.4   Triangular factorization

Consider an elementary row operation of replacing the kth row of an $m \times n$ matrix A by its sum with the lth row multiplied by m_{kl}, where $k > l$. Observe that this operation is equivalent to performing a matrix multiplication $E_{kl}^{(m_{kl})} A$, where $E_{kl}^{(m_{kl})}$ is an $m \times m$ matrix, which is obtained from the identity matrix I_m by replacing zero on the intersection of kth row and l^{th} column with m_{kl}. As an illustration, consider

$$M^{(1)} = \underbrace{\begin{bmatrix} 1 & 2 & -4 \\ 5 & 11 & -21 \\ 3 & -2 & 3 \end{bmatrix}}_{A^{(1)}} \left.\underbrace{\begin{bmatrix} -4 \\ -22 \\ 11 \end{bmatrix}}_{b^{(1)}}\right.$$

from Example 3.5 (page 61). In this example, we replaced the second row of $M^{(1)}$ by its sum with the first row multiplied by -5 to obtain

$$C = \left[\begin{array}{ccc|c} 1 & 2 & -4 & -4 \\ 0 & 1 & -1 & -2 \\ 3 & -2 & 3 & 11 \end{array}\right].$$

It is easy to check that the same outcome results from performing the following matrix multiplication:

$$\underbrace{\begin{bmatrix} 1 & 0 & 0 \\ -5 & 1 & 0 \\ 0 & 0 & 1 \end{bmatrix}}_{E_{21}^{(-5)}} \underbrace{\left[\begin{array}{ccc|c} 1 & 2 & -4 & -4 \\ 5 & 11 & -21 & -22 \\ 3 & -2 & 3 & 11 \end{array}\right]}_{M^{(1)}} = \underbrace{\left[\begin{array}{ccc|c} 1 & 2 & -4 & -4 \\ 0 & 1 & -1 & -2 \\ 3 & -2 & 3 & 11 \end{array}\right]}_{C}.$$

Next, in Example 3.5, we replaced the third row of the current matrix C by its sum with the first row of C multiplied by -3 to obtain $M^{(2)}$, that is,

$$M^{(2)} = E_{31}^{(-3)} C = E_{31}^{(-3)} E_{21}^{(-5)} M^{(1)}.$$

Finally, we replaced the third row of $M^{(2)}$ by its sum with the second row of $M^{(2)}$ multiplied by 8 to obtain $M^{(3)}$:

$$M^{(3)} = E_{32}^{(8)} M^{(2)} = E_{32}^{(8)} E_{31}^{(-3)} E_{21}^{(-5)} M^{(1)}.$$

In the process of obtaining $M^{(3)} = [A^{(3)}|b^{(3)}]$, the coefficient matrix A has undergone the following transformations to become an upper triangular matrix $U \equiv A^{(3)}$:

$$U = E_{32}^{(8)} E_{31}^{(-3)} E_{21}^{(-5)} A.$$

Hence,

$$A = \left(E_{21}^{(-5)}\right)^{-1} \left(E_{31}^{(-3)}\right)^{-1} \left(E_{32}^{(8)}\right)^{-1} U = LU,$$

where $L = \left(E_{21}^{(-5)} \right)^{-1} \left(E_{31}^{(-3)} \right)^{-1} \left(E_{32}^{(8)} \right)^{-1}$. Next we show that L is a lower triangular matrix. Note that $\left(E_{kl}^{(m_{kl})} \right)^{-1} = E_{kl}^{(-m_{kl})}$ and

$$E_{21}^{(-m_{21})} E_{31}^{(-m_{31})} E_{32}^{(-m_{32})} = \begin{bmatrix} 1 & 0 & 0 \\ -m_{21} & 1 & 0 \\ -m_{31} & -m_{32} & 1 \end{bmatrix}.$$

Hence,

$$L = \left(E_{21}^{(-5)} \right)^{-1} \left(E_{31}^{(-3)} \right)^{-1} \left(E_{32}^{(8)} \right)^{-1} = E_{21}^{(5)} E_{31}^{(3)} E_{32}^{(-8)} = \begin{bmatrix} 1 & 0 & 0 \\ 5 & 1 & 0 \\ 3 & -8 & 1 \end{bmatrix}.$$

Thus, we obtained a representation of A as the product of a lower and upper triangular matrices, $A = LU$. This process is called the *triangular (LU) factorization* of A.

In the above description of triangular factorization, we have not exchanged rows while performing Gaussian elimination. Next we discuss the case involving row exchanges. As an example, we solve the same system as above using Gaussian elimination with the partial pivoting strategy. We have

$$M^{(1)} = \left[\begin{array}{ccc|c} 1 & 2 & -4 & -4 \\ 5 & 11 & -21 & -22 \\ 3 & -2 & 3 & 11 \end{array} \right],$$

$$\underbrace{\phantom{\begin{array}{ccc} 1 & 2 & -4 \end{array}}}_{A} \underbrace{\phantom{\begin{array}{c} -4 \end{array}}}_{b}$$

and we first exchange the first two rows of $M^{(1)}$. The permutation corresponding to this row exchange is $p_1 = \begin{pmatrix} 1 & 2 & 3 \\ 2 & 1 & 3 \end{pmatrix}$. For each permutation p we can define the corresponding *permutation matrix* $P = [p_{ij}]_{n \times n}$ such that

$$p_{ij} = \begin{cases} 1, & \text{if } p(i) = j, \\ 0, & \text{otherwise.} \end{cases}$$

Note that $P^{-1} = P^T$ for a permutation matrix P. The matrix P_1 corresponding to our permutation p_1 above is given by

$$P_1 = \begin{bmatrix} 0 & 1 & 0 \\ 1 & 0 & 0 \\ 0 & 0 & 1 \end{bmatrix}.$$

Note that pre-multiplying $M^{(1)}$ by P_1 yields

$$P_1 M^{(1)} = \begin{bmatrix} 0 & 1 & 0 \\ 1 & 0 & 0 \\ 0 & 0 & 1 \end{bmatrix} \left[\begin{array}{ccc|c} 1 & 2 & -4 & -4 \\ 5 & 11 & -21 & -22 \\ 3 & -2 & 3 & 11 \end{array} \right] = \left[\begin{array}{ccc|c} 5 & 11 & -21 & -22 \\ 1 & 2 & -4 & -4 \\ 3 & -2 & 3 & 11 \end{array} \right],$$

which is exactly $M^{(1)}$ after exchange of its first two rows. Next, we eliminate the below-diagonal elements in the first column:

$$E_{31}^{(-3/5)} E_{21}^{(-1/5)} P_1 M^{(1)} = \begin{bmatrix} 5 & 11 & -21 & -22 \\ 0 & -1/5 & 1/5 & 2/5 \\ 0 & -43/5 & 78/5 & 121/5 \end{bmatrix} = M^{(2)}.$$

Exchanging the second and third rows of $M^{(2)}$, we obtain:

$$P_2 E_{31}^{(-3/5)} E_{21}^{(-1/5)} P_1 M^{(1)} = \begin{bmatrix} 5 & 11 & -21 & -22 \\ 0 & -43/5 & 78/5 & 121/5 \\ 0 & -1/5 & 1/5 & 2/5 \end{bmatrix},$$

where

$$P_2 = \begin{bmatrix} 1 & 0 & 0 \\ 0 & 0 & 1 \\ 0 & 1 & 0 \end{bmatrix}.$$

Finally,

$$E_{32}^{(-1/43)} P_2 E_{31}^{(-3/5)} E_{21}^{(-1/5)} P_1 M^{(1)} = \underbrace{\begin{bmatrix} 5 & 11 & -21 & -22 \\ 0 & -43/5 & 78/5 & 121/5 \\ 0 & 0 & -7/43 & -7/43 \end{bmatrix}}_{U}.$$

Since $M^{(1)} = [A|b]$, we have

$$\left(E_{32}^{(-1/43)} P_2 E_{31}^{(-3/5)} E_{21}^{(-1/5)} P_1 \right) A = U.$$

Multiplying both sides by $\left(E_{32}^{(-1/43)} \right)^{-1} = E_{32}^{(1/43)}$, we have:

$$\left(P_2 E_{31}^{(-3/5)} E_{21}^{(-1/5)} P_1 \right) A = E_{32}^{(1/43)} U.$$

Since $P_2^{-1} = P_2^T = P_2$, we have:

$$\left(E_{31}^{(-3/5)} E_{21}^{(-1/5)} P_1 \right) A = P_2 E_{32}^{(1/43)} U = \begin{bmatrix} 1 & 0 & 0 \\ 0 & 1/43 & 1 \\ 0 & 1 & 0 \end{bmatrix} U.$$

Next,

$$P_1 A = P_2 E_{32}^{(1/43)} U = \left(E_{21}^{(-1/5)} \right)^{-1} \left(E_{31}^{(-3/5)} \right)^{-1} \begin{bmatrix} 1 & 0 & 0 \\ 0 & 1/43 & 1 \\ 0 & 1 & 0 \end{bmatrix} U,$$

so

$$P_1 A = \begin{bmatrix} 1 & 0 & 0 \\ 1/5 & 1 & 0 \\ 3/5 & 0 & 1 \end{bmatrix} \begin{bmatrix} 1 & 0 & 0 \\ 0 & 1/43 & 1 \\ 0 & 1 & 0 \end{bmatrix} U = \begin{bmatrix} 1 & 0 & 0 \\ 1/5 & 1/43 & 1 \\ 3/5 & 1 & 0 \end{bmatrix} U.$$

Multiplying both sides by P_2 from the left, we obtain:

$$P_2 P_1 A = \underbrace{\begin{bmatrix} 1 & 0 & 0 \\ 3/5 & 1 & 0 \\ 1/5 & 1/43 & 1 \end{bmatrix}}_{L} U.$$

Thus, for a permutation matrix

$$P = P_2 P_1 = \begin{bmatrix} 0 & 1 & 0 \\ 0 & 0 & 1 \\ 1 & 0 & 0 \end{bmatrix},$$

we have

$$PA = LU \iff A = P^T LU.$$

Consider a system $Ax = b$ and suppose that a triangular factorization $A = LU$ of A is known. Then

$$Ax = b \iff LUx = b \iff Ly = b, \text{ where } Ux = y.$$

Hence, we can solve $Ly = b$ for y using forward substitution. Having computed y, we can find x by solving the system $Ux = y$ via backward substitution.

Thus, using a triangular factorization $A = LU$, we solve the system $Ax = b$ in two steps:

1. compute y that solves $Ly = b$;

2. compute x that solves $Ux = y$.

Similarly, if we have $PA = LU$, then

$$Ax = b \iff (PA)x = Pb \iff LUx = b',$$

where $b' = Pb$. Then the system $Ax = b$ can be solved by first finding the solution y of $Ly = b$ and then computing the solution x of the system $Ux = y$.

3.2 Iterative Methods for Solving Systems of Linear Equations

Given a system $Ax = b$ of n linear equations with n unknowns, an iterative method starts with an initial guess $x^{(0)}$ and produces a sequence of vectors $x^{(1)}, x^{(2)}, \ldots$, such that

$$x^{(k)} = x^{(k-1)} + C(b - Ax^{(k-1)}), \quad k = 1, 2 \ldots, \tag{3.3}$$

where C is an appropriately selected matrix. We are interested in a choice of C, such that the above iteration produces a sequence $\{x^{(k)} : k \geq 0\}$ such that $x^{(k)} \to x^*, k \to \infty$, where x^* is the solution of the system, that is, $Ax^* = b$. We have

$$x^{(k)} - x^* = x^{(k-1)} + C(b - Ax^{(k-1)}) - x^* = x^{(k-1)} + Cb - CAx^{(k-1)} - x^*.$$

Replacing b with Ax^* in the last expression, we obtain

$$x^{(k)} - x^* = x^{(k-1)} + CAx^* - CAx^{(k-1)} - x^* = (I_n - CA)(x^{(k-1)} - x^*),$$

where I_n is the $n \times n$ identity matrix. Consider a matrix norm $\| \cdot \|_M$ and a vector norm $\| \cdot \|_V$ which are compatible, that is,

$$\|Ax\|_V \leq \|A\|_M \cdot \|x\|_V, \quad \forall A \in \mathbb{R}^{n \times n}, x \in \mathbb{R}^n.$$

Then the absolute error after k iterations of the scheme expressed in terms of the norm $\| \cdot \|_V$ satisfies the following inequality:

$$\|x^{(k)} - x^*\|_V = \|(I_n - CA)(x^{(k-1)} - x^*)\|_V \leq \|(I_n - CA)\|_M \cdot \|(x^{(k-1)} - x^*)\|_V.$$

Since

$$\|x^{(k-1)} - x^*\|_V \leq \|(I_n - CA)\|_M \cdot \|(x^{(k-2)} - x^*)\|_V,$$

we have

$$\|x^{(k)} - x^*\|_V \leq \|(I_n - CA)\|_M^2 \cdot \|(x^{(k-2)} - x^*)\|_V.$$

Continuing likewise and noting that the error bound is multiplied by the factor of $\|I_n - CA\|_M$ each time, we obtain the following upper bound on the error:

$$\|x^{(k)} - x^*\|_V \leq \|(I_n - CA)\|_M^k \cdot \|(x^{(0)} - x^*)\|_V.$$

Therefore, if $\|I_n - CA\|_M < 1$, then $\|x^{(k)} - x^*\|_V \to 0, k \to \infty$, and the method converges. A matrix C such that $\|I_n - CA\|_M < 1$ for some matrix norm is called an *approximate inverse* of A.

Example 3.9 *If we choose* $C = A^{-1}$, *then for any initial* $x^{(0)}$ *we have:*

$$\begin{aligned} x^{(1)} &= x^{(0)} + A^{-1}(b - Ax^{(0)}) \\ &= A^{-1}b, \end{aligned}$$

thus, we obtain the exact solution in one iteration.

3.2.1   Jacobi method

Note that $A = [a_{ij}]_{n \times n}$ can be easily represented as the sum of three matrices, $A = L + U + D$, where L, U, and D are a lower triangular, an upper triangular, and a diagonal matrix, respectively, such that L and U both have zero diagonals.

In the *Jacobi method*, we choose C in (3.3) to be

$$C = D^{-1} = \begin{bmatrix} \frac{1}{a_{11}} & 0 & \cdots & 0 \\ 0 & \frac{1}{a_{22}} & \cdots & 0 \\ \vdots & \vdots & \ddots & \vdots \\ 0 & 0 & \cdots & \frac{1}{a_{nn}} \end{bmatrix}.$$

Thus, an iteration of the Jacobi method is given by the following expression:

$$x^{(k+1)} = x^{(k)} + D^{-1}(b - Ax^{(k)}), \quad k = 0, 1, \ldots \tag{3.4}$$

Recall that when $\|I_n - D^{-1}A\|_M < 1$ for some matrix norm, the method is guaranteed to converge. We have

$$I_n - D^{-1}A = \begin{bmatrix} 0 & \frac{a_{12}}{a_{11}} & \cdots & \frac{a_{1n}}{a_{11}} \\ \frac{a_{21}}{a_{22}} & 0 & \cdots & \frac{a_{2n}}{a_{22}} \\ \vdots & \vdots & \ddots & \vdots \\ \frac{a_{n1}}{a_{nn}} & \frac{a_{n2}}{a_{nn}} & \cdots & 0 \end{bmatrix}.$$

A matrix $A \in \mathbb{R}^{n \times n}$ is called *strictly diagonally dominant by rows* if

$$|a_{ii}| > \sum_{\substack{j=1 \\ j \neq i}}^{n} |a_{ij}|, \quad i = 1, 2, \ldots, n.$$

Similarly, $A \in \mathbb{R}^{n \times n}$ is called *strictly diagonally dominant by columns* if

$$|a_{jj}| > \sum_{\substack{i=1 \\ i \neq j}}^{n} |a_{ij}|, \quad j = 1, 2, \ldots, n.$$

If A is strictly diagonally dominant by rows, then

$$\|I_n - D^{-1}A\|_\infty = \max_{1 \leq i \leq n} \sum_{\substack{j=1 \\ j \neq i}}^{n} \frac{|a_{ij}|}{|a_{ii}|} < 1,$$

and the Jacobi method converges.

Likewise, if A is strictly diagonally dominant by columns, then

$$\|I_n - D^{-1}A\|_1 = \max_{1 \leq i \leq n} \sum_{\substack{i=1 \\ i \neq j}}^{n} \frac{|a_{ij}|}{|a_{ii}|} < 1,$$

and the method converges in this case as well.

Example 3.10 *Consider a system $Ax = b$, where*

$$A = \begin{bmatrix} 2 & 1 \\ 3 & 5 \end{bmatrix}, \ b = \begin{bmatrix} 3 \\ 8 \end{bmatrix}.$$

The true solution is $x^* = [1, 1]^T$. We apply the Jacobi method three times with $x^{(0)} = [0, 0]^T$ to find an approximate solution of this system. We have:

$$D^{-1} = \begin{bmatrix} \frac{1}{2} & 0 \\ 0 & \frac{1}{5} \end{bmatrix}.$$

Hence, using the Jacobi iteration (3.4) we obtain:

$$x^{(1)} = \begin{bmatrix} 1/2 & 0 \\ 0 & 1/5 \end{bmatrix} \begin{bmatrix} 3 \\ 8 \end{bmatrix} = \begin{bmatrix} 3/2 \\ 8/5 \end{bmatrix};$$

$$x^{(2)} = \begin{bmatrix} 3/2 \\ 8/5 \end{bmatrix} + \begin{bmatrix} 1/2 & 0 \\ 0 & 1/5 \end{bmatrix} \left(\begin{bmatrix} 3 \\ 8 \end{bmatrix} - \begin{bmatrix} 23/5 \\ 25/2 \end{bmatrix} \right) = \begin{bmatrix} 7/10 \\ 7/10 \end{bmatrix};$$

$$x^{(3)} = \begin{bmatrix} 7/10 \\ 7/10 \end{bmatrix} + \begin{bmatrix} 1/2 & 0 \\ 0 & 1/5 \end{bmatrix} \begin{bmatrix} 9/10 \\ 12/5 \end{bmatrix} = \begin{bmatrix} 23/20 \\ 59/50 \end{bmatrix}.$$

Notice how the error decreases:

$$\|x^{(1)} - x^*\|_\infty = 0.60;$$
$$\|x^{(2)} - x^*\|_\infty = 0.30;$$
$$\|x^{(3)} - x^*\|_\infty = 0.18.$$

3.2.2   Gauss-Seidel method

Again, we use the same representation of A as in the Jacobi method, $A = L + U + D$. To derive an iteration of the Gauss-Seidel method, we can use $C = (L + D)^{-1}$ in the general scheme (3.3):

$$\begin{aligned} x^{(k+1)} &= x^{(k)} + (L + D)^{-1}(b - Ax^{(k)}) \Rightarrow \\ (L + D)x^{(k+1)} &= (L + D - A)x^{(k)} + b \Rightarrow \\ Dx^{(k+1)} &= -Lx^{(k+1)} - Ux^{(k)} + b. \end{aligned}$$

For example, for $n = 3$ we have:

$$\begin{bmatrix} a_{11} & 0 & 0 \\ 0 & a_{22} & 0 \\ 0 & 0 & a_{33} \end{bmatrix} x^{(k+1)} = - \begin{bmatrix} 0 & 0 & 0 \\ a_{21} & 0 & 0 \\ a_{31} & a_{32} & 0 \end{bmatrix} x^{(k+1)} - \begin{bmatrix} 0 & a_{12} & a_{13} \\ 0 & 0 & a_{23} \\ 0 & 0 & 0 \end{bmatrix} x^{(k)} + \begin{bmatrix} b \\ b \\ b \end{bmatrix}$$

The above is equivalent to

$$\begin{aligned} a_{11}x_1^{(k+1)} &= -0 - (a_{12}x_2^{(k)} + a_{13}x_3^{(k)}) + b_1 \\ a_{22}x_2^{(k+1)} &= -(a_{21}x_1^{(k+1)}) - (a_{23}x_3^{(k)}) + b_2 \\ a_{33}x_3^{(k+1)} &= -(a_{31}x_1^{(k+1)} + a_{32}x_2^{(k+1)}) - 0 + b_3. \end{aligned}$$

In general,

$$x_i^{(k+1)} = \frac{-\sum_{j<i} a_{ij} x_j^{(k+1)} - \sum_{j>i} a_{ij} x_j^{(k)} + b_i}{a_{ii}}, \quad i = 1, 2, \ldots, n. \quad (3.5)$$

Hence, computing $x_i^{(k+1)}$ requires the values of $x_1^{(k+1)}, x_2^{(k+1)}, \ldots, x_{i-1}^{(k+1)}$.

Example 3.11 *Given the system $Ax = b$ from Example 3.10, where*

$$A = \begin{bmatrix} 2 & 1 \\ 3 & 5 \end{bmatrix}, b = \begin{bmatrix} 3 \\ 8 \end{bmatrix},$$

we apply three iterations of the Gauss-Seidel method starting with $x^{(0)} = [0, 0]^T$.

We have

$$Dx^{(k+1)} = -Lx^{(k+1)} - Ux^{(k)} + b,$$

or, equivalently,

$$2x_1^{(k+1)} = -x_2^{(k)} + 3,$$
$$5x_2^{(k+1)} = -3x_1^{(k+1)} + 8.$$

For $k = 0$ we have:

$$2x_1^{(1)} = 3 \qquad \Rightarrow \quad x_1^{(1)} = 1.5,$$
$$5x_2^{(1)} = -9/2 + 8 \quad \Rightarrow \quad x_2^{(1)} = 0.7.$$

For $k = 1$ we obtain:

$$2x_1^{(2)} = -7/10 + 3 \quad \Rightarrow \quad x_1^{(2)} = 23/20 = 1.15,$$
$$5x_2^{(2)} = -69/20 + 8 \quad \Rightarrow \quad x_2^{(2)} = 91/100 = 0.91.$$

Finally, for $k = 2$:

$$2x_1^{(3)} = -91/100 + 3 \quad \Rightarrow \quad x_1^{(3)} = 209/200 = 1.045,$$
$$5x_2^{(3)} = -627/200 + 8 \quad \Rightarrow \quad x_2^{(3)} = 973/1000 = 0.973.$$

Recall that the true solution of the considered system is $x^* = [1, 1]^T$, therefore the errors for each step are

$$\|x^{(1)} - x^*\|_\infty = 0.500;$$
$$\|x^{(2)} - x^*\|_\infty = 0.150;$$
$$\|x^{(3)} - x^*\|_\infty = 0.045.$$

We can observe that for this system, the Gauss-Seidel method converges faster than the Jacobi method.

3.2.3   Application: input-output models in economics

The input-output analysis is a quantitative methodology used to study the interdependencies between the sectors of an economy. It was developed by Leontief [21], who was awarded the Nobel Prize in Economics in 1973 for this work.

Consider a system consisting of n industries, each producing a different product. By a unit of input or output of the ith product, we mean one dollar's worth of the product (that is, $1/w_i$, where w_i is the worth of the ith product). Assume that $a_{ij} \geq 0$ units of product i are required to produce one unit of product j, $i = 1, \ldots, n$. Then the cost of production of one unit of j will include $\sum_{i=1}^{n} a_{ij}$, therefore production of the jth product will be profitable only if

$$\sum_{i=1}^{n} a_{ij} < 1. \tag{3.6}$$

In the *open model* there is also the open sector demand for $d_i \geq 0$ units of product i, whereas in the *closed model* the open sector is ignored ($d = 0$). Denote by x_i the output of the ith product needed to meet the total demand. Then the output of x_j units of product j will require the input of $a_{ij}x_j$ units of product i. Since the volume of production of the ith product should be equal to the total demand for the product i, we have:

$$x_i = a_{i1}x_1 + a_{i2}x_2 + \ldots + a_{in}x_n + d_i \quad \text{for } i = 1, 2, \ldots, n.$$

Thus, we obtain the following system:

$$
\begin{array}{rcrcccrcl}
(1 - a_{11})x_1 & + & (-a_{12})x_2 & +\ldots+ & (-a_{1n})x_n & = & d_1 \\
(-a_{21})x_1 & + & (1 - a_{22})x_2 & +\ldots+ & (-a_{2n})x_n & = & d_2 \\
\vdots & & \vdots & & \vdots & & \vdots \\
(-a_{n1})x_1 & + & (-a_{n2})x_2 & +\ldots+ & (1 - a_{nn})x_n & = & d_n,
\end{array}
$$

or, in the matrix form,

$$(I_n - A)x = d. \tag{3.7}$$

We are interested in finding a nonnegative solution of the above system. We can use the Jacobi method to show that such a solution always exists.

From (3.6) we have

$$\sum_{\substack{i=1 \\ i \neq j}}^{n} a_{ij} < 1 - a_{jj},$$

and, taking into account that $a_{ij} > 0$ for $i, j = 1, 2, \ldots, n$, we conclude that $I_n - A$ is strictly diagonally dominant by columns. Hence, for any initial guess $x^{(0)}$, the Jacobi method applied to (3.7) will converge to the unique solution of this system. The kth iteration of the Jacobi method is

$$x_i^{(k+1)} = \frac{d_i + \sum\limits_{\substack{j=1 \\ j \neq i}}^{n} a_{ij} x_j^{(k)}}{(1 - a_{ii})} \quad \text{for } i = 1, 2, \ldots, n,$$

where $0 \leq a_{ij} < 1$, $d_i \geq 0$ for all $i, j = 1, 2, \ldots, n$. Therefore, if $x^{(k)} \geq 0$ then $x_i^{(k+1)} \geq 0$ for $i = 1, 2, \ldots, n$, so $x^{(k+1)} \geq 0$. Thus, choosing $x^{(0)} \geq 0$, the method will generate a sequence of nonnegative vectors converging to the unique solution, which is also nonnegative.

3.3   Overdetermined Systems and Least Squares Solution

Consider a system $Ax = b$, where A is an $m \times n$ matrix with entries $a_{ij}, i = 1, \ldots m, j = 1, \ldots, n$, $x = [x_1, \ldots, x_n]^T$ is the vector of unknowns, and $b = [b_1, \ldots, b_m]^T$ is the vector of right-hand sides. If $m > n$, that is, the number of equations is greater than the number of variables, the system $Ax = b$ is called *overdetermined*. An overdetermined system is typically not expected to have a solution x such that $Ax = b$, and one looks for an answer which is the "closest," in some sense, to satisfying the system. A common way of measuring how close a given vector $x \in \mathbb{R}^n$ is to satisfying the overdetermined system $Ax = b$ is to look at the norm $\|r\|_2 = \sqrt{r^T r}$ of the *residual vector* $r = b - Ax$.

Definition 3.1 (Least Squares Solution) *A vector x^* is called the least squares solution of the system $Ax = b$ if it minimizes $\|b - Ax\|_2$, that is, $\|b - Ax^*\|_2 \leq \|b - Ax\|_2$ for any $x \in \mathbb{R}^n$.*

The following theorem yields a method for finding a least squares solution of an overdetermined system of linear equations.

Theorem 3.2 *If x^* solves the system*

$$A^T A x = A^T b,$$

then x^* is a least squares solution of the system $Ax = b$.

Proof. We need to show that $\|b - Ax^*\|_2 \leq \|b - Ax\|_2$ for any $x \in \mathbb{R}^n$. For an arbitrary $x \in \mathbb{R}^n$, we have:

$$
\begin{aligned}
\|b - Ax\|_2^2 &= \|(b - Ax^*) + (Ax^* - Ax)\|_2^2 \\
&= \|b - Ax^*\|_2^2 + 2(x^* - x)^T \underbrace{(A^T b - A^T Ax^*)}_{=0} + \underbrace{\|A(x^* - x)\|^2}_{\geq 0} \\
&\geq \|b - Ax^*\|_2^2,
\end{aligned}
$$

which proves that x^* is a least squares solution of the system $Ax = b$. $\square$

Example 3.12 *Find the least squares solution of the system*

$$
\begin{aligned}
x_1 + x_2 &= 1 \\
x_1 - x_2 &= 3 \\
x_1 + 2x_2 &= 2.
\end{aligned}
$$

First, we show that the system has no exact solution. Solving for the first two equations, we find that $x_1 = 2$, $x_2 = -1$. Putting these values in the third equation we obtain $0 = 2$, implying that the system has no solution.

To find the least squares solution, we solve the system $A^T Ax = A^T b$, where

$$
A = \begin{bmatrix} 1 & 1 \\ 1 & -1 \\ 1 & 2 \end{bmatrix}, \quad b = \begin{bmatrix} 1 \\ 3 \\ 2 \end{bmatrix}.
$$

We have

$$
A^T Ax = A^T b \iff \begin{bmatrix} 3 & 2 \\ 2 & 6 \end{bmatrix} \begin{bmatrix} x_1 \\ x_2 \end{bmatrix} = \begin{bmatrix} 6 \\ 2 \end{bmatrix}.
$$

This system has a unique solution

$$
\begin{bmatrix} x_1^* \\ x_2^* \end{bmatrix} = \begin{bmatrix} 16/7 \\ -3/7 \end{bmatrix}.
$$

3.3.1 Application: linear regression

The following problem frequently arises in statistics. Given n points $[x_1, y_1]^T, \ldots, [x_n, y_n]^T \in \mathbb{R}^2$ on the plane, find the line $l(x) = ax + b$ (defined by the coefficients a and b) that minimizes the sum of squared residuals

$$
\sum_{i=1}^n (ax_i + b - y_i)^2.
$$

Such a line is called the *linear regression line* for the points $[x_1, y_1]^T, \ldots, [x_n, y_n]^T$.

Clearly, this problem is equivalent to the problem of finding a least squares solution of the following overdetermined linear system with n equations and two unknowns (a and b):

$$
\begin{aligned}
ax_1 + b &= y_1 \\
&\cdots \\
ax_n + b &= y_n.
\end{aligned}
$$

The solution can be obtained by solving the system $A^T A z = A^T y$, where $z = \begin{bmatrix} a \\ b \end{bmatrix}$, $y = [y_1, \ldots, y_n]^T$, and $A^T = \begin{bmatrix} x_1 & \cdots & x_n \\ 1 & \cdots & 1 \end{bmatrix}$. We have:

$$A^T A z = A^T y \iff \begin{aligned} \left(\sum_{i=1}^n x_i^2 \right) a &+ \left(\sum_{i=1}^n x_i \right) b &= \sum_{i=1}^n x_i y_i \\ \left(\sum_{i=1}^n x_i \right) a &+ nb &= \sum_{i=1}^n y_i. \end{aligned}$$

Expressing b via a from the second equation, we obtain:

$$b = \frac{\sum_{i=1}^n y_i - \left(\sum_{i=1}^n x_i \right) a}{n}.$$

Hence, from the first equation,

$$a = \frac{n \sum_{i=1}^n x_i y_i - \sum_{i=1}^n x_i \sum_{i=1}^n y_i}{n \sum_{i=1}^n x_i^2 - \left(\sum_{i=1}^n x_i \right)^2}.$$

3.4 Stability of a Problem

Consider the following pair of linear systems:

$$\begin{aligned} x_1 &- x_2 &= 1 \\ x_1 &- 1.001 x_2 &= 0 \end{aligned}$$

and

$$\begin{aligned} x_1 &- x_2 &= 1 \\ x_1 &- 0.999 x_2 &= 0. \end{aligned}$$

The solution to the first system is $[x_1, x_2] = [1001, 1000]$, whereas the solution to the second system is $[x_1, x_2] = [-999, -1000]$. A key observation here is that despite a high similarity between the coefficient matrices of the two systems, their solutions are very different. We call an instance of a problem *unstable* if a small variation in the values of input parameters may result in a significant change in the solution to the problem.

Given a nonsingular matrix $A \in \mathbb{R}^{n \times n}$ and a vector $b \in \mathbb{R}^n$, let $\bar{x} = A^{-1}b$ be the solution of the system $Ax = b$. Let $\tilde{A}$ be a nonsingular matrix obtained from A by perturbing its entries, and let $\tilde{x} = \tilde{A}^{-1}b$ be the solution of the system $\tilde{A}x = b$. Then $A(\tilde{x} - \bar{x}) = (A - \tilde{A})\tilde{x}$, hence, $\tilde{x} - \bar{x} = A^{-1}(A - \tilde{A})\tilde{x}$, and for any given compatible vector and matrix norms we have

$$\|\tilde{x} - \bar{x}\| \leq \|A^{-1}\| \cdot \|\tilde{A} - A\| \cdot \|\tilde{x}\|.$$

The last inequality can be rewritten as

$$\frac{\|\tilde{x} - \bar{x}\|}{\|\tilde{x}\|} \leq k(A)\frac{\|\tilde{A} - A\|}{\|A\|},$$

where

$$k(A) = \|A\| \cdot \|A^{-1}\|$$

is referred to as the *condition number* of A with respect to the given norm. When $k(A)$ is large then we say that A is *ill-conditioned*.

Note that

$$\|x\| = \|A^{-1}Ax\| \leq \|A^{-1}\| \cdot \|Ax\|,$$

implying

$$\|A^{-1}\| \geq \frac{\|x\|}{\|Ax\|}, \quad \forall x \neq 0.$$

Hence,

$$k(A) = \|A^{-1}\| \cdot \|A\| \geq \frac{\|x\|}{\|Ax\|}\|A\| \quad \forall x \neq 0. \tag{3.8}$$

This lower bound on $k(A)$ is easier to compute and can be used as a rough estimate of $k(A)$.

Example 3.13 *Consider the coefficient matrix of the unstable system discussed in the beginning of this section:*

$$A = \begin{bmatrix} 1 & -1 \\ 1 & -1.001 \end{bmatrix}.$$

Its inverse is given by

$$A^{-1} = \begin{bmatrix} 1001 & -1000 \\ 1000 & -1000 \end{bmatrix}.$$

The condition number of this matrix with respect to the infinity norm is

$$k(A) = \|A\|_\infty \|A^{-1}\|_\infty = 2.001 \cdot 2001 = 4004.001.$$

For $x = [1,1]^T$, the lower bound in (3.8) is given by

$$\frac{\|x\|_\infty}{\|Ax\|_\infty}\|A\|_\infty = 2001.$$

3.5   Computing Eigenvalues and Eigenvectors

In this section we briefly discuss the problem of computing the eigenvalues and eigenvectors of a square matrix. Recall that an eigenvalue λ_i and the corresponding eigenvector v_i of an $n \times n$ matrix A satisfy the system

$$(A - \lambda I_n)v_i = 0, \ v_i \neq 0. \tag{3.9}$$

The eigenvalues are the n roots of the characteristic polynomial $p(\lambda) = \det(A - \lambda I_n)$, which is of degree n. Given λ_i, the corresponding eigenvector v_i can be computed by solving the linear system (3.9). Also, if an eigenvector v of A is known, pre-multiplying $Av = \lambda v$ by v^T and dividing both sides by $v^T v \neq 0$, we obtain

$$\lambda = \frac{v^T A v}{v^T v}. \tag{3.10}$$

In theory, one could compute the coefficients of the characteristic polynomial and then find its roots using numerical methods (such as those discussed in Chapter 4). However, due to a high sensitivity of roots of a polynomial to perturbations in its coefficients, this would not be practical in the presence of round-off errors.

We will restrict our discussion of numerical methods for computing eigenvalues and eigenvectors to the *power method* described next.

3.5.1 The power method

Given a matrix $A \in \mathbb{R}^{n \times n}$ with eigenvalues $\lambda_1, \lambda_2, \ldots, \lambda_n$, by its *dominant eigenvalue* we mean an eigenvalue λ_i such that $|\lambda_i| > |\lambda_j|$ for all $j \neq i$ (if such an eigenvalue exists).

The power method discussed here is suitable for computation of the dominant eigenvalue and the corresponding eigenvector. Assume that $A \in \mathbb{R}^{n \times n}$ has n linearly independent eigenvectors $v_1, \ldots, v_n$ and a unique dominant eigenvalue, that is,

$$|\lambda_1| > |\lambda_2| \geq |\lambda_3| \geq \ldots \geq |\lambda_n|.$$

Starting with an arbitrary initial guess $x^{(0)}$ of the dominant eigenvector, we construct a sequence of vectors $\{x^{(k)} : k \geq 0\}$ determined by

$$x^{(k)} = A x^{(k-1)} = A^k x^{(0)}, \ k \geq 1.$$

Let $x^{(0)} = \sum_{i=1}^{n} c_i v_i$ be the representation of the initial guess as a linear combination of the eigenvectors, then

$$x^{(k)} = A^k \sum_{i=1}^{n} c_i v_i = \sum_{i=1}^{n} \lambda_i^k c_i v_i = \lambda_1^k \left(c_1 v_1 + \sum_{i=2}^{n} \left(\frac{\lambda_i}{\lambda_1} \right)^k c_i v_i \right), \ k \geq 1.$$

Since $|\lambda_i/\lambda_1| < 1$, the direction of $x^{(k)}$ tends to that of v_1 as $k \to \infty$ (assuming $c_1 \neq 0$). Also, for

$$\mu^{(k)} = \frac{(x^{(k)})^T A x^{(k)}}{(x^{(k)})^T x^{(k)}},$$

we have $\mu^{(k)} \to \lambda_1, k \to \infty$ according to (3.10).

To ensure the convergence of $x^{(k)}$ to a nonzero vector of bounded length, we scale $x^{(k)}$ at each iteration:

$$v^{(k)} = \frac{x^{(k)}}{\|x^{(k)}\|_2}, \quad x^{(k+1)} = Av^{(k)}, \ k \geq 0. \tag{3.11}$$

Also,

$$\mu^{(k)} = \frac{(v^{(k)})^T A v^{(k)}}{(v^{(k)})^T v^{(k)}} = (v^{(k)})^T x^{(k+1)}. \tag{3.12}$$

In summary, starting with an initial guess $x^{(0)}$, we proceed by computing $\{v^{(k)} : k \geq 1\}$ and $\{\mu^{(k)} : k \geq 1\}$ using (3.11)–(3.12) to find approximations of v_1 and λ_1, respectively.

3.5.2 Application: ranking methods

A notable application of the power method for computing the dominant eigenpair of a matrix can be found in ranking methods. We discuss two examples, the celebrated PageRank algorithm and the Analytic Hierarchy Process.

PageRank algorithm. The PageRank method was originally developed by Brin and Page [7] for the purpose of ranking webpages in Google's search engine. Its effectiveness and simplicity made the method a popular choice in many other applications. PageRank scores each webpage by summing up the scores of webpages that point to the page being scored via a hyperlink. This ranking system can be interpreted as a democracy where hyperlinks can be thought of as votes in favor of the linked webpages [8]. By linking to multiple webpages, a website's "voting score" is split evenly among the set of webpages to which it links.

Assume that there are a total of n webpages, and denote by w_i the importance score (same as the "voting score") of the ith webpage, $i = 1, \ldots, n$. Let $N^+(i)$ denote the set of webpages that have a hyperlink to the ith webpage, and let n_i be the number of pages to which the ith page points. Then we have

$$w_i = \sum_{j \in N^+(i)} \frac{1}{n_j} w_j, \quad i = 1, \ldots, n. \tag{3.13}$$

This system can be written as

$$w = Aw, \tag{3.14}$$

where $A = [a_{ij}]_{n \times n}$ is given by

$$a_{ij} = \begin{cases} 1/n_j, & \text{if } j \in N^+(i); \\ 0, & \text{otherwise.} \end{cases} \tag{3.15}$$

Note that A is a column-stochastic matrix (that is, all its elements are nonnegative and each column sums up to 1), and as such is guaranteed to have an eigenvalue $\lambda_1 = 1$. Thus, the problem of computing the importance scores reduces to finding the eigenvector of A corresponding to the eigenvalue $\lambda_1 = 1$. Then the webpages are ranked according to a nonincreasing order of their importance scores given by the corresponding components of the eigenvector.

Analytic Hierarchy Process. Another example of using eigenvalue computations for ranking alternatives is the *Analytic Hierarchy Process (AHP)*. AHP was first introduced by Saaty in 1980 [30], and has since been developed into a powerful decision-making tool.

Assume that there are n elements (alternatives, options) to be ranked. At some stage of the AHP, a matrix P is created, which is called the *preference matrix*, and whose elements are

$$p_{ij} = \frac{w_i}{w_j}, \ i, j = 1, \ldots, n.$$

Here numbers w_i and w_j are used to compare the alternatives i and j. To compare two options, a 10-point scale is often used, in which w_i, $i = 1, \ldots, n$ is assigned values from $\{0, 1, 2, \ldots, 9\}$ as follows. If $i = j$, or i and j are equal alternatives, then $w_i = w_j = 1$. Otherwise,

$$w_i = \begin{cases} 3 & \text{moderately} \\ 5 & \text{strongly} \\ 7 & \text{very strongly} \\ 9 & \text{extremely} \end{cases} \text{if } i \text{ is} \quad \text{preferable over } j.$$

The numbers $2, 4, 6, 8$ are used for levels of preference between two of the specified above. In all these cases, w_j is set equal to 1. For example, if element i is strongly preferable over element j, we have $p_{ij} = 5$ and $p_{ji} = 1/5$. Zeros are used when there is not enough information to compare two elements, in which case the diagonal element in each row is increased by the number of zeros in that row.

As soon as the preference matrix is constructed, one of the techniques used to rank the elements is the following *eigenvalue method*. Denote by $w = [w_i]_{i=1}^n$ the eigenvector of P corresponding to its largest eigenvalue. Then element i is assigned the value w_i, and the elements are ranked according to the nonincreasing order of the components of vector w. This method is based on the observation that if the numbers $w_i, i = 1, \ldots, n$ used to construct the preference matrix P were the true preference scores of the corresponding alternatives, then for the vector $w = [w_i]_{i=1}^n$ we would have $Pw = nw$, that is, w is the eigenvector of P corresponding to its largest eigenvalue $\lambda = n$.

Exercises

3.1. Solve the following systems of linear algebraic equations using

82   *Numerical Methods and Optimization: An Introduction*

(a) the Gaussian elimination method;

(b) the Gauss-Jordan method.

(i)
$$\begin{aligned}
x_1 + x_2 + 2x_3 &= 1 \\
2x_1 + x_2 - 3x_3 &= 0 \\
-3x_1 - x_2 + x_3 &= 1
\end{aligned}$$

(ii)
$$\begin{aligned}
x_1 + \tfrac{1}{2}x_2 + \tfrac{1}{3}x_3 &= 1 \\
\tfrac{1}{2}x_1 + \tfrac{1}{3}x_2 + \tfrac{1}{4}x_3 &= 0 \\
\tfrac{1}{3}x_1 + \tfrac{1}{4}x_2 + \tfrac{1}{5}x_3 &= 0
\end{aligned}$$

(iii)
$$\begin{aligned}
2x_1 + 4x_2 - 4x_3 &= 12 \\
x_1 + 5x_2 - 5x_3 - 3x_4 &= 18 \\
2x_1 + 3x_2 + x_3 + 3x_4 &= 8 \\
x_1 + 4x_2 - 2x_3 + 2x_4 &= 8.
\end{aligned}$$

3.2. Consider the system $Ax = b$, where

(a)
$$A = \begin{bmatrix} 2 & 0 & 1 \\ 1 & 3 & -1 \\ -1 & 2 & 4 \end{bmatrix}, \quad b = \begin{bmatrix} 0 \\ -1 \\ 0 \end{bmatrix};$$

(b)
$$A = \begin{bmatrix} 1 & 0 & \tfrac{1}{2} \\ 1 & 2 & 0 \\ 0 & 1 & 3 \end{bmatrix}, \quad b = \begin{bmatrix} 1 \\ 2 \\ 3 \end{bmatrix}.$$

Find A^{-1}. Then solve the system by computing $x = A^{-1}b$.

3.3. Use Gaussian elimination to factorize the matrix

(a)
$$A = \begin{bmatrix} 1 & 2 & 1 \\ 3 & 4 & 2 \\ 2 & 5 & 1 \end{bmatrix};$$
(b)
$$A = \begin{bmatrix} 1 & 4 & 2 \\ 4 & 4 & 1 \\ 2 & 6 & 1 \end{bmatrix}.$$

That is, compute a lower triangular matrix L and an upper triangular matrix U such that $PA = LU$, where P is the permutation matrix corresponding to the final permutation p used for making the pivots.

3.4. How many multiplication and division operations are required to solve a linear system using

(a) Gaussian elimination;

(b) Gauss-Jordan;

(c) triangular factorization

methods? What procedure would you follow to solve k systems of linear equations $Ax = b_i, i = 1, \ldots, k$?

3.5. Let

$$A = \begin{bmatrix} 3 & 2 \\ 1 & 2 \end{bmatrix}, \quad b = \begin{bmatrix} 1 \\ 0 \end{bmatrix}.$$

(a) Verify that A is strictly diagonally dominant by rows.

(b) Verify that $\|I_2 - D^{-1}A\|_\infty < 1$, where the matrices involved are defined as in the description of the Jacobi method.

(c) Apply the Jacobi method twice to get an approximate solution $x^{(2)}$ to the system $Ax = b$. Use $x^{(0)} = [1, -1]^T$ as the initial guess.

(d) Will the method converge? Explain.

3.6. For the systems in Exercise 3.2, use $x^{(0)} = [-1, 1, 2]^T$ as the initial guess and do the following:

(a) Compute $\|x^{(0)} - x^*\|_\infty$, where x^* is the solution found in Exercise 3.2.

(b) Apply the Jacobi method twice and compute $\|x^{(k)} - x^*\|_\infty, k = 1, 2$.

(c) Apply the Gauss-Seidel method twice and compute $\|x^{(k)} - x^*\|_\infty, k = 1, 2$.

(d) How many steps k of the Jacobi and Gauss-Seidel methods will guarantee that $\|x^{(k)} - x^*\|_\infty \leq 10^{-8}$?

3.7. Let

$$A = \begin{bmatrix} 3 & -1 & 1 \\ 0 & 2 & 1 \\ -1 & 1 & 4 \end{bmatrix}; \quad b = \begin{bmatrix} -1 \\ 0 \\ 1 \end{bmatrix}.$$

(a) Use Gaussian elimination to find the solution x^* of the system $Ax = b$.

(b) Compute $det(A)$ and A^{-1} above. Then find $x = A^{-1}b$.

(c) Use the Jacobi method twice, starting with $x^{(0)} = [1, 1, 0]^T$, to find an approximation to the solution. Report the error $\|x^{(k)} - x^*\|_\infty, k = 1, 2$.

(d) Use the Gauss-Seidel method twice, starting with $x^{(0)} = [1, 1, 0]^T$, to find an approximation to the solution. Compare the corresponding errors to those obtained in (c) for the Jacobi method.

(e) After how many steps will the Jacobi method find an approximation to the solution with the error not exceeding 10^{-8}?

(f) After how many steps will the Gauss-Seidel method find an approximation to the solution with the error not exceeding 10^{-8}?

3.8. Given

(i)

$$A = \begin{bmatrix} -3 & 1 & 1 \\ 1 & 2 & 0 \\ 1 & -1 & 3 \end{bmatrix}, \quad b = \begin{bmatrix} 1 \\ -1 \\ 0 \end{bmatrix}, \quad \text{and} \quad x^{(0)} = \begin{bmatrix} 1 \\ 1 \\ 1 \end{bmatrix};$$

(ii)

$$A = \begin{bmatrix} 2 & 0 & 1 \\ -1 & 3 & 1 \\ 0 & 1 & -2 \end{bmatrix}, \quad b = \begin{bmatrix} -1 \\ 0 \\ 1 \end{bmatrix}, \quad \text{and} \quad x^{(0)} = \begin{bmatrix} 1 \\ 2 \\ -1 \end{bmatrix};$$

(iii)

$$A = \begin{bmatrix} 3 & 1 & 1 \\ -1 & 2 & 0 \\ 2 & -1 & -4 \end{bmatrix}, \quad b = \begin{bmatrix} -1 \\ 0 \\ 2 \end{bmatrix}, \quad \text{and} \quad x^{(0)} = \begin{bmatrix} 1 \\ 2 \\ 3 \end{bmatrix};$$

(iv)

$$A = \begin{bmatrix} 4 & 3 & -2 \\ -2 & 5 & 0 \\ 1 & -1 & -4 \end{bmatrix}, \quad b = \begin{bmatrix} 1 \\ 2 \\ 0 \end{bmatrix}, \quad \text{and} \quad x^{(0)} = \begin{bmatrix} 1 \\ 0 \\ -1 \end{bmatrix};$$

(v)

$$A = \begin{bmatrix} 5 & 2 & -1 & 0 \\ 1 & 7 & 2 & -3 \\ 0 & 3 & -6 & 1 \\ -2 & 4 & -2 & 9 \end{bmatrix}, \quad b = \begin{bmatrix} 4 \\ 0 \\ -5 \\ 5 \end{bmatrix}, \quad \text{and} \quad x^{(0)} = \begin{bmatrix} 0 \\ 0 \\ 0 \\ 0 \end{bmatrix};$$

find an approximate solution of $Ax = b$ using two steps of

(a) the Jacobi method;

(b) the Gauss-Seidel method.

Use $x^{(0)}$ as an initial guess. What can you say about the convergence of the Jacobi and Gauss-Seidel methods in each case?

3.9. For each system below, do the following:

(a) Prove that the system has *no* solution.

(b) Find the least squares solution, x^*.

(c) Compute $\|Ax^* - b\|_2$, where A is the matrix of coefficients and b is the right-hand-side vector.

$$\begin{array}{llll}
\text{(i)} & 2x_1 & + & x_2 & = & 1 \\
& x_1 & - & x_2 & = & 2 \\
& x_1 & + & 2x_2 & = & 1
\end{array}$$

$$\begin{array}{llll}
\text{(ii)} & x_1 & + & x_2 & = & 1 \\
& 2x_1 & + & 3x_2 & = & 0 \\
& x_1 & + & 2x_2 & = & 2
\end{array}$$

$$\begin{array}{llll}
\text{(iii)} & -x_1 & + & x_2 & = & 1 \\
& x_1 & + & 2x_2 & = & 0 \\
& 2x_1 & + & x_2 & = & 1
\end{array}$$

$$\begin{array}{llll}
\text{(iv)} & -3x_1 & + & x_2 & = & -2 \\
& 4x_1 & + & 7x_2 & = & 11 \\
& x_1 & - & x_2 & = & 3
\end{array}$$

$$\begin{array}{llll}
\text{(v)} & 2x_1 & - & x_2 & + & x_3 & = & 1 \\
& -x_1 & + & x_2 & - & x_3 & = & 2 \\
& x_1 & + & x_2 & + & 2x_3 & = & -1 \\
& 2x_1 & + & x_2 & + & 2x_3 & = & 3.
\end{array}$$

3.10. Consider the following two matrices:

$$A = \begin{bmatrix} 101 & -90 \\ 110 & -98 \end{bmatrix}, \quad B = \begin{bmatrix} 100.999 & -90.001 \\ 110 & -98 \end{bmatrix}.$$

Find the eigenvalues of A and B. What do you observe?

3.11. Given the matrix $A = \begin{bmatrix} 4 & 0 & 1 \\ -2 & 1 & 0 \\ -2 & 0 & 1 \end{bmatrix}$,

(a) Find the eigenvalues and corresponding eigenvectors of A.
(b) Use your answer from (a) to compute $\det(A)$.
(c) What are the eigenvalues of A^{-1}?

3.12. Let a polynomial $p(x) = c_0 + c_1 x + c_2 x^2 + \ldots + c_{n-1} x^{n-1} + x^n$ be given. Prove that $p(x)$ is the characteristic polynomial of matrix $A = [a_{ij}]_{n \times n}$ defined as follows:

$$a_{i,j} = \begin{cases} 1, & \text{if } j = i+1; \\ -c_{j-1} & \text{if } i = n, j = 1, 2, \ldots, n; \\ 0 & \text{otherwise.} \end{cases}$$

i.e.,

$$A = \begin{bmatrix} 0 & 1 & 0 & \cdots & 0 \\ 0 & 0 & 1 & \cdots & 0 \\ \vdots & \vdots & \vdots & \ddots & \vdots \\ 0 & 0 & 0 & \cdots & 1 \\ -c_0 & -c_1 & -c_2 & \cdots & -c_{n-1} \end{bmatrix}.$$

This matrix is called the *companion* matrix of $p(x)$.

3.13. Given the polynomial $p(x) = (x-1)(x+2)(x-3)$, find a *non-diagonal* 3×3 matrix M whose eigenvalues are the roots of $p(x)$.

3.14. Find two different matrices A and B whose eigenvalues are the roots of the polynomial

$$P(\lambda) = \lambda^3 - 3\lambda^2 - \lambda + 3.$$

Chapter 4

Solving Equations

In this chapter, we discuss several numerical methods for solving a nonlinear equation in the form

$$F(x) = 0, \qquad (4.1)$$

where $F(x)$ is a given real function.

> **Definition 4.1** *A number x^* such that $F(x^*) = 0$ is called a* root *of the function $F(x)$ or of the equation $F(x) = 0$.*

The problem of finding a root of a function (equation) arises frequently in science, engineering and everyday life.

Example 4.1 *John saves money by making regular monthly deposits in the amount p to his account. Assume that the annual interest rate is r. Then the total amount, A, that John will have after n payments is*

$$A = p + p\left(1 + \frac{r}{12}\right) + p\left(1 + \frac{r}{12}\right)^2 + \ldots + p\left(1 + \frac{r}{12}\right)^{n-1}. \qquad (4.2)$$

In this expression, the term $p\left(1 + \frac{r}{12}\right)^{i-1}$ is the contribution of the $(n - i)$-th deposit toward the total sum, that is, the contribution of the last, month n, payment is p, the contribution of the payment made the month before is $p\left(1 + \frac{r}{12}\right)$ (the deposit plus the interest for one month), etc.; and the contribution of the first payment is $p\left(1 + \frac{r}{12}\right)^{n-1}$ (the deposit plus the interest for $n - 1$ months). The right-hand side of Equation (4.2) is the sum of n terms of a geometric series, hence

$$A = \sum_{i=1}^{n} p\left(1 + \frac{r}{12}\right)^{i-1} = p\frac{\left(1 + \frac{r}{12}\right)^n - 1}{\left(1 + \frac{r}{12}\right) - 1},$$

which simplifies to the equation

$$A = \frac{p}{r/12}\left(\left(1 + \frac{r}{12}\right)^n - 1\right), \qquad (4.3)$$

called the annuity-due equation. Suppose that John has n months left until his retirement, and the target amount of money he wants to have in his account *upon retirement is A. Then in order to find an interest rate that would be high enough to achieve his target, John needs to solve the following equation with respect to r:*

$$F(r) = \frac{p}{r/12}\left(\left(1 + \frac{r}{12}\right)^n - 1\right) - A = 0. \qquad (4.4)$$

There are only a few types of equations that can be solved by simple analytical methods. These include linear and quadratic equations

$$ax + b = 0,$$

$$ax^2 + bx + c = 0.$$

In general, the problem of finding a solution of a nonlinear equation is not easy. For example, for a polynomial of degree n

$$p(x) = a_n x^n + a_{n-1} x^{n-1} + a_{n-2} x^{n-2} + \ldots + a_1 x + a_0,$$

where $n > 4$ there are no general formulas to solve the equation $p(x) = 0$. Moreover, in 1823 a Norwegian mathematician Niels Henrik Abel (1802–1829) had shown that no such formulas can be developed for general polynomials of degree higher than four.

When direct solution methods are not available, numerical (iterative) techniques are used, which start with an initial solution estimate x_0 and proceed by recursively computing improved estimates $x_1, x_2, \ldots, x_n$ until a certain stopping criterion is satisfied.

Numerical methods typically give only an approximation to the exact solution. However, this approximation can be of a very good (predefined) accuracy, depending on the amount of computational effort one is willing to invest. One of the advantages of numerical methods is their simplicity: they can be concisely expressed in algorithmic form and can be easily implemented on a computer. The main drawback is that in some cases numerical methods fail. We will illustrate this point when discussing specific methods.

4.1   Fixed Point Method

In this section, we discuss the *fixed point method*, which is also known as *simple one-point iteration* or *method of successive substitutions*.

We consider an equation in the form

$$F(x) = 0, \qquad (4.5)$$

where $F : [a, b] \to \mathbb{R}$, and suppose that the interval (a, b) is known to contain a root of this equation. We can always reduce (4.5) to an equation in the form

$$x = f(x), \qquad (4.6)$$

such that (4.6) is equivalent to (4.5) at the interval (a, b), i.e., such that $x^* \in (a, b)$ solves (4.5) if and only if x^* solves (4.6).

Example 4.2 *Put $f(x) = x - \alpha(x)F(x)$, where $\alpha(x) \neq 0$ if $x \in (a, b)$. Then, obviously, $f(x) = x \Leftrightarrow \alpha(x)F(x) = 0 \Leftrightarrow F(x) = 0$.*

To find a root $x^* \in (a, b)$ of Equation (4.6), we take some starting point $x_0 \in (a, b)$ and sequentially calculate $x_1, x_2, \ldots$ using the formula

$$x_{k+1} = f(x_k), k = 0, 1, \ldots. \tag{4.7}$$

In the above, we assume that $f(x_k) \in (a, b)$ for all k, so x_k is always in the domain of F.

Definition 4.2 *A* fixed point *of a function f is a real number x^* such that $f(x^*) = x^*$.*

Clearly, if x^* is a fixed point of f, then $F(x^*) = 0$, and thus x^* is a root of F.

Definition 4.3 *The iteration $x_{k+1} = f(x_k)$ is called a* fixed point iteration.

Next we show that if the fixed point method converges, it converges to a fixed point of f, which is also a root of F.

Theorem 4.1 *Let f be a continuous function, and let the sequence $\{x_k : k \geq 0\}$ be generated by the fixed point iteration, $x_{k+1} = f(x_k)$. If $\lim_{k \to \infty} x_k = x^*$, then x^* is a fixed point of f and hence x^* solves $F(x) = 0$.*

Proof. Since f is continuous, we have

$$\lim_{k \to \infty} f(x_k) = f(\lim_{k \to \infty} x_k) = f(x^*),$$

and since $x_{k+1} = f(x_k)$,

$$x^* = \lim_{k \to \infty} x_k = \lim_{k \to \infty} f(x_{k-1}) = f(x^*). \qquad \square$$

Example 4.3 *For $f(x) = \exp(-x)$ and $x_0 = -1$, four fixed point iterations produce the following points:*

$$x_1 = f(x_0) \approx 2.718282,$$
$$x_2 = f(x_1) \approx 0.065988,$$
$$x_3 = f(x_2) \approx 0.936142,$$
$$x_4 = f(x_3) \approx 0.392138.$$

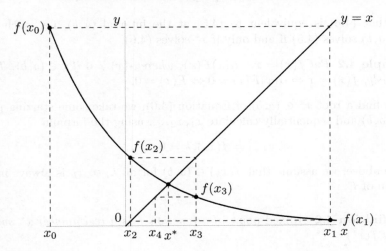

FIGURE 4.1: Four fixed point iterations applied to $f(x) = \exp(-x)$ starting with $x_0 = -1$.

The iterations are illustrated in Figure 4.1. Observe that applying the fixed point iteration recursively will produce a sequence converging to the fixed point $x^* \approx 0.567143$.

The following example shows that a repeated application of the fixed point iteration may not produce a convergent sequence.

Example 4.4 *For $f(x) = 1/x$ and $x_0 = 4$, we have*

$$x_1 = f(x_0) = 1/4,$$
$$x_2 = f(x_1) = 4 = x_0,$$

hence, if we continue, we will cycle between the same two points, x_0 and x_1. See Figure 4.2 for an illustration.

Theorem 4.2 *Assume that $f : (a, b) \to \mathbb{R}$ is differentiable on (a, b) and there exists a constant q, such that $0 \leq q < 1$ and $|f'(x)| \leq q$ for any $x \in (a, b)$. If there exists a solution $x^* \in (a, b)$ of the fixed point equation $f(x) = x$, then this solution is unique on (a, b). Moreover, there exists $\epsilon > 0$, such that for any x_0 satisfying $|\hat{x} - x_0| < \epsilon$, the sequence defined by the fixed point iteration, $x_{k+1} = f(x_k)$, satisfies the following inequality:*

$$|x_k - x^*| \leq q^k |x_0 - x^*|, k \geq 0. \tag{4.8}$$

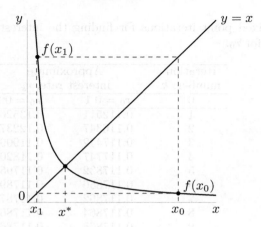

FIGURE 4.2: Fixed point iterations applied to $f(x) = 1/x$ starting with $x_0 = 4$ lead to a cycle $x_0, x_1, x_0, x_1, \ldots$, where $x_1 = 1/4$.

Proof. Let $x^* \in (a, b)$ be an arbitrary fixed point of f, and let $x_k \in (a, b)$. According to the mean value theorem, there exists c_k between x_k and x^* in (a, b) such that $f(x_k) - f(x^*) = f'(c_k)(x_k - x^*)$, therefore

$$|x_k - x^*| = |f(x_{k-1}) - f(x^*)| = |f'(c_{k-1})(x_{k-1} - x^*)| \le q|x_{k-1} - x^*|.$$

Hence,

$$|x_k - x^*| \le q|x_{k-1} - x^*| \le q^2|x_{k-2} - x^*| \le \ldots \le q^k|x_0 - x^*| \to 0, k \to \infty,$$

and $\lim_{k \to \infty} x_k = x^*$. Since x^* is an arbitrary fixed point of f on (a, b) and $x_k \to x^*, k \to \infty$, this implies that for any other fixed point $\hat{x} \in (a, b)$ of f we must have $x_k \to \hat{x}, k \to \infty$, meaning that $\hat{x} = x^*$. $\qquad \square$

Example 4.5 *Assume that John from Example 4.1 can make the monthly payment of \$300, and we want to find an interest rate r which would yield \$1,000,000 in 30 years. We will use the annuity-due equation given in (4.4):*

$$\frac{p}{r/12} \left(\left(1 + \frac{r}{12}\right)^n - 1 \right) - A = 0. \tag{4.9}$$

Rearrange the annuity-due equation in the following way:

$$\left(1 + \frac{r}{12}\right)^n - 1 - \frac{Ar}{12p} = 0, \tag{4.10}$$

so,

$$r = 12 \left(\frac{Ar}{12p} + 1 \right)^{1/n} - 12.$$

TABLE 4.1: Fixed point iterations for finding the interest rate with two different choices for r_0.

Iteration number, k	Approximate interest rate, r_k	
0	$r_0 = 0.1$	$r_0 = 0.2$
1	0.112511	0.135264
2	0.116347	0.122372
3	0.117442	0.119091
4	0.117747	0.118203
5	0.117832	0.117958
6	0.117856	0.117891
7	0.117862	0.117872
8	0.117864	0.117867
9	0.117865	0.117865
10	0.117865	0.117865

We have $p = 300$, $A = 1,000,000$, and $n = 12 \cdot 30 = 360$, that is,

$$r = 12 \left(\frac{10^6 r}{3600} + 1 \right)^{1/360} - 12 = f(r).$$

As a plausible initial approximation, we may take the value of $r_0 = 0.1$. Then the first approximation becomes

$$r_1 = f(r_0) = f(0.1) = 0.112511.$$

Likewise, the second approximation is

$$r_2 = f(r_1) = f(0.116347) = 0.116347.$$

The results of 10 fixed point iterations, as well as 10 iterations with another initial guess $r_0 = 0.2$ are shown in Table 4.1, which illustrates the convergence of the method. We conclude that the target interest rate is $\approx 11.79\%$

4.2 Bracketing Methods

The bracketing methods are based on the observation that if $f : [a, b] \to \mathbb{R}$ is continuous and if $f(a)f(b) \leq 0$, then there exists $x^* \in [a, b]$ such that $f(x^*) = 0$, which follows from the intermediate value theorem. To find a root x^*, we start with the interval $[a_0, b_0] = [a, b]$ and identify a smaller subinterval $[a_1, b_1]$ of $[a_0, b_0]$ containing a root (i.e., such that $f(a_1)f(b_1) \leq 0$). Then the search is continued on $[a_1, b_1]$. This procedure is carried out recursively, until the search interval is reduced to a small enough size.

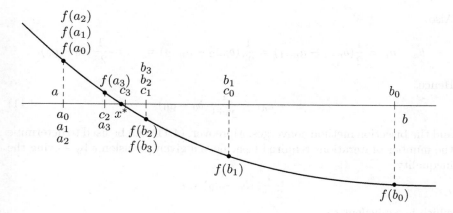

FIGURE 4.3: An illustration of the bisection method.

4.2.1 Bisection method

The idea of the bisection method, which is also called the *half-interval method*, is to split the search space at each iteration in half by choosing the middle point of the interval as the point of separation. One can determine which of the two halves contains the root using the intermediate value theorem as described above. Namely, we choose the half where the function has opposite-sign values at the endpoints. The algorithm terminates when the search interval is smaller than a given precision ϵ.

The method is illustrated in Figure 4.3 and proceeds as follows. We start with an interval $[a, b]$ such that $f(a)f(b) < 0$, and set $a_0 = a$, $b_0 = b$. Let $c_0 = (a_0 + b_0)/2$ be the mid-point of the interval $[a_0, b_0]$. Since $f(a_0) > 0$, $f(b_0) < 0$, and $f(c_0) < 0$, out of the two subintervals, $[a_0, c_0]$ and $[c_0, b_0]$, the first one is guaranteed to contain a root of f. Hence, we set $a_1 = a_0$, $b_1 = c_0$, and continue the search with the interval $[a_1, b_1]$. For the mid-point c_1 of this interval we have $f(c_1) < 0$, therefore there must be a root of f between a_1 and c_1, and we continue with $[a_2, b_2] = [a_1, c_1]$. After one more step, we obtain $[a_3, b_3] = [c_2, b_2]$ as the new search interval. If $b_3 - a_3 < \epsilon$, then $c_3 = (a_3 + b_3)/2$ is output as an approximation of the root x^*.

4.2.1.1 Convergence of the bisection method

The bisection algorithm produces a set of *centers* $c_1, \ldots, c_k, \ldots$ such that $\lim_{k \to \infty} c_k = x^*$, where $x^* \in [a_0, b_0]$ is a root of f. Since at iteration n of the bisection method the root x^* is in the interval $[a_n, b_n]$ and c_n is the mid-point of this interval, we have

$$|x^* - c_n| \leq \frac{1}{2}(b_n - a_n).$$

Also,

$$b_n - a_n = \frac{1}{2}(b_{n-1} - a_{n-1}) = \frac{1}{2^2}(b_{n-2} - a_{n-2}) = \ldots = \frac{1}{2^n}(b_0 - a_0).$$

Hence,

$$|x^* - c_n| \leq \frac{1}{2^{n+1}}(b_0 - a_0), \tag{4.11}$$

and the bisection method converges. Moreover, (4.11) can be used to determine the number of iterations required to achieve a given precision ϵ by solving the inequality

$$\frac{1}{2^{n+1}}(b_0 - a_0) < \epsilon,$$

which is equivalent to

$$n > \log_2\left(\frac{b_0 - a_0}{\epsilon}\right) - 1 = \frac{\ln\left(\frac{b_0 - a_0}{\epsilon}\right)}{\ln 2} - 1. \tag{4.12}$$

For example, if $[a_0, b_0] = [0, 1]$ and $\epsilon = 10^{-5}$, we have

$$n > \frac{5\ln(10)}{\ln 2} - 1 \approx 15.6,$$

and the number of iterations guaranteeing the precision ϵ is 16.

A pseudo-code for the bisection method is presented in Algorithm 4.1. We stop if $f(c_{k-1}) = 0$ at some iteration (line 6 of the algorithm), meaning that the algorithm found the exact root of f, or when the search interval is small enough to guarantee that $|c_n - x^*| < \epsilon$ (line 14). We could also add the requirement that $|f(\bar{x})| < \delta$ to the stopping criterion to make sure that not only our approximate solution is near the real root, but also that the corresponding function value is very close to 0.

Example 4.6 *We use 4 iterations of Algorithm 4.1 to approximate a root $x^* \in [0, 1]$ of $f(x) = x^3 + x - 1$. First we need to make sure that there is a root in the interval $[0, 1]$. We have*

$$f(0) = -1, \ f(1) = 1, \ f(0)f(1) < 0,$$

therefore we can say that there exists at least one root within $[0, 1]$. In fact, we can show that this interval contains exactly one root of f. Indeed, suppose that there are at least two roots $\hat{x}, x^* \in [0, 1]$, such that $f(\hat{x}) = f(x^*) = 0$. Then by the mean value theorem there exists γ between $\hat{x}$ and x^* such that $f'(\gamma) = 0$. But $f'(x) = 3x^2 + 1 \geq 1$ for all x in $[0, 1]$, leading to a contradiction. Therefore, the assumption that f has more than one root in $[0, 1]$ is not correct.

Using Algorithm 4.1 we obtain:

$$a_0 = 0, b_0 = 1, c_0 = (0 + 1)/2 = 0.5, f(c_0) = (0.5)^3 + 0.5 - 1 = -0.375 < 0,$$

Algorithm 4.1 Bisection method for solving $f(x) = 0$.

Input: f, ϵ, a, b such that $f(a)f(b) < 0$
Output: $\bar{x}$ such that $|\bar{x} - x^*| < \epsilon$, where $f(x^*) = 0$

1: $a_0 = a, b_0 = b, c_0 = (a_0 + b_0)/2$
2: $n = \left\lceil \ln\left(\frac{b_0 - a_0}{\epsilon}\right)/\ln 2 \right\rceil$
3: **for** $k = 1, \ldots, n$ **do**
4:     **if** $f(a_{k-1})f(c_{k-1}) \leq 0$ **then**
5:         **if** $f(c_{k-1}) = 0$ **then**
6:             **return** $\bar{x} = c_{k-1}$
7:         **end if**
8:         $a_k = a_{k-1}, b_k = c_{k-1}$
9:     **else**
10:         $a_k = c_{k-1}, b_k = b_{k-1}$
11:     **end if**
12:     $c_k = (a_k + b_k)/2$
13: **end for**
14: **return** $\bar{x} = c_n$

---

and the first 4 iterations give the following results.

$k = 1:$ $f(b_0)f(c_0) < 0 \Rightarrow [a_1, b_1] = [0.5, 1],$
        $c_1 = (0.5 + 1)/2 = 0.75,$
        $f(c_1) = (0.75)^3 + 0.75 - 1 = 0.172 > 0;$

$k = 2:$ $f(a_1)f(c_1) < 0 \Rightarrow [a_2, b_2] = [0.5, 0.75],$
        $c_2 = (0.5 + 0.75)/2 = 0.625,$
        $f(c_2) = (0.625)^3 + 0.625 - 1 = -0.131 < 0;$

$k = 3:$ $f(b_2)f(c_2) < 0 \Rightarrow [a_3, b_3] = [0.625, 0.75],$
        $c_3 = (0.625 + 0.75)/2 = 0.6875,$
        $f(c_3) = (0.6875)^3 + 0.6875 - 1 = 0.01245 > 0;$

$k = 4:$ $f(a_3)f(c_3) < 0 \Rightarrow [a_4, b_4] = [0.625, 0.6875],$
        $c_4 = (0.625 + 0.6875)/2 = 0.65625,$
        $f(c_4) = (0.65625)^3 + 0.65625 - 1 \approx -0.06113 < 0.$

4.2.1.2 Intervals with multiple roots

If $[a, b]$ is known to contain multiple roots of a continuous function f, then:

1. if $f(a)f(b) > 0$, then there is an *even* number of roots of f in $[a, b]$;

2. if $f(a)f(b) < 0$, then there is an *odd* number of roots of f in $[a, b]$.

Example 4.7 *Consider $f(x) = x^6 + 4x^4 + x^2 - 6$, where $x \in [-2, 2]$. Since $f(-2) = f(2) = 126$ and $f(-2)f(2) > 0$, we can say that if the interval $[-2, 2]$ contains a root of f, then it contains an even number of roots. Using* *the bisection method we have:*

$$a_0 = -2, b_0 = 2, c_0 = (-2+2)/2 = 0, f(c_0) = f(0) = -6 < 0,$$

and since $f(a_0)f(c_0) < 0$ and $f(b_0)f(c_0) < 0$, both $[-2,0]$ and $[0,2]$ contain at least one root of f. Hence, we can apply the bisection method for each of these intervals to find the corresponding roots. For $[a_1, b_1] = [0, 2]$ we have

$$c_1 = (0+2)/2 = 1, f(c_1) = 0,$$

and we found a root $\hat{x} = 1$. Similarly, for $[a_1, b_1] = [-2, 0]$ we obtain

$$c_1 = (-2+0)/2 = -1, f(c_1) = 0,$$

and we found another root $x^* = -1$. Since $f'(x) = 6x^5 + 16x^3 + 2x$, we have $f'(x) < 0$ for all $x < 0$ and $f'(x) > 0$ for all $x > 0$, implying that the function is decreasing on $(-\infty, 0)$ and increasing on $(0, +\infty)$. Hence, f can have at most one negative and one positive root, and it has no real roots other than $x^* = -1$ and $\hat{x} = 1$.

4.2.2   Regula-falsi method

The *regula-falsi method*, also known as the *false-position method*, is similar to the bisection method, but instead of the mid-point at the kth iteration we take the point c_k defined by an intersection of the line segment joining the points $(a_k, f(a_k))$ and $(b_k, f(b_k))$ with the x-axis (see Figure 4.4). The line passing through $(a, f(a))$ and $(b, f(b))$ is given by

$$y - f(b) = \frac{f(b) - f(a)}{b - a}(x - b),$$

so for $y = 0$,

$$\frac{f(b) - f(a)}{b - a} = -\frac{f(b)}{x - b} = \frac{f(b)}{b - x},$$

implying

$$x = b - f(b)\left(\frac{b - a}{f(b) - f(a)}\right) = \frac{af(b) - bf(a)}{f(b) - f(a)}.$$

Thus, we have the following expression for c_k:

$$c_k = \frac{a_k f(b_k) - b_k f(a_k)}{f(b_k) - f(a_k)}.$$

A pseudo-code for the regula-falsi method is given in Algorithm 4.2.

Note that in this case we stop when $|f(c_k)| < \epsilon$, since, unlike the bisection method, the regula-falsi method does not guarantee that the length of the interval containing the root of $f(x)$ tends to zero. Indeed, as can be seen from the illustration in Figure 4.4, it is possible that one of the endpoints of the

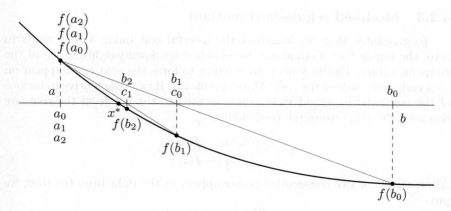

FIGURE 4.4: An illustration of the regula-falsi method.

Algorithm 4.2 Regula-falsi method for solving $f(x) = 0$.

Input: f, ϵ, a, b such that $f(a)f(b) < 0$
Output: $\bar{x}$ such that $|f(\bar{x})| < \epsilon$

1: $k = 0, a_0 = a, b_0 = b, c_0 = \frac{a_0 f(b_0) - b_0 f(a_0)}{f(b_0) - f(a_0)}$
2: **repeat**
3:    $k = k + 1$
4:    **if** $f(a_{k-1})f(c_{k-1}) \leq 0$ **then**
5:       **if** $f(c_{k-1}) = 0$ **then**
6:          **return** $\bar{x} = c_{k-1}$
7:       **end if**
8:       $a_k = a_{k-1}, b_k = c_{k-1}$
9:    **else**
10:       $a_k = c_{k-1}, b_k = b_{k-1}$
11:    **end if**
12:    $c_k = \frac{a_k f(b_k) - b_k f(a_k)}{f(b_k) - f(a_k)}$
13: **until** $|f(c_k)| < \epsilon$
14: **return** $\bar{x} = c_k$

---

search interval converges to the root x^*, whereas the other endpoint of the search interval always remains the same. For example, if we apply the regula-falsi method to $f(x) = 2x^3 - 4x^2 + 3x$ on $[-1, 1]$, we observe that the left endpoint is always -1. At the same time, the right endpoint approaches 0, which is the root (Exercise 4.4). Thus, the length of the interval is always at least 1 in this case.

4.2.3   Modified regula-falsi method

To guarantee that the length of the interval containing a root tends to zero, the regula-falsi method can be modified by down-weighting one of the endpoint values. This is done in an attempt to force the next c_k to appear on that endpoint's side of the root. More specifically, if two consecutive iterations of the regula-falsi method yield points located to the left from the root, we down-weight $f(b_k)$ using the coefficient of $\frac{1}{2}$:

$$c_k = \frac{\frac{1}{2}f(b_k)a_k - f(a_k)b_k}{\frac{1}{2}f(b_k) - f(a_k)}.$$

Alternatively, if two consecutive points appear to the right from the root, we put

$$c_k = \frac{f(b_k)a_k - \frac{1}{2}f(a_k)b_k}{f(b_k) - \frac{1}{2}f(a_k)}.$$

Figure 4.5 provides an illustration using the same example as the one used in the previous figure for the regula-falsi method.

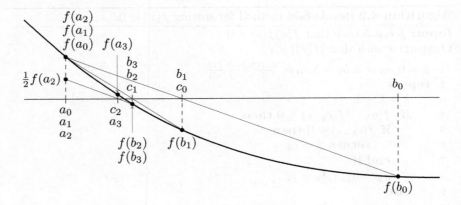

FIGURE 4.5: An illustration of the modified regula-falsi method.

This simple modification guarantees that the width of the bracket tends to zero. Moreover, the modified regula-falsi method often converges faster than the original regula-falsi method. A pseudo-code for the modified regula-falsi method is given in Algorithm 4.3.

Example 4.8 *Let $f(x) = x^3 + x - 1$. We use the modified regula-falsi method to compute an approximate root of f on $[0,1]$.*
We have $a_0 = 0$, $b_0 = 1$, $f(a_0) = -1$, $f(b_0) = 1$, and $f(a_0)f(b_0) < 0$. We compute

$$c_0 = \frac{a_0 f(b_0) - b_0 f(a_0)}{f(b_0) - f(a_0)} = \frac{1}{2} = 0.5, \quad f(c_0) = -0.375.$$

Algorithm 4.3 Modified regula-falsi method for solving $f(x) = 0$.

Input: f, ϵ, a, b such that $f(a)f(b) < 0$
Output: $\bar{x}$ such that $|f(\bar{x})| < \epsilon$

1: $k = 0, a_0 = a, b_0 = b, c_0 = \frac{a_0 f(b_0) - b_0 f(a_0)}{f(b_0) - f(a_0)}$
2: **repeat**
3:      $k = k + 1$
4:      $\alpha = 1, \beta = 1$
5:      **if** $f(a_{k-1})f(c_{k-1}) \leq 0$ **then**
6:          **if** $f(c_{k-1}) = 0$ **then**
7:             **return** $\bar{x} = c_{k-1}$
8:          **end if**
9:          $a_k = a_{k-1}, b_k = c_{k-1}$
10:         **if** $k > 1$ and $f(c_{k-1})f(c_k) > 0$ **then**
11:            $\alpha = 1/2$
12:         **end if**
13:      **else**
14:          $a_k = c_{k-1}, b_k = b_{k-1}$
15:         **if** $k > 1$ and $f(c_{k-1})f(c_k) > 0$ **then**
16:            $\beta = 1/2$
17:         **end if**
18:      **end if**
19:      $c_k = \frac{a_k \beta f(b_k) - b_k \alpha f(a_k)}{\beta f(b_k) - \alpha f(a_k)}$
20: **until** $|f(c_k)| < \epsilon$
21: **return** $\bar{x} = c_k$

$k = 1:$ $\alpha = \beta = 1; f(b_0)f(c_0) < 0 \Rightarrow [a_1, b_1] = [0.5, 1],$
         $c_1 = \frac{a_1 f(b_1) - b_1 f(a_1)}{f(b_1) - f(a_1)} = \frac{0.5 + 0.375}{1 + 0.375} \approx 0.63636, f(c_1) \approx -0.10594.$

$k = 2:$ $\alpha = \beta = 1; f(b_1)f(c_1) < 0 \Rightarrow [a_2, b_2] = [0.63636, 1],$
         $f(c_0)f(c_1) > 0 \Rightarrow \beta = 0.5,$
         $c_2 = \frac{a_2 0.5 f(b_2) - b_2 f(a_2)}{0.5 f(b_2) - f(a_2)} \approx 0.699938, f(c_2) \approx 0.042847.$

4.3   Newton's Method

Similarly to the regula-falsi method, *Newton's method* (also called *Newton-Raphson method*) uses a linear approximation of the function to obtain an estimate of a root. However, in the case of Newton's method, the linear approximation at each iteration is given by the tangent line to $f(x)$ at x_k, the current estimate of the root.

To derive Newton's method iteration, we assume that $f(x)$ is continuously differentiable and consider the first-order Taylor's series approximation of $f(x)$

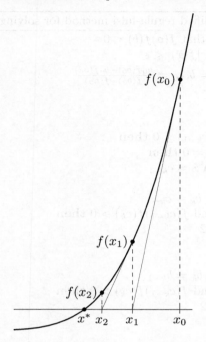

FIGURE 4.6: An illustration of Newton's method.

about x_k:

$$f(x) \approx f(x_k) + f'(x_k)(x - x_k).$$

Instead of solving $f(x) = 0$, we solve the linear equation

$$f(x_k) + f'(x_k)(x - x_k) = 0$$

for x to obtain the next approximation x_{k+1} of a root. If $f'(x_k) \neq 0$, the solution of this equation is

$$x_{k+1} = x_k - \frac{f(x_k)}{f'(x_k)}.$$

So, the kth iteration of Newton's method can be written as

$$x_k = x_{k-1} - \frac{f(x_{k-1})}{f'(x_{k-1})}, k \geq 1. \tag{4.13}$$

Figure 4.6 illustrates the steps of Newton's method geometrically. For the function f in this illustration, the method quickly converges to the root x^* of

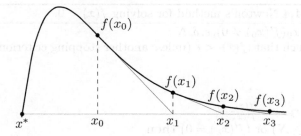

FIGURE 4.7: Newton's method produces a sequence of points moving away from the unique root x^* of f.

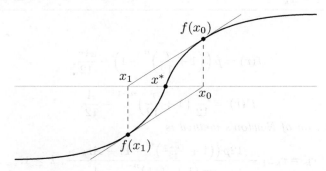

FIGURE 4.8: Newton's method cycles.

f. However, as can be seen from examples in Figures 4.7 and 4.8, sometimes Newton's method fails to converge to a root.

A pseudo-code of Newton's method is presented in Algorithm 4.4. Besides terminating when we reach a point x_k with $|f(x_k)| < \epsilon$ or when the step size becomes very small ($|x_k - x_{k-1}| < \delta$), we included additional stopping criteria, such as allowing no more than N iterations and stopping whenever we encounter a point x_k with $f'(x_k) = 0$. This makes sure that the algorithm always terminates, however, as illustrated in Figures 4.7 and 4.8, it may still produce an erroneous output. On the other hand, if the method does converge to a root, it typically has a very good speed of convergence.

Example 4.9 *Consider the annuity-due equation (4.4) derived in Example 4.1:*

$$\frac{p}{r/12}\left(\left(1 + \frac{r}{12}\right)^n - 1\right) - A = 0. \tag{4.14}$$

Multiplying both sides of this equation by $r/12$, we obtain

$$p\left(\left(1 + \frac{r}{12}\right)^n - 1\right) - \frac{Ar}{12} = 0.$$

Algorithm 4.4 Newton's method for solving $f(x) = 0$.

Input: $f, f', x_0(f'(x_0) \neq 0), \epsilon, \delta, N$
Output: $\bar{x}$ such that $|f(\bar{x})| < \epsilon$ (unless another stopping criterion is satisfied)

1: $k = 0$
2: **repeat**
3:     $k = k + 1$
4:     $x_k = x_{k-1} - \frac{f(x_{k-1})}{f'(x_{k-1})}$
5:     **if** $(k \geq N)$ or $(f'(x_k) = 0)$ **then**
6:         STOP
7:     **end if**
8: **until** $(|f(x_k)| < \epsilon)$ or $(|x_k - x_{k-1}| < \delta)$
9: **return** $\bar{x} = x_k$

Denote by

$$f(r) = p\left(\left(1 + \frac{r}{12}\right)^n - 1\right) - \frac{Ar}{12},$$

then

$$f'(r) = \frac{pn}{12}\left(1 + \frac{r}{12}\right)^{n-1} - \frac{A}{12},$$

and an iteration of Newton's method is

$$r_k = r_{k-1} - \frac{12p\left(\left(1 + \frac{r_{k-1}}{12}\right)^n - 1\right) - Ar_{k-1}}{pn\left(1 + \frac{r_{k-1}}{12}\right)^{n-1} - A}, \; k \geq 1. \qquad (4.15)$$

Like in Example 4.5, assume that the monthly payment is $300, and we want to find an interest rate r which would yield $1,000,000 in 30 years. We have $p = 300$, $A = 1,000,000$, and $n = 12 \cdot 30 = 360$. We also need to define the starting point r_0. Table 4.2 contains the results of applying 7 iterations of Newton's method with two different initial guesses: $r_0 = 0.1$ and $r_0 = 0.2$. In both cases the method converges to the same solution $r \approx 0.117865$, or 11.79%.

Example 4.10 *We can use Newton's method for the equation*

$$f(x) = x^2 - \alpha = 0 \qquad (4.16)$$

to approximate the square root $x = \sqrt{\alpha}$ of a positive number α. Newton's step for the above equation is given by

$$
\begin{aligned}
x_k &= x_{k-1} - \frac{f(x_{k-1})}{f'(x_{k-1})} \\
&= x_{k-1} - \frac{x_{k-1}^2 - \alpha}{2x_{k-1}} \\
&= \frac{2x_{k-1}^2 - x_{k-1}^2 + \alpha}{2x_{k-1}} \\
&= \frac{1}{2}\left(x_{k-1} + \frac{\alpha}{x_{k-1}}\right).
\end{aligned}
$$

TABLE 4.2: Newton's iterations for finding the interest rate with two different choices of r_0.

Iteration number, k	Approximate interest rate, r_k	
0	$r_0 = 0.1$	$r_0 = 0.2$
1	0.128616	0.170376
2	0.119798	0.145308
3	0.117938	0.127727
4	0.117865	0.119516
5	0.117865	0.117919
6	0.117865	0.117865
7	0.117865	0.117865

For example, to find $\sqrt{7}$, we use $f(x) = x^2 - 7 = 0$ and the iteration

$$x_k = \frac{1}{2}\left(x_{k-1} + \frac{7}{x_{k-1}}\right).$$

Starting with $x_0 = 3$, we obtain

$$x_1 = \frac{1}{2}\left(3 + \frac{7}{3}\right) = 8/3 \approx 2.66667,$$

$$x_2 = \frac{1}{2}\left(8/3 + \frac{7}{8/3}\right) = 127/48 \approx 2.64583,$$

$$x_3 = \frac{1}{2}\left(127/48 + \frac{7}{127/48}\right) = 32,257/12,192 \approx 2.64575.$$

More generally, to approximate $\sqrt[p]{a}$ for a positive number α, we can use the equation $f(x) = x^p - \alpha = 0$ and Newton's iteration in the form:

$$
\begin{aligned}
x_k &= x_{k-1} - \frac{x_{k-1}^p - \alpha}{px_{k-1}^{p-1}} \\
&= \frac{1}{p}\left((p-1)x_{k-1} + \frac{\alpha}{x_{k-1}^{p-1}}\right), \quad k \geq 1.
\end{aligned}
$$

4.3.1 Convergence rate of Newton's method

Assume that Newton's iteration produces a sequence $\{x_k : k \geq 1\}$ converging to a root x^* of function f. We have

$$x_{k+1} = x_k - \frac{f(x_k)}{f'(x_k)},$$

and x^* is a fixed point of

$$\phi(x) = x - \frac{f(x)}{f'(x)}.$$

Hence, $x^* = \phi(x^*)$, and by the mean value theorem

$$x_{k+1} - x^* = \phi(x_k) - \phi(x^*) = \phi'(\xi_k)(x_k - x^*)$$

for some ξ_k between x_k and x^*. We have

$$\phi'(x) = 1 - \frac{(f'(x))^2 - f''(x)f(x)}{(f'(x))^2} = \frac{f''(x)f(x)}{(f'(x))^2},$$

therefore

$$x_{k+1} - x^* = \frac{f''(\xi_k)f(\xi_k)}{(f'(\xi_k))^2}(x_k - x^*). \tag{4.17}$$

Since $f(x^*) = 0$, by the mean value theorem we get

$$|f(\xi_k)| = |f(\xi_k) - f(x^*)| = |f'(\nu_k)| \cdot |\xi_k - x^*| \tag{4.18}$$

for some ν_k lying between ξ_k and x^*. Recall that ξ_k is located between x_k and x^*, thus

$$|\xi_k - x^*| \leq |x_k - x^*|$$

and from (4.18),

$$|f(\xi_k)| \leq |f'(\nu_k)| \cdot |x_k - x^*|. \tag{4.19}$$

Combining (4.17) and (4.19), we obtain

$$|x_{k+1} - x^*| \leq \frac{|f''(\xi_k)f'(\nu_k)|}{(f'(\xi_k))^2}|x_k - x^*|^2. \tag{4.20}$$

If $f(x)$ is such that for some constant C

$$\frac{|f''(\xi)f'(\nu)|}{(f'(\xi))^2} \leq C$$

for all ξ and ν in some neighborhood of x^*, then

$$|x_{k+1} - x^*| \leq C|x_k - x^*|^2.$$

Hence, if certain conditions are satisfied, Newton's method is quadratically convergent.

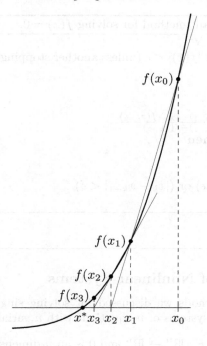

FIGURE 4.9: An illustration of the secant method.

4.4 Secant Method

The secant method can be viewed as a modification of Newton's method that avoids calculating the derivative by replacing it with a difference-based approximation:

$$f'(x_k) \approx \frac{f(x_k) - f(x_{k-1})}{x_k - x_{k-1}}.$$

Hence, an iteration of the secant method is given by

$$x_{k+1} = x_k - f(x_k) \frac{x_k - x_{k-1}}{f(x_k) - f(x_{k-1})}, \quad k \ge 1. \qquad (4.21)$$

Geometrically, this change in Newton's iteration corresponds to replacing the tangent line at x_k with the secant line passing through $(x_{k-1}, f(x_{k-1}))$ and $(x_k, f(x_k))$, as illustrated in Figure 4.9. Algorithm 4.5 provides a pseudo-code of the secant method.

Algorithm 4.5 Secant method for solving $f(x) = 0$.

Input: $f, x_0, x_1, \epsilon, \delta, N$
Output: $\bar{x}$ such that $|f(\bar{x})| < \epsilon$ (unless another stopping criterion is satisfied)

1: $k = 1$
2: **repeat**
3:     $k = k + 1$
4:     $x_k = \frac{x_{k-2}f(x_{k-1}) - x_{k-1}f(x_{k-2})}{f(x_{k-1}) - f(x_{k-2})}$
5:     **if** $(k \geq N)$ **then**
6:         STOP
7:     **end if**
8: **until** $(|f(x_k)| < \epsilon)$ or $(|x_k - x_{k-1}| < \delta)$
9: **return** $\bar{x} = x_k$

4.5 Solution of Nonlinear Systems

Some of the methods we discussed for solving single-variable equations extend to nonlinear systems of n equations with n variables in the form

$$F(x) = 0, \text{ where } F : \mathbb{R}^n \to \mathbb{R}^n \text{ and } 0 \text{ is an } n\text{-dimensional zero-vector.}$$

Then $x^* \in \mathbb{R}^n$ is a solution (root) of the above system if $F(x^*) = 0$. While the methods we discuss in this section apply to $n \times n$ systems for an arbitrary dimension n, for simplicity of presentation we will focus on the case of $n = 2$, corresponding to the systems in the form

$$\begin{cases} F_1(x, y) & = & 0 \\ F_2(x, y) & = & 0. \end{cases} \tag{4.22}$$

In this case, a solution is given by a pair of numbers, $[x^*, y^*]^T$ such that $F_1(x^*, y^*) = 0$ and $F_2(x^*, y^*) = 0$.

4.5.1 Fixed point method for systems

Suppose that we know that the root $[x^*, y^*]^T$ we are looking for is such that $x^* \in (a, b)$ and $y^* \in (c, d)$. Assume that we can reduce (4.22) to an equivalent in the region $(a, b) \times (c, d) = \{[x, y]^T \in \mathbb{R}^2 : x \in (a, b), y \in (c, d)\}$ system of the form

$$\begin{cases} x & = & f_1(x, y) \\ y & = & f_2(x, y). \end{cases}$$

Example 4.11 *Put*

$$f_1(x, y) = x - \alpha F_1(x, y),$$
$$f_2(x, y) = y - \alpha F_2(x, y),$$

where $\alpha \neq 0$. Then, obviously

$$\begin{cases} f_1(x,y) & = & x \\ f_2(x,y) & = & y \end{cases} \Leftrightarrow \begin{cases} \alpha F_1(x,y) & = & 0 \\ \alpha F_2(x,y) & = & 0. \end{cases}$$

To find a root $(x^*, y^*) \in (a, b) \times (c, d)$, we take some $[x_0, y_0]^T \in (a, b) \times (c, d)$ and then recursively calculate (x_{k+1}, y_{k+1}) by the formula

$$\begin{cases} x_{k+1} & = & f_1(x_k, y_k) \\ y_{k+1} & = & f_2(x_k, y_k) \end{cases} \tag{4.23}$$

for $k \geq 0$. In the above, we assume that $[f_1(x_k, y_k), f_2(x_k, y_k)]^T \in (a, b) \times (c, d)$ for all $[x_k, y_k]^T$, so the next vector $[x_{k+1}, y_{k+1}]^T$ in the sequence is in the domain of $f_i(x, y), i = 1, 2$.

Similar to the one-dimensional case, we call $[x^*, y^*]^T$ a *fixed point* of a function $f(x,y) = \begin{bmatrix} f_1(x,y) \\ f_2(x,y) \end{bmatrix}$ if $f(x^*, y^*) = \begin{bmatrix} x^* \\ y^* \end{bmatrix}$, that is, $f_1(x^*, y^*) = x^*$ and $f_2(x^*, y^*) = y^*$. Obviously, if $[x^*, y^*]^T$ is a fixed point of $f(x,y)$ then it is a root of the system (4.22). The iteration (4.23) is called a *fixed point iteration*.

4.5.2 Newton's method for systems

Recall that for $F(x, y) = \begin{bmatrix} F_1(x,y) \\ F_2(x,y) \end{bmatrix}$, its first-order derivative at point $[x, y]^T$ is given by the Jacobian matrix

$$J_F(x, y) = \begin{bmatrix} \frac{\partial F_1(x,y)}{\partial x} & \frac{\partial F_1(x,y)}{\partial y} \\ \frac{\partial F_2(x,y)}{\partial x} & \frac{\partial F_2(x,y)}{\partial y} \end{bmatrix}.$$

Assume that we are given $[x_k, y_k]^T \in \mathbb{R}^2$. Similar to the one-dimensional case, we use the Taylor's approximation of $F(x, y)$ about $[x_k, y_k]^T$,

$$\begin{bmatrix} F_1(x,y) \\ F_2(x,y) \end{bmatrix} \approx \begin{bmatrix} F_1(x_k, y_k) \\ F_2(x_k, y_k) \end{bmatrix} + \begin{bmatrix} \frac{\partial F_1(x_k,y_k)}{\partial x} & \frac{\partial F_1(x_k,y_k)}{\partial y} \\ \frac{\partial F_2(x_k,y_k)}{\partial x} & \frac{\partial F_2(x_k,y_k)}{\partial y} \end{bmatrix} \begin{bmatrix} x - x_k \\ y - y_k \end{bmatrix},$$

instead of $F(x, y)$ in order to find the next approximation $[x_{k+1}, y_{k+1}]^T$ of a root $[x^*, y^*]^T$ of $F(x, y)$. Assume that the matrix $J_F(x_k, y_k)$ is nonsingular. Solving the system

$$\begin{bmatrix} F_1(x_k, y_k) \\ F_2(x_k, y_k) \end{bmatrix} + \begin{bmatrix} \frac{\partial F_1(x_k,y_k)}{\partial x} & \frac{\partial F_1(x_k,y_k)}{\partial y} \\ \frac{\partial F_2(x_k,y_k)}{\partial x} & \frac{\partial F_2(x_k,y_k)}{\partial y} \end{bmatrix} \begin{bmatrix} x - x_k \\ y - y_k \end{bmatrix} = \begin{bmatrix} 0 \\ 0 \end{bmatrix}$$

with respect to x and y, and denoting the solution by $[x_{k+1}, y_{k+1}]^T$, we obtain

$$\begin{bmatrix} x_{k+1} \\ y_{k+1} \end{bmatrix} = \begin{bmatrix} x_k \\ y_k \end{bmatrix} - \begin{bmatrix} \frac{\partial F_1(x_k,y_k)}{\partial x} & \frac{\partial F_1(x_k,y_k)}{\partial y} \\ \frac{\partial F_2(x_k,y_k)}{\partial x} & \frac{\partial F_2(x_k,y_k)}{\partial y} \end{bmatrix}^{-1} \begin{bmatrix} F_1(x_k, y_k) \\ F_2(x_k, y_k) \end{bmatrix}.$$

Assuming that the Jacobian $J_F(x, y)$ is always nonsingular, the $(k+1)$-st iteration of Newton's method, where $k \geq 0$, is

$$
\begin{bmatrix} x_{k+1} \\ y_{k+1} \end{bmatrix} = \begin{bmatrix} x_k \\ y_k \end{bmatrix} - \begin{bmatrix} \frac{\partial F_1(x_k, y_k)}{\partial x} & \frac{\partial F_1(x_k, y_k)}{\partial y} \\ \frac{\partial F_2(x_k, y_k)}{\partial x} & \frac{\partial F_2(x_k, y_k)}{\partial y} \end{bmatrix}^{-1} \begin{bmatrix} F_1(x_k, y_k) \\ F_2(x_k, y_k) \end{bmatrix}. \quad (4.24)
$$

Example 4.12 *Consider the system of nonlinear equations*

$$
\begin{aligned}
(x-1)^2 + (y-1)^2 - 1 &= 0 \\
x + y - 2 &= 0.
\end{aligned}
$$

We apply 2 iterations of Newton's method starting with $[x_0, y_0]^T = [0, 2]^T$. We have

$$
F(x, y) = \begin{bmatrix} F_1(x, y) \\ F_2(x, y) \end{bmatrix} = \begin{bmatrix} (x-1)^2 + (y-1)^2 - 1 \\ x + y - 2 \end{bmatrix},
$$

and the Jacobian of F is

$$
J_F(x, y) = \begin{bmatrix} 2(x-1) & 2(y-1) \\ 1 & 1 \end{bmatrix}.
$$

We obtain

$$
\begin{bmatrix} x_1 \\ y_1 \end{bmatrix} = \begin{bmatrix} 0 \\ 2 \end{bmatrix} - J_F^{-1}(0, 2) \begin{bmatrix} F_1(0, 2) \\ F_2(0, 2) \end{bmatrix} = \begin{bmatrix} 1/4 \\ 7/4 \end{bmatrix};
$$

$$
\begin{bmatrix} x_2 \\ y_2 \end{bmatrix} = \begin{bmatrix} 1/4 \\ 7/4 \end{bmatrix} - J_F^{-1}(1/4, 7/4) \begin{bmatrix} F_1(1/4, 7/4) \\ F_2(1/4, 7/4) \end{bmatrix} = \begin{bmatrix} 7/24 \\ 41/24 \end{bmatrix} \approx \begin{bmatrix} 0.2917 \\ 1.7083 \end{bmatrix}.
$$

　　Note that geometrically the first equation is described by a circle, and the second equation is given by a line (see Figure 4.12). The roots are at points

$$
\hat{x} = \begin{bmatrix} 1 - 1/\sqrt{2} \\ 1 + 1/\sqrt{2} \end{bmatrix} \approx \begin{bmatrix} 0.2929 \\ 1.7071 \end{bmatrix} \text{ and } \tilde{x} = \begin{bmatrix} 1 + 1/\sqrt{2} \\ 1 - 1/\sqrt{2} \end{bmatrix} \approx \begin{bmatrix} 1.7071 \\ 0.2929 \end{bmatrix},
$$

which are located at the intersections of the line and the circle. The steps of Newton's method in this example correspond to movement from the point $[0, 2]^T$ along the line given by the second equation toward the root $\hat{x}$.

Exercises

4.1. Let $f(x) = 5 - \frac{6}{x}$.

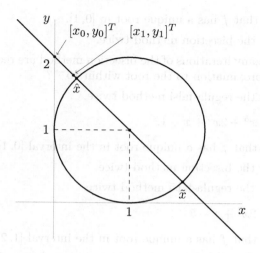

FIGURE 4.10: An illustration of Example 4.12.

(a) Find all fixed points of f by solving a quadratic equation.

(b) Apply the fixed point method twice with $x_0 = 4$.

4.2. Let $f(x) = \frac{1}{4}(x^2 - 4x + 7)$.

(a) Verify that $x^* = 1$ is a fixed point of f.

(b) Apply the fixed point method twice with $x_0 = 3$.

(c) Perform 3 iterations of the fixed point method with $x_0 = 5$. What do you observe?

(d) Graph $f(x)$ and illustrate your computations in (b) and (c) geometrically.

4.3. Consider $f(x) = x^3 + 2x^2 + 10x - 20$.

(a) Prove that f has a unique root x^* in the interval $[1, 2]$.

(b) Apply the bisection method twice.

(c) How many iterations of the bisection method are required to guarantee an approximation of the root within 10^{-10}? (Note: In 1225 Leonardo of Pisa computed the root $x^* \approx 1.368808107$. This was a remarkable result for his time.)

4.4. Plot $f(x) = 2x^3 - 4x^2 + 3x$ and illustrate three steps of the regula-falsi method applied to f on the interval $[-1, 1]$ graphically. What do you observe?

4.5. Consider $f(x) = x^5 + 2x - 1$.

    (a) Prove that f has a unique root in $[0, 1]$.

    (b) Apply the bisection method twice.

    (c) How many iterations of the bisection method are required to obtain an approximation to the root within 10^{-8}?

    (d) Apply the regula-falsi method twice.

4.6. Let $f(x) = x^5 + 2x^3 + x - 1$.

    (a) Prove that f has a unique root in the interval $[0, 1]$.

    (b) Apply the bisection method twice.

    (c) Apply the regula-falsi method twice.

4.7. Let $f(x) = 2x^7 + x - 5$.

    (a) Prove that f has a unique root in the interval $[1, 2]$.

    (b) Apply three iterations of the bisection method to find an approximation of the root.

4.8. Consider the function $f(x) = e^{-2x} - x^3$.

    (a) Prove that f has a unique root x^* in the interval $[0, 1]$.

    (b) Apply the bisection method three times to obtain an approximation c_3 of the root.

    (c) How many iterations k of the bisection method are needed to ensure that the absolute error $|c_k - x^*| \leq 10^{-5}$?

4.9. Consider the function $f(x) = e^x + x - 4$.

    (a) Prove that f has a unique root in the interval $[1, 2]$.

    (b) Perform three iterations of the regula-falsi method.

    (c) Perform three iterations of the modified regula-falsi method.

    (d) Compare the results you obtained in (b) and (c).

4.10. Let $f(x) = x^5 + 3x - 1$.

    (a) Prove that $f(x) = 0$ has a unique solution in $[0, 1]$.

    (b) Apply two iterations of the bisection method to compute an approximation c_2 of the root.

    (c) Apply Newton's method twice using c_2 computed in (b) as the initial guess.

4.11. Let $f(x) = x^3 + 2x - 2$.

    (a) Prove that $f(x)$ has a unique root in $[0, 1]$.

    (b) Apply the regula-falsi method twice.

(c) Use Newton's method twice using your answer in (b) as the starting point.

4.12. Approximate $\sqrt[3]{2}$ as follows.

(a) Use the bisection method twice for $f(x) = x^3 - 2$ in the interval $[1, 2]$.

(b) Use Newton's method twice starting with the point obtained using bisection in (a).

4.13. Find an approximation of $\sqrt[3]{9}$ by finding the root of the function $f(x) = x^3 - 9$ using two steps of the modified regula-falsi method. Choose an appropriate initial interval with integer endpoints.

4.14. Use three iterations of Newton's method with the initial guess $x_0 = 2$ to approximate

(a) $\sqrt{3}$;

(b) $\sqrt{5}$;

(c) $3^{2/5}$.

4.15. Let $f(x) = x^6 - x - 1$.

(a) Use 7 iterations of Newton's method with $x_0 = 2$ to get an approximate root for this equation.

(b) Use 7 iterations of the secant method with $x_0 = 2, x_1 = 1$ to get an approximate root for this equation.

Compare your results in (a) and (b).

4.16. Consider the equation $e^{100x}(x - 2) = 0$. Apply Newton's method several times with $x^0 = 1$. What do you observe?

4.17. Propose a method to solve the equation $e^{100x}(x - 2)(x^5 + 2x - 1) = 0$.

4.18. Consider the system of equations

$$\begin{aligned} (x - 1)^2 + (y - 1)^2 - 1 &= 0 \\ x + y - 1 &= 0. \end{aligned}$$

(a) Draw the sets of points on the plane that satisfy each equation, and indicate the solutions of the system.

(b) Solve this system exactly.

(c) Apply Newton's method twice with $[x_0, y_0]^T = [1/2, 1/2]^T$. Illustrate the corresponding steps geometrically.

4.19. Consider the system of equations

$$x^2 + y^2 - 9 = 0$$
$$x + y - 1 = 0.$$

Apply Newton's method twice with starting point $[1, 0]^T$. Show the exact solutions and the steps of Newton's method graphically.

4.20. Use Newton's method twice with initial guess $[x_0, y_0]^T = [1/2, 0]^T$ to solve the nonlinear system

$$x^2 - x - y = 0$$
$$x + y - 2 = 0.$$

Illustrate your steps graphically.

4.21. Consider the system of equations

$$x^4 + y^4 - 3 = 0$$
$$x^3 - 3xy^2 + 1 = 0.$$

Apply Newton's method twice with the starting point $[1, 1]^T$.

Chapter 5

Polynomial Interpolation

Many problems arising in engineering require evaluation of real-valued functions. Evaluation of some of the commonly used functions, such as $\cos(x)$, on a computer usually utilizes the so-called lookup tables, which store a finite set of numbers describing the function of interest in a way that allows one to approximate the value of $f(x)$ for any given x with a high level of accuracy. For example, a finite set of coefficients in the power series decomposition of $f(x)$ could be used for this purpose.

Example 5.1 *Consider $f(x) = \cos(x), x \in [-1, 1]$. The Taylor series approximation of this function about the origin is*

$$\cos(x) = 1 - \frac{x^2}{2!} + \frac{x^4}{4!} - \frac{x^6}{6!} + \ldots + (-1)^n \frac{x^{2n}}{(2n)!} + \ldots,$$

and for a given n, the set of $2n+1$ coefficients

$$1, 0, -\frac{1}{2!}, 0, \frac{1}{4!}, 0, \ldots, (-1)^n \frac{1}{(2n)!}$$

could be used as the table representing $f(x)$ for the values of x close to 0. Given such a table, the approximation to $f(x)$ is computed as

$$\cos(x) \approx 1 - \frac{x^2}{2!} + \frac{x^4}{4!} - \frac{x^6}{6!} + \ldots + (-1)^n \frac{x^{2n}}{(2n)!}.$$

In this formula, a better approximation is obtained for a larger n.

However, in many practical situations, the function $f(x)$ of interest is not known explicitly, and the only available information about the function is a table of measured or computed values $y_0 = f(x_0), y_1 = f(x_1), \ldots, y_n = f(x_n)$, for a set of points $x_0 < x_1 < \ldots < x_n$ in some interval $[a, b]$:

x_0	x_1	$\cdots$	x_n
y_0	y_1	$\cdots$	y_n

Given such a table, our goal is to approximate $f(x)$ with a function, say a polynomial $p(x)$, such that $p(x_i) = f(x_i)$, $i = 0, 1, \ldots, n$. Then for any $x \in [a, b]$ we can approximate the value $f(x)$ with $p(x)$: $f(x) \approx p(x)$. If $x_0 < x < x_n$, the approximation $p(x)$ is called an *interpolated value*, otherwise it is an *extrapolated value*.

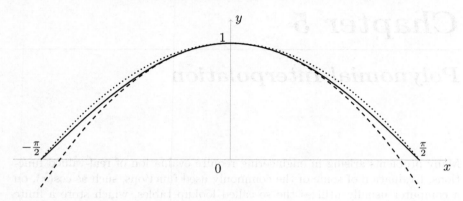

FIGURE 5.1: Two polynomials approximating $y = f(x) = \cos(x)$ over $[-\pi/2, \pi/2]$ (solid line): the Taylor polynomial $p(x) = 1 - \frac{x^2}{2}$ (dashed) and the quadratic polynomial $q(x) = 1 - \frac{4}{\pi^2}x^2$ passing through points $(-\pi/2, 0), (0, 1)$ and $(\pi/2, 0)$ (dotted).

Example 5.2 *Figure 5.1 shows two approximations to $f(x) = \cos(x)$ over $[-\pi/2, \pi/2]$: the Taylor polynomial $p(x) = 1 - \frac{x^2}{2}$ of degree 2 (computed as in Example 5.1), and the quadratic polynomial $q(x) = 1 - \frac{4}{\pi^2}x^2$ passing through points given by the table*

$-\pi/2$	0	$\pi/2$
0	1	0

Note that $|f(x) - p(x)|$ increases significantly when x moves away from the origin. At the same time, the polynomial $q(x)$ passing through the three equidistant points in $[-\pi/2, \pi/2]$ provides a better approximation of $f(x)$.

Polynomials represent a natural choice for approximation of more complicated functions, since their simple structure facilitates easy manipulation (differentiation, integration, etc.) In this chapter we will discuss several basic methods for polynomial interpolation. We start by introducing various forms of representation for polynomials. Then we show how these forms can be used to compute the unique interpolating polynomial in different ways. Finally, in Section 5.3 we discuss upper bounds for the error of interpolation and introduce Chebyshev polynomials, which ensure the minimum possible worst-case error bound.

5.1 Forms of Polynomials

Depending on the objective, the same polynomial can be represented in several different forms. In particular, a polynomial $p(x)$ in the *power form* is written as follows:

$$p(x) = a_0 + a_1 x + a_2 x^2 + \cdots + a_n x^n.$$

If $a_n \neq 0$, then $p(x)$ has degree n.

A polynomial $p(x)$ is in the *shifted power form* with the *center* c if

$$p(x) = b_0 + b_1(x - c) + b_2(x - c)^2 + \cdots + b_n(x - c)^n.$$

A polynomial $p(x)$ is in the *Newton form* if

$$
\begin{aligned}
p(x) &= a_0 + a_1(x - c_1) + a_2(x - c_1)(x - c_2) + \cdots + a_n(x - c_1)\cdots(x - c_n) \\
&= a_0 + \sum_{k=1}^{n} a_k \prod_{i=1}^{k}(x - c_i).
\end{aligned}
$$

Note that for $c_1 = c_2 = \cdots = c_n = c$ we obtain the shifted power form, while for $c_1 = c_2 = \cdots = c_n = 0$ we have the power form.

Evaluation of $p(x)$ in the Newton form can trivially be done using $n + n(n+1)/2$ additions and $n(n+1)/2$ multiplications. However, this can be done much more efficiently by expressing $p(x)$ in the *nested form* as follows:

$$
\begin{aligned}
p(x) &= a_0 + (x - c_1)\left(a_1 + \sum_{k=2}^{n} a_k \prod_{i=2}^{k}(x - c_i) \right) \\
&= a_0 + (x - c_1)\left(a_1 + (x - c_2)\left(a_2 + \sum_{k=3}^{n} a_k \prod_{i=3}^{k}(x - c_i) \right) \right) \\
&\;\;\vdots \\
&= a_0 + (x - c_1)(a_1 + (x - c_2)(a_2 + (x - c_3)(a_3 + \cdots \\
&\qquad + (x - c_{n-1})(a_{n-1} + (x - c_n)a_n)\cdots))).
\end{aligned}
$$

We can start evaluating $p(x)$ by first computing $a'_{n-1} = a_{n-1} + (x - c_n)a_n$, then going backward and finding $a'_{n-2} = a_{n-2} + (x - c_{n-1})a'_{n-1}$, $a'_{n-3} = a_{n-3} + (x - c_{n-2})a'_{n-2}$, ..., to eventually compute $a_0 + (x - c_1)a'_1 = p(x)$. This evaluation procedure, which is summarized in Algorithm 5.1, requires only $2n$ additions and n multiplications.

Example 5.3 *Find* $p(5)$ *for* $p(x) = 1 + 2(x - 1) + 3(x - 1)(x - 2) + 4(x - 1)(x - 2)(x - 3) + 5(x - 1)(x - 2)(x - 3)(x - 4).$

Algorithm 5.1 Evaluating a polynomial $p(x)$ in the Newton form.

1: **Input:** $a_0, a_1, \ldots, a_n; c_1, \ldots, c_n; x$
2: **Output:** $p(x) = a_0 + \sum_{k=1}^{n} a_k \prod_{i=1}^{k} (x - c_i)$
3: $a'_n := a_n$
4: **for** $i = n-1, n-2, \ldots, 0$ **do**
5: $\quad a'_i := a_i + (x - c_{i+1})a'_{i+1}$
6: **end for**
7: **return** a'_0

We have $a_0 = 1, a_1 = 2, a_2 = 3, a_3 = 4, a_4 = 5; c_1 = 1, c_2 = 2, c_3 = 3, c_4 = 4$, and $x = 5$. Applying Algorithm 5.1 we obtain:

$$
\begin{aligned}
i = 4 : a_4' &= 5 \\
i = 3 : a_3' &= 4 + (5-4)5 = 9 \\
i = 2 : a_2' &= 3 + (5-3)9 = 21 \\
i = 1 : a_1' &= 2 + (5-2)21 = 65 \\
i = 0 : a_0' &= 1 + (5-1)65 = 261.
\end{aligned}
$$

Thus, $p(5) = 261$.

5.2 Polynomial Interpolation Methods

Given $n+1$ pairs $(x_0, y_0), \ldots, (x_n, y_n)$ of numbers, where $x_i \neq x_j$ if $i \neq j$, our goal is to find a polynomial $p(x)$ such that $p(x_i) = y_i, i = 0, \ldots, n$.

For $x_0, \ldots, x_n$, we define the following n^{th} degree polynomials for $j = 0, \ldots, n$:

$$
\begin{aligned}
l_j(x) &= \frac{(x - x_0)(x - x_1) \cdots (x - x_{j-1})(x - x_{j+1}) \cdots (x - x_n)}{(x_j - x_0)(x_j - x_1) \cdots (x_j - x_{j-1})(x_j - x_{j+1}) \cdots (x_j - x_n)} \\
&= \prod_{\substack{i=0 \\ i \neq j}}^{n} \frac{(x - x_i)}{(x_j - x_i)},
\end{aligned}
$$

which are referred to as *Lagrange polynomials*. It is easy to verify that $l_j(x_j) = 1$, whereas $l_j(x_i) = 0$ if $i \neq j, \forall i, j = 0, \ldots, n$. Hence, the following polynomial has the desired property, $p(x_i) = y_i, i = 0, \ldots, n$:

$$
p(x) = y_0 l_0(x) + y_1 l_1(x) + \cdots + y_n l_n(x) = \sum_{j=0}^{n} y_j \prod_{\substack{i=0 \\ i \neq j}}^{n} \frac{(x - x_i)}{(x_j - x_i)}. \tag{5.1}
$$

Now, assume that there are two polynomials of degree $\leq n$, $p(x)$ and $q(x)$, such that $p(x_i) = q(x_i) = y_i, i = 0, \ldots, n$. Then $r(x) = p(x) - q(x)$ is a polynomial of degree $\leq n$, which has $n + 1$ distinct roots $x_0, x_1, \ldots, x_n$. We obtain a contradiction with the fundamental theorem of algebra. Thus, we proved the following theorem.

Theorem 5.1 *Given* $(x_0, y_0), \ldots, (x_n, y_n)$, *where* $x_i \neq x_j$ *if* $i \neq j$, *there exists a unique interpolating polynomial* $p(x)$ *of degree no greater than* n, *such that* $p(x_i) = y_i, i = 0, \ldots, n$.

5.2.1 Lagrange method

The proof of the theorem above was constructive and provided a method for determining the interpolating polynomial, called the *Lagrange method* in (5.1):

$$p(x) = y_0 l_0(x) + y_1 l_1(x) + \cdots + y_n l_n(x) = \sum_{j=0}^{n} y_j \prod_{\substack{i=0 \\ i \neq j}}^{n} \frac{(x - x_i)}{(x_j - x_i)}.$$

Example 5.4 *Use the Lagrange method to find the second-degree polynomial* $p(x)$ *interpolating the data*

x	-1	0	1
y	3	5	2

We have:

$$l_0(x) = \frac{(x-0)(x-1)}{(-1-0)(-1-1)} = \frac{x(x-1)}{2},$$

$$l_1(x) = \frac{(x+1)(x-1)}{(0+1)(0-1)} = 1 - x^2,$$

$$l_2(x) = \frac{(x+1)(x-0)}{(1+1)(1-0)} = \frac{x(x+1)}{2}.$$

Hence, the interpolating polynomial is

$$p(x) = \sum_{j=0}^{2} y_j l_j(x) = 3 \cdot \frac{x(x-1)}{2} + 5 \cdot (1 - x^2) + 2 \cdot \frac{x(x+1)}{2}$$

$$= -\frac{5}{2}x^2 - \frac{1}{2}x + 5.$$

5.2.2   The method of undetermined coefficients

Alternatively, we could use the power form for the interpolating polynomial,

$$p(x) = a_0 + a_1 x + a_2 x^2 + \cdots + a_n x^n,$$

and use the fact that $p(x_i) = y_i, i = 0, \ldots, n$, to find its coefficients by solving the following linear system for $a_0, \ldots, a_n$:

$$
\begin{aligned}
a_0 + a_1 x_0 + a_2 x_0^2 + \cdots + a_n x_0^n &= y_0 \\
a_0 + a_1 x_1 + a_2 x_1^2 + \cdots + a_n x_1^n &= y_1 \\
&\vdots \\
a_0 + a_1 x_n + a_2 x_n^2 + \cdots + a_n x_n^n &= y_n.
\end{aligned}
$$

The same system in the matrix form is given by $Va = y$, where

$$
V = \begin{bmatrix}
1 & x_0 & x_0^2 & \cdots & x_0^n \\
1 & x_1 & x_1^2 & \cdots & x_1^n \\
\vdots & \vdots & \vdots & \ddots & \vdots \\
1 & x_n & x_n^2 & \cdots & x_n^n
\end{bmatrix}, \quad
a = \begin{bmatrix}
a_0 \\
a_1 \\
\vdots \\
a_n
\end{bmatrix}, \quad
y = \begin{bmatrix}
y_0 \\
y_1 \\
\vdots \\
y_n
\end{bmatrix}.
$$

The matrix V is a *Vandermonde matrix* and is known to have the determinant

$$det(V) = \prod_{0 \le i < j \le n} (x_i - x_j) \ne 0.$$

This implies that there exists a unique solution $a = V^{-1}y$ to the system.

Example 5.5 *For the same data as in the previous example,*

x	-1	0	1
y	3	5	2

we have the following system:

$$
\begin{bmatrix}
1 & -1 & 1 \\
1 & 0 & 0 \\
1 & 1 & 1
\end{bmatrix}
\begin{bmatrix}
a_0 \\
a_1 \\
a_2
\end{bmatrix}
=
\begin{bmatrix}
3 \\
5 \\
2
\end{bmatrix}.
$$

The solution to this system is $a_0 = 5, a_1 = -\frac{1}{2}, a_2 = -\frac{5}{2}$, and the corresponding interpolating polynomial is $p(x) = -\frac{5}{2}x^2 - \frac{1}{2}x + 5$.

5.2.3   Newton's method

Assume that we are given the interpolating polynomial $p_n(x)$ for data points $(x_i, y_i), i = 0, 1, \ldots, n$. Suppose that a new data point (x_{n+1}, y_{n+1}) is added, then a natural question is, how can we use p_n in order to construct the interpolating polynomial p_{n+1} for points $(x_i, y_i), i = 0, 1, \ldots, n + 1$? Unfortunately, the Lagrange method and method of undetermined coefficients do not utilize the fact of having $p_n(x)$ when computing $p_{n+1}(x)$. Consider, however, the Newton form of $p_{n+1}(x)$:

$$p_{n+1}(x) = \underbrace{\sum_{j=0}^{n} a_j \prod_{i=0}^{j-1} (x - x_i)}_{q(x)} + \underbrace{a_{n+1} \prod_{i=0}^{n} (x - x_i)}_{r(x)} = q(x) + r(x),$$

where

$$q(x) = \sum_{j=0}^{n} a_j \prod_{i=0}^{j-1} (x - x_i) = a_0 + a_1(x - x_0) + \ldots + a_n(x - x_0) \cdot \ldots \cdot (x - x_{n-1})$$

and

$$r(x) = a_{n+1} \prod_{i=0}^{n} (x - x_i) = a_{n+1}(x - x_0) \cdot \ldots \cdot (x - x_n).$$

We can show that $q(x) = p_k(x)$. Indeed, by definition of $p_n(x)$ and $p_{n+1}(x)$,

$$p_{n+1}(x_j) = p_n(x_j) = y_j \text{ for } j = 0, \ldots, n.$$

Since $r(x_j) = a_{n+1} \prod_{i=0}^{n} (x_j - x_i) = 0$ for $j = 0, \ldots, n$, we have

$$q(x_j) = p_{n+1}(x_j) - r(x_j) = p_{n+1}(x_j) = y_j \text{ for } j = 0, \ldots, n,$$

and due to the uniqueness of the interpolating polynomial, $q(x) = p_n(x)$. As a result, we obtain the following representation for $p_{n+1}(x)$:

$$p_{n+1}(x) = p_n(x) + a_{n+1} \prod_{i=0}^{n} (x - x_i).$$

In view of the fact that $p_{n+1}(x_{n+1}) = y_{n+1}$, this implies that

$$y_{n+1} = p_{n+1}(x_{n+1}) = p_n(x_{n+1}) + a_{n+1} \prod_{i=0}^{n} (x_{n+1} - x_i) \Rightarrow$$

$$a_{n+1} = \frac{y_{n+1} - p_n(x_{n+1})}{\prod_{i=0}^{n} (x_{n+1} - x_i)}.$$

Therefore, $p_{n+1}(x)$ can be computed efficiently using $p_n(x)$.

Example 5.6 *Assume that a new data point $(x_3, y_3) = (2, 4)$ is added to the data in Example 5.4:*

x	-1	0	1	2
y	3	5	2	4

From Example 5.4 we know that the quadratic interpolating polynomial for the first three data points is given by $p_2(x) = -\frac{5}{2}x^2 - \frac{1}{2}x + 5$. We use Newton's method to find the 3rd degree interpolating polynomial for the given data. We have:

$$p_3(x) = p_2(x) + a_3(x+1)(x)(x-1), \; p_2(2) = -6, \; and \; p_3(2) = 4.$$

Hence, $a_3 = \frac{4+6}{3\cdot2\cdot1} = \frac{5}{3}$, and

$$p_3(x) = -\frac{5}{2}x^2 - \frac{1}{2}x + 5 + \frac{5}{3}(x+1)(x)(x-1) = \frac{5}{3}x^3 - \frac{5}{2}x^2 - \frac{13}{6}x + 5.$$

5.3 Theoretical Error of Interpolation and Chebyshev Polynomials

Assume that we interpolate a given function f using the data points $(x_i, f(x_i)), i = 0, \ldots, n$, where $x_0 < x_1 < \cdots < x_n$. The following theorem describes the error of interpolation over the interval $[x_0, x_n]$.

Theorem 5.2 *Let $f(x) \in C^{(n+1)}([x_0, x_n])$. Then there exists $\xi \in [x_0, x_n]$ such that for all $x \in [x_0, x_n]$, the error*

$$e_n(x) = f(x) - p_n(x) = \frac{f^{(n+1)}(\xi)}{(n+1)!} \prod_{i=0}^{n}(x - x_i). \tag{5.2}$$

Note that ξ depends on x and its value is not available explicitly. However, even if $f^{(n+1)}(\xi)$ is not known, it may be possible to obtain an *upper bound* c_n such that

$$|f^{(n+1)}(x)| \le c_n \;\; \forall x \in [x_0, x_n].$$

Then

$$|e_n(x)| \le \frac{c_n}{(n+1)!} \prod_{i=0}^{n} |x - x_i| \;\; \forall x \in [x_0, x_n]. \tag{5.3}$$

Example 5.7 *Consider the quadratic interpolation of $f(x) = \cos x$ using 3 points, $x_0 = -\pi/2, x_1 = 0$, and $x_2 = \pi/2$ shown in Figure 5.1 at page 114. We will estimate the error of interpolation at $x = \pi/4$.*

In this case,

$$|f^{(3)}(x)| \le 1 \;\; \forall x,$$

hence,

$$|e_2(x)| = |f(x) - p_2(x)| \le \frac{1}{3!} \left| \left(x + \frac{\pi}{2} \right) x \left(x - \frac{\pi}{2} \right) \right| = \frac{1}{6} \left| x^3 - \frac{\pi^2}{4} x \right|,$$

and for $x = \pi/4$ we get the bound

$$|e_2(x)| \leq \frac{\pi^3}{128} \approx 0.2422365.$$

The actual error of interpolation is $e_2(\pi/4) = 1/\sqrt{2} - 3/4 \approx -0.0428932$.

Note that the upper bound on the error in (5.3) depends on the choice of the interpolation nodes $x_1, \ldots, x_{n-1}$. Our next goal is to try to select these nodes in such a way that the largest possible value of the upper bound in (5.3) is as small as possible.

Without loss of generality, we can assume that $[x_0, x_n] = [-1, 1]$. Indeed, consider an arbitrary interval $[x_0, x_n]$ within the domain of $f(x)$. Denoting by

$$F(x) = f\left(\frac{x(x_n - x_0) + x_0 + x_n}{2}\right),$$

we transform the function $f(x)$ defined over $[x_0, x_n]$ into the function $F(x)$ with the domain $[-1, 1]$.

Denote by

$$R_{n+1}(x) = \prod_{k=0}^{n}(x - x_k) = (x - x_0)(x - x_1) \cdot \ldots \cdot (x - x_n).$$

Then, from (5.3),

$$|e_n(x)| \leq \frac{c_n}{(n+1)!} \max_{-1 \leq x \leq 1} |R_{n+1}(x)|. \tag{5.4}$$

We can control the error bound in (5.4) by choosing a set of nodes x_i, $i = 0, 1, \ldots, n$ which would minimize the largest deviation of $R_{n+1}(x)$ from zero over $[-1, 1]$. That is, our objective is to minimize $\max_{-1 \leq x \leq 1} |R_{n+1}(x)|$ by selecting an appropriate set of interpolating nodes. This leads to a discussion on *Chebyshev polynomials*. Let us introduce the following function:

$$T_n(x) = \cos(n \arccos x), \quad n = 0, 1, \ldots. \tag{5.5}$$

Theorem 5.3 *For any $n = 0, 1, \ldots$, function $T_n(x)$ is a polynomial of degree n (called the nth Chebyshev polynomial), moreover,*

$$T_0(x) = 1, \; T_1(x) = x,$$

and for $n \geq 2$, $T_n(x)$ can be found recursively from the following relation:

$$T_{n+1}(x) = 2xT_n(x) - T_{n-1}(x). \tag{5.6}$$

Proof. We have

$$T_0(x) = \cos 0 = 1; \quad T_1(x) = \cos(\arccos x) = x.$$

We will use the following trigonometric identity to prove (5.6):

$$\cos((n+1)\alpha) = 2\cos\alpha\cos(n\alpha) - \cos((n-1)\alpha), \quad n = 1, 2, \ldots.$$

With $\alpha = \arccos x$, this identity is transformed into (5.6).

To complete the proof, we need to show that $T_n(x)$ is a polynomial of degree n. We will use relation (5.6) and induction for the proof. We have already shown that for $n = 0$ and $n = 1$, $T_n(x)$ is a polynomial of degree 0 and 1, respectively. Assume that T_k is a polynomial of degree k for all $k = 0, 1, \ldots, n$. Then

$$T_{n+1}(x) = 2xT_n(x) - T_{n-1}(x)$$

is a polynomial of degree that is equal to the degree of $2xT_n(x)$, which is $n+1$. Therefore, by induction the statement is correct for any integer $n \geq 0$. $\square$

Table 5.1 shows the first six Chebyshev polynomials.

TABLE 5.1: Chebyshev polynomials $T_0(x)$ through $T_5(x)$.

$T_0(x) = 1$
$T_1(x) = x$
$T_2(x) = 2x^2 - 1$
$T_3(x) = 4x^3 - 3x$
$T_4(x) = 8x^4 - 8x^2 + 1$
$T_5(x) = 16x^5 - 20x^3 + 5x$

5.3.1 Properties of Chebyshev polynomials

Chebyshev polynomials have the following properties.

1. If $n \geq 1$, then the coefficient of x^n in $T_n(x)$ is 2^{n-1}.

2. $T_n(x)$ has n distinct roots $x_0, x_1, \ldots, x_{n-1} \in [0, 1]$ called *Chebyshev nodes*, which are given by

$$x_k = \cos\left(\frac{\pi + 2\pi k}{2n}\right), \quad k = 0, 1, \ldots, n - 1. \tag{5.7}$$

3. The deviation of T_n from zero on $[-1, 1]$ is bounded by

$$\max_{-1 \leq x \leq 1} |T_n(x)| = 1. \tag{5.8}$$

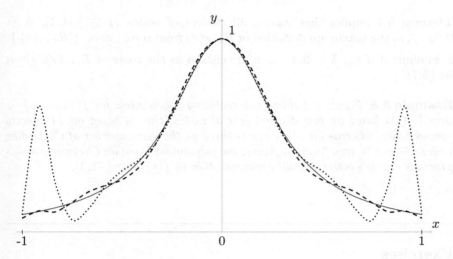

FIGURE 5.2: Polynomial approximations to $y = f(x) = \frac{1}{1+10x^2}$ over $[-1, 1]$ (shown with a solid line) based on 11 equally spaced nodes (dotted) and based on 11 Chebyshev nodes (dashed).

4. $T_n(x)$ is an even function for $n = 2k$, and an odd function for $n = 2k+1$, that is,

$$T_{2k}(-x) = T_{2k}(x) \ \text{ and } \ T_{2k+1}(-x) = -T_{2k+1}(x), \ \ k = 0, 1, \ldots.$$

It appears, that the normalized Chebyshev polynomial of degree $n + 1$, $T_{n+1}(x)/2^n$ has a property useful for polynomial approximation: it has the smallest deviation from zero over $[-1, 1]$ among all normalized polynomials of degree $n + 1$. This is expressed by the following theorem, which we state without a proof.

Theorem 5.4 (Minimax) *There is a unique polynomial of degree $n+1$, $T(x) = T_{n+1}(x)/2^n$ such that for any polynomial $R_{n+1}(x)$ of degree $n+1$ (with the coefficient for x^{n+1} equal to 1), the following property holds:*

$$\max_{-1 \leq x \leq 1}\{|T(x)|\} \leq \max_{-1 \leq x \leq 1}\{|R_{n+1}(x)|\},$$

Since any polynomial of degree $n + 1$ with leading coefficient 1 can be represented as

$$R_{n+1}(x) = (x - x_0)(x - x_1) \cdot \ldots \cdot (x - x_n),$$

Theorem 5.4 implies that among all choices of nodes $x_k \in [-1, 1]$, $k = 0, 1, \ldots, n$, the maximum deviation of $R_{n+1}(x)$ from zero, $\max\limits_{-1 \le x \le 1} \{|R_{n+1}(x)|\}$, is minimized if x_k, $k = 0, 1, \ldots, n$ are chosen as the roots of $T_{n+1}(x)$, given by (5.7).

Example 5.8 *Figure 5.2 shows interpolating polynomials for $f(x) = \frac{1}{1+10x^2}$ over $[-1, 1]$ based on two distinct sets of nodes. One is based on 11 equally spaced nodes, whereas the other one is based on the same number of Chebyshev nodes. As can be seen from the figure, the polynomial based on Chebyshev nodes provides a much better overall approximation of $f(x)$ over $[-1, 1]$.*

Exercises

5.1. Use Algorithm 5.1 to evaluate $p(3)$, where

(a) $p(x) = 4 + 5(x + 1) - 7(x + 1)(x - 2) + 8(x + 1)(x - 2)(x - 4)$;

(b) $p(x) = 7 + 3(x - 2) + 4(x - 2)(x - 4) - 10(x - 2)(x - 4)(x - 5) + 6(x - 2)(x - 4)(x - 5)x$.

5.2. Consider the following polynomial given in the shifted power form: $p(x) = 1 + (x - 5555.5)^2$. The power form representation of $p(x)$ is $p(x) = 30863581.25 - 11.111x + x^2$. For each of the two representations, evaluate $p(5555)$ using 6-digit floating-point arithmetic.

5.3. Find the quadratic polynomial $p_2(x)$ that interpolates the data

(i)
x	1	2	4
y	-1	-1	2

(ii)
x	-1	2	5
y	1	0	-2

(iii)
x	-2	0	2
y	9	7	9

(iv)
x	0	3	5
y	-1	8	-4

(a) using the Lagrange method;

(b) using the method of undetermined coefficients.

5.4. Add the data point $(x_3, y_3) = (6, 1)$ to each table in Exercise 5.3.

(a) Use Newton's method to find the cubic interpolating polynomial $p_3(x)$ for the resulting data.

(b) Find $p_3(x)$ using the Lagrange method.

(c) Find $p_3(x)$ using the method of undetermined coefficients.

5.5. Find the trigonometric function $T(x) = a_0 + a_1 \sin x + a_2 \sin 2x$ that interpolates the following data.

x	0	$\pi/2$	$\pi/3$
y	-1	2	1

5.6. Find the cubic interpolating polynomial $p_3(x)$ for $f(x) = \cos x$ using $x_0 = 0.3, x_1 = 0.4, x_2 = 0.5$, and $x_3 = 0.6$. For $x = 0.44$,

    (a) compute $e_3(x) = f(x) - p_3(x)$;

    (b) estimate the error $e_3(x)$ using the bound in Equation (5.3) at page 120.

5.7. Find the quartic interpolating polynomial $p_4(x)$ for $f(x) = e^{3x}$ using $x_0 = -1, x_1 = -0.5, x_2 = 0, x_3 = 0.5$, and $x_4 = 1$. For $x = 0.8$,

    (a) compute $e_4(x) = f(x) - p_4(x)$;

    (b) estimate the error $e_4(x)$ using the bound in Equation (5.3) at page 120.

Chapter 6

Numerical Integration

In this chapter we will deal with the problem of integration of a function over an interval,

$$\int_a^b f(x)dx,$$

which arises in many situations.

Example 6.1 *A manufacturer produces 2-inch wafers. Due to the production process variability, the actual diameter of wafers made is normally distributed with a mean of 2 inches and a standard deviation of 0.01 inch. Specifications require that the diameter is between 1.985 and 2.02 inches. We need to estimate the fraction of the produced wafers that will be acceptable.*

Denote by ξ the random variable corresponding to the diameter of a wafer. We know that ξ has the normal probability density function (pdf)

$$p(x) = \frac{1}{0.01\sqrt{2\pi}} \exp\left(\frac{-(x-2)^2}{2 \cdot 0.01^2}\right).$$

Then to answer our question, we need to estimate the probability $P(1.985 \le \xi \le 2.02)$ of ξ being in the range $[1.985, 2.02]$. From probability theory, it is known that this probability is given by the following integral:

$$P(1.985 \le \xi \le 2.02) = \int_{1.985}^{2.02} p(x)dx = \frac{1}{0.01\sqrt{2\pi}} \int_{1.985}^{2.02} \exp\left(\frac{-(x-2)^2}{2 \cdot 0.01^2}\right) dx,$$

which is equal to the area enclosed between the x-axis and the plot of $p(x)$, where $x \in [1.985, 2.02]$, as shown in Figure 6.1. This integral cannot be computed analytically, and numerical methods are required to approximate its value.

The idea behind the numerical methods to be discussed in this chapter is simple: we replace the integrated function (*integrand*) with a polynomial approximating this function. Recall the Lagrange form of a polynomial:

$$p_n(x) = a_0 l_0(x) + a_1 l_1(x) + \cdots + a_n l_n(x) = \sum_{i=0}^{n} a_i l_i(x),$$

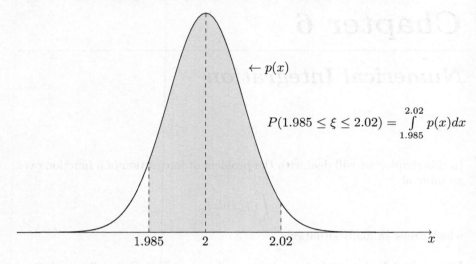

FIGURE 6.1: Computing the probability using integral.

where

$$l_k(x) = \prod_{\substack{i=0 \\ i \neq k}}^{n} \frac{(x - x_i)}{(x_k - x_i)}, \quad k = 0, 1, \ldots, n.$$

For a given function $f(x)$ and n distinct points $x_0, \ldots, x_n$, the interpolating polynomial is

$$p_n(x) = \sum_{k=0}^{n} f(x_k) l_k(x). \tag{6.1}$$

Consider now the problem of approximating $\int_a^b f(x) dx$ for some "complicated" $f : [a, b] \to \mathbb{R}$, e.g., such that the integral cannot be computed analytically. Then we could replace the complicated $f(x)$ with an interpolating polynomial $p_n(x)$ based on evaluating f in n equally spaced points $x_0, \ldots, x_n$ on $[a, b]$, where $x_i = a + \frac{b-a}{n} i, i = 0, \ldots, n$, and compute $\int_a^b p_n(x) dx$ instead. Note that the operation of integration is linear, that is

$$\int_a^b (f(x) + g(x)) dx = \int_a^b f(x) dx + \int_a^b g(x) dx,$$

and

$$\int_a^b \alpha f(x) dx = \alpha \int_a^b f(x) dx,$$

where $g(x) : [a, b] \to \mathbb{R}$ is a function and α is a scalar. Therefore, if $f(x) = p_n(x) + e_n(x)$, where $e_n(x)$ is the error of approximation of f by p_n, then

$$\int_a^b f(x) dx = \int_a^b p_n(x) dx + \int_a^b e_n(x) dx,$$

and the error of approximation of $\int_a^b f(x)dx$ by $\int_a^b p_n(x)dx$ equals $\int_a^b e_n(x)dx$.

Consider the interpolating polynomial of $f(x)$ in the form (6.1). Since the operation of integration is linear, we have

$$\int_a^b p_n(x)dx = \int_a^b \left(\sum_{i=0}^n f(x_i)l_i(x) \right) dx = \sum_{i=0}^n f(x_i) \int_a^b l_i(x)dx, \qquad (6.2)$$

where $\int_a^b l_i(x)dx$ does not depend on f for $i = 0, \ldots, n$. Thus, for a function $f : [a, b] \to \mathbb{R}$ the integral $\int_a^b f(x)dx$ is approximated by

$$Q_n(f) = \sum_{i=0}^n A_i f(x_i), \qquad (6.3)$$

where $A_i = \int_a^b l_i(x)dx$. $Q_n(f)$ is called the *quadrature formula* or *numerical integration formula*.

Taking $n = 1$ and $n = 2$ in (6.2) we obtain the trapezoidal and Simpson's rules, respectively, which are discussed next.

6.1 Trapezoidal Rule

Given a function $f(x)$ and points a and b, in the trapezoidal rule we want to compute the integral $\int_a^b f(x)dx$ using the linear interpolating polynomial passing through the points $(a, f(a))$ and $(b, f(b))$. We take $x_0 = a$, $x_1 = b$, and approximate f by p_1:

$$
\begin{aligned}
p_1(x) &= f(a)l_0(x) + f(b)l_1(x) \\
&= f(a)\frac{x-b}{a-b} + f(b)\frac{x-a}{b-a}.
\end{aligned}
$$

Next, we use p_1 to approximate the integral of f over $[a, b]$:

$$
\begin{aligned}
\int_a^b f(x)dx &\approx f(a)\int_a^b \frac{x-b}{a-b}dx + f(b)\int_a^b \frac{x-a}{b-a}dx \\
&= \frac{b-a}{2}(f(a) + f(b)).
\end{aligned}
$$

Thus, we obtain the formula for approximately computing the integral known as the *trapezoidal rule*:

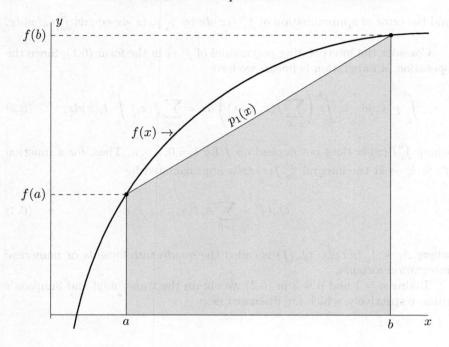

FIGURE 6.2: An illustration of the trapezoidal rule.

Trapezoidal Rule:

$$\int_a^b f(x)dx \approx \frac{b-a}{2}(f(a) + f(b)). \qquad (6.4)$$

Note that (6.4) is just the area of the trapezoid defined by the points $(a, 0), (a, f(a)), (b, f(b))$, and $(b, 0)$, as shown in Figure 6.2.

Example 6.2 *Using the trapezoidal rule to approximate the integral* $\int_0^1 \exp(x^2)$, *we obtain:*

$$\int_0^1 \exp(x^2)dx \approx \frac{1}{2}(\exp(0) + \exp(1)) = \frac{1}{2}(1 + e) \approx 1.85914.$$

For $f(x) = \frac{\sin x}{x}$ *and* $[a, b] = [1, 5]$ *the trapezoidal rule gives*

$$\int_1^5 \frac{\sin x}{x}dx \approx \frac{5-1}{2}\left(\frac{\sin 1}{1} + \frac{\sin 5}{5}\right) = 2\left(\sin 1 + \frac{\sin 5}{5}\right) \approx 1.29937.$$

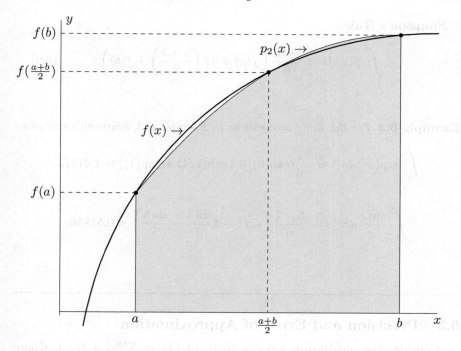

FIGURE 6.3: An illustration for Simpson's rule.

6.2   Simpson's Rule

To derive Simpson's rule, we approximate f with a quadratic interpolating polynomial p_2 based on the points $x_0 = a, x_1 = \frac{a+b}{2}$ and $x_2 = b$. Then we have

$$p_2(x) = f(a)\frac{(x - \frac{a+b}{2})(x - b)}{(x_0 - \frac{a+b}{2})(x_0 - b)} + f\left(\frac{a+b}{2}\right)\frac{(x - a)(x - b)}{(x_1 - a)(x_1 - b)}$$
$$+ f(b)\frac{(x - a)(x - \frac{a+b}{2})}{(x_2 - a)(x_2 - x_1)}.$$

Integrating p_2 over $[a, b]$, we obtain

$$\int_a^b p_2(x)dx = \frac{b-a}{6}\left(f(a) + 4f\left(\frac{a+b}{2}\right) + f(b)\right). \tag{6.5}$$

This yields Simpson's rule:

Simpson's Rule:

$$\int_a^b f(x)dx \approx \frac{b-a}{6}\left(f(a) + 4f\left(\frac{a+b}{2}\right) + f(b)\right). \tag{6.6}$$

Example 6.3 *For the same integrals as in Example 6.2, Simpson's rule gives*

$$\int_0^1 \exp(x^2)dx \approx \frac{1}{6}(\exp(0) + 4\exp(1/4) + \exp(1)) \approx 1.47573,$$

$$\int_1^5 \frac{\sin x}{x}dx \approx \frac{5-1}{6}\left(\sin 1 + 4\frac{\sin 3}{3} + \frac{\sin 5}{5}\right) \approx 0.55856.$$

6.3 Precision and Error of Approximation

Consider the quadrature formula (6.3), $Q_n(f) = \sum_{i=0}^n A_i f(x_i)$, where $A_i = \int_a^b l_i(x)dx$.

Definition 6.1 (Precision of a Quadrature Formula) *If $Q_n(f) = \int_a^b f(x)dx$ for all polynomials f of degree $\leq m$, and if there exists a polynomial $\hat{f}$ of degree $m+1$ for which $Q_n(\hat{f}) \neq \int_a^b \hat{f}(x)dx$, then we say that the formula has precision m.*

Clearly $m \geq n$ when $n+1$ points are used, since in this case $p_n(x) = f(x)$ for any polynomial $f(x)$ of degree at most n. For example, for the trapezoidal rule we have

$$f(x) = 1: \quad \int_a^b 1dx = (b-a) = Q_1(f);$$

$$f(x) = x: \quad \int_a^b xdx = \frac{b^2 - a^2}{2} = \frac{b-a}{2}(b+a) = Q_1(f).$$

To find the precision of a quadrature, we use it to compute the integrals $\int_a^b x^k dx$ for $k \geq n+1$ until we encounter k for which the error of approximation is nonzero. Using the trapezoidal rule for $f(x) = x^2$, we obtain:

$$\int_a^b x^2 dx = \frac{b^3 - a^3}{3},$$

$$Q_1(f) = \frac{b-a}{2}(a^2 + b^2) = \frac{1}{2}(b^3 + a^2b - ab^2 - a^3),$$

and the error of approximation is

$$\int_a^b x^2 dx - Q_1(f) = \frac{1}{3}(b^3 - a^3) - \frac{1}{2}(b^3 + a^2b - ab^2 - a^3) = -\frac{1}{6}(b-a)^3,$$

thus $\int_a^b f(x)dx \neq Q_1(f)$ for $f(x) = x^2$ whenever $a \neq b$. Hence the trapezoidal rule has the precision 1.

Since $n = 2$ for Simpson's rule, it is precise for all polynomials of degree up to 2. For $f(x) = x^3$ we have

$$\int_a^b x^3 dx = \frac{b^4 - a^4}{4},$$

$$Q_2(f) = \frac{b-a}{6}\left(a^3 + 4((a+b)/2)^3 + b^3\right) = \frac{b^4 - a^4}{4},$$

and Simpson's rule is exact. For $f(x) = x^4$, we can use $[a,b] = [0,1]$ to show that Simpson's rule does not give the exact answer. Indeed,

$$\int_0^1 x^4 dx = \frac{1}{5} \neq \frac{5}{24} = \frac{1}{6}(0 + 4(1/2)^4 + 1) = Q_2(f).$$

Therefore, the precision of Simpson's rule is 3.

Next we discuss the error of approximation of the trapezoidal and Simpson's rules. Recall from Theorem 5.2 that for $f(x) \in C^{(n+1)}([a,b])$ the error of interpolating polynomial $p_n(x)$ based on $x_0, \ldots, x_n \in [a,b]$ at any $x \in [a,b]$ is

$$e_n(x) = f(x) - p_n(x) = \frac{f^{(n+1)}(\xi)}{(n+1)!}\prod_{i=0}^n (x - x_i),$$

for some $\xi \in [a,b]$. In particular, for $n = 1$ the error is

$$e_1(x) = f(x) - p_1(x) = \frac{f''(\xi)}{2}(x-a)(x-b).$$

Therefore, the error of approximation of the trapezoidal rule is

$$e_T = \int_a^b e_1(x)dx = \frac{f''(\xi)}{2}\int_a^b (x-a)(x-b)dx = -\frac{f''(\xi)(b-a)^3}{12}.$$

For Simpson's rule, it can be shown that there exists $\xi \in [a,b]$ such that its error of approximation is

$$e_S = -\frac{f^{(4)}(\xi)}{90}\left(\frac{b-a}{2}\right)^5.$$

In summary, we have

$$e_T = -\frac{f''(\xi)}{12}(b-a)^3 \quad \text{for some } \xi \in [a,b] \tag{6.7}$$

for the trapezoidal rule, and

$$e_S = -\frac{f^{(4)}(\xi)}{90}\left(\frac{b-a}{2}\right)^5 \quad \text{for some } \xi \in [a,b] \tag{6.8}$$

for Simpson's rule.

6.4   Composite Rules

As can be seen from the error terms (6.7) and (6.8) above, the trapezoidal and Simpson's rules above may not work well if the interval $[a,b]$ is large. To reduce the error in this case, we can divide the interval $[a,b]$ into N smaller subintervals using points $a = x_0, x_1, \ldots, x_N = b$, and since

$$\int_a^b f(x)dx = \sum_{i=0}^{N-1} \int_{x_i}^{x_{i+1}} f(x)dx,$$

we can apply the rules to the integrals defined on the N smaller intervals.

6.4.1   The composite trapezoidal rule

Divide $[a,b]$ into N equal intervals, so we have

$$h = \frac{b-a}{N}, \quad x_i = a + ih, i = 0, \ldots, N$$

(see Figure 6.4). Therefore,

$$a = x_0 < x_1 < \cdots < x_N = b,$$

and for each interval we have

$$\int_{x_i}^{x_{i+1}} f(x)dx = \frac{h}{2}(f(x_i) + f(x_{i+1})) - \frac{f''(\xi_i)}{12}h^3$$

for some $\xi_i \in [x_i, x_{i+1}]$. Hence, we obtain the following formulas for the composite trapezoidal rule:

$$\int_a^b f(x)dx \approx T_N(f),$$

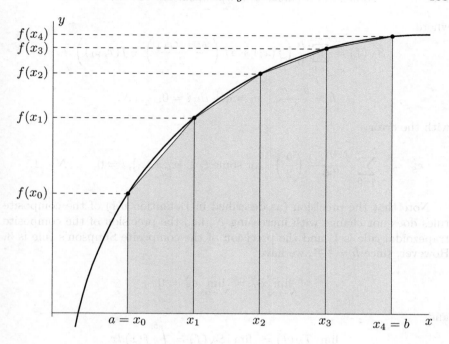

FIGURE 6.4: An illustration of the composite trapezoidal rule with $N = 4$.

where

$$T_N(f) = \sum_{i=0}^{N-1} \frac{h}{2} \left(f(x_i) + f(x_{i+1}) \right) = h \sum_{i=1}^{N-1} f(x_i) + \frac{h}{2} \left(f(x_0) + f(x_N) \right),$$

$$h = \frac{b-a}{N}, \quad x_i = a + ih, i = 0, \ldots, N,$$

with the error

$$e_T^N = -\sum_{i=0}^{N-1} \frac{f''(\xi_i)h^3}{12} \text{ for some } \xi_i \in [x_i, x_{i+1}], i = 0, \ldots, N-1.$$

6.4.2 Composite Simpson's rule

Similarly, for the composite version of Simpson's rule we have the following formulas:

$$\int_a^b f(x)dx \approx S_N(f),$$

where

$$S_N(f) = \sum_{i=0}^{N-1} \frac{h}{6}\left(f(x_i) + 4f\left(\frac{x_i + x_{i+1}}{2}\right) + f(x_{i+1})\right),$$

$$h = \frac{b-a}{N}, \quad x_i = a + ih, i = 0, \ldots, N,$$

with the error

$$e_S^N = -\sum_{i=0}^{N-1} \frac{f^{(4)}(\xi_i)}{90}\left(\frac{h}{2}\right)^5 \text{ for some } \xi_i \in [x_i, x_{i+1}], i = 0, \ldots, N-1.$$

Note that the precision (as described in Definition 6.1) of the composite rules does not change with increasing N, i.e., the precision of the composite trapezoidal rule is 1, and the precision of the composite Simpson's rule is 3. However, since $h = \frac{b-a}{N}$, we have

$$\lim_{N\to\infty} e_T^N = \lim_{N\to\infty} e_S^N = 0,$$

and

$$\lim_{N\to\infty} T_N(f) = \lim_{N\to\infty} S_N(f) = \int_a^b f(x)dx.$$

Example 6.4 *Estimate the integral*

$$\int_1^4 \frac{1}{x}dx$$

using the composite trapezoidal rule and the composite Simpson's rule with $N = 3$.

Applying the composite trapezoidal rule, we obtain

$$\begin{aligned}
\int_1^4 \frac{1}{x}dx &= \sum_{i=1}^3 \int_i^{i+1} \frac{1}{x}dx \\
&\approx \sum_{i=1}^3 \frac{1}{2}\left(\frac{1}{i} + \frac{1}{i+1}\right) \\
&= \frac{1}{2}\left[\left(1 + \frac{1}{2}\right) + \left(\frac{1}{2} + \frac{1}{3}\right) + \left(\frac{1}{3} + \frac{1}{4}\right)\right] \\
&= \frac{1}{2} + \frac{1}{3} + \frac{1}{2}\left(1 + \frac{1}{4}\right) = \frac{35}{24} \approx 1.3863.
\end{aligned}$$

Note, that the exact value of the integral is $\int_1^4 \frac{1}{x}dx = \ln 4$, therefore the absolute error is $1.3863 - \ln 4 \approx 0.0720$.

Using the composite Simpson's rule, we have:

$$\int_1^4 \frac{1}{x}dx = \sum_{i=1}^{3} \int_i^{i+1} \frac{1}{x}dx$$

$$\approx \sum_{i=1}^{3} \frac{1}{6}\left(\frac{1}{i} + 4\frac{2}{2i+1} + \frac{1}{i+1}\right)$$

$$= \frac{1}{6}\left[\left(1 + \frac{8}{3} + \frac{1}{2}\right) + \left(\frac{1}{2} + \frac{8}{5} + \frac{1}{3}\right) + \left(\frac{1}{3} + \frac{8}{7} + \frac{1}{4}\right)\right]$$

$$= \frac{3497}{2520} \approx 1.3877.$$

The absolute error is $1.3877 - \ln 4 \approx 0.0014$ *in this case.*

6.5 Using Integrals to Approximate Sums

If the value of an integral can be computed analytically, the numerical methods for computing integrals can be used to approximate the sums appearing in the corresponding computations. For example, consider the integral

$$\int_1^N \frac{1}{x}dx.$$

We can use the composite trapezoidal rule with step size $h = 1$ to estimate this integral.

$$\int_1^N \frac{1}{x}dx = \sum_{i=1}^{N-1} \int_i^{i+1} \frac{1}{x}dx$$

$$\approx \frac{1}{2}\sum_{i=1}^{N-1}\left(\frac{1}{i} + \frac{1}{i+1}\right)$$

$$= \frac{1}{2}\left(\sum_{i=1}^{N-1}\frac{1}{i} + \sum_{i=2}^{N}\frac{1}{i}\right)$$

$$= \sum_{i=1}^{N}\frac{1}{i} - \frac{1}{2}\left(1 + \frac{1}{N}\right)$$

$$= 1 + \frac{1}{2} + \cdots + \frac{1}{N} - \left(\frac{1}{2} + \frac{1}{2N}\right).$$

Thus,

$$\sum_{i=1}^{N}\frac{1}{i} \approx \int_1^N \frac{1}{x}dx + \frac{1}{2} - \frac{1}{2N} = \ln N + \frac{1}{2} - \frac{1}{2N}.$$

Example 6.5 *Consider the integral* $\int\limits_{1}^{5000} \frac{1}{x+10}dx$. *Use the composite trape-zoidal rule to estimate* $\sum\limits_{n=1}^{5000} \frac{1}{n+10}$.

Applying the composite trapezoidal rule, and denoting by $x = \sum\limits_{n=1}^{5000} \frac{1}{n+10}$, *we have*

$$\int\limits_{1}^{5000} \frac{1}{x+10}dx \approx \frac{1}{2} \sum\limits_{i=1}^{4999} \left(\frac{1}{i+10} + \frac{1}{(i+1)+10} \right) =$$

$$\frac{1}{2} \left(\sum\limits_{i=1}^{4999} \frac{1}{i+10} + \sum\limits_{i=2}^{5000} \frac{1}{i+10} \right) =$$

$$= \frac{1}{2} \left(x - \frac{1}{5000+10} + x - \frac{1}{1+10} \right) = x - \frac{1}{2} \left(\frac{1}{5010} + \frac{1}{11} \right).$$

On the other hand, $\int\limits_{1}^{5000} \frac{1}{x+10}dx = \ln(x+10)|_{1}^{5000} = \ln(5010) - \ln(11)$, *which gives* $x - \frac{1}{2} \left(\frac{1}{5010} + \frac{1}{11} \right) \approx \ln(5010) - \ln(11)$, *or* $x \approx \ln(5010) - \ln(11) + \frac{1}{2} \left(\frac{1}{5010} + \frac{1}{11} \right) \approx 6.1669$.

Exercises

6.1. Estimate the integral $\int\limits_{-1}^{1} x^5 \cos x\, dx$ using

    (a) the trapezoidal rule;

    (b) Simpson's rule.

6.2. Estimate the integral $\int_{0}^{\pi} \exp(x) \cos x\, dx$ using

    (a) the trapezoidal rule;

    (b) Simpson's rule.

6.3. Evaluate the integral $\int_{0}^{1} x \exp(x)dx$ using the trapezoidal and Simpson's rule. Find lower and upper bounds on the approximation error in both cases.

6.4. Consider the integral

$$I = \int\limits_{-4}^{4} \frac{1}{1+x^2} dx$$

with the exact value given by $2\arctan(4)$.

(a) Compute the integral by the trapezoidal rule and the composite trapezoidal rule with $N = 2$ and $N = 4$, respectively. Compute the absolute error of approximation.

(b) Compute the integral by Simpson's rule and the composite Simpson's rule with $N = 2$ and $N = 4$, respectively. Compute the absolute error of approximation.

(c) Summarize the results you obtained in (a) and (b) in a table. Which method is more accurate?

6.5. Consider the integral $\int\limits_{1}^{2001} \frac{1}{x+11} dx$. Use the composite trapezoidal rule to estimate $\sum\limits_{n=1}^{2000} \frac{1}{n+11}$.

6.6. Find N such that $\sum\limits_{n=1}^{N} \frac{1}{n} \approx 2015$ using the composite trapezoidal rule.

Chapter 7

Numerical Solution of Differential Equations

Differential equations relate a function f to its derivatives. They play a very important role in science and engineering and arise in a wide range of applications.

Example 7.1 *In finance, a bond is a fixed-income security that can be described using the following attributes [22]: current price P; coupon payment rate c; the number of coupon payments per year m; the number of years to maturity n; yield to maturity λ–the annual interest rate that is implied by the current price; and the modified duration D_M, which is computed by the following formula:*

$$D_M = D_M(c, m, n, \lambda) = \frac{1}{\lambda} - \frac{1 + nc - (mn-1)\lambda/m}{(1+\lambda/m)(c(1+\lambda/m)^{mn} - c + \lambda)}.$$

The price and the yield to maturity are related by the following differential equation, referred to as the price sensitivity formula:

$$\frac{dP}{d\lambda} = -PD_M, \quad \text{with } P(\lambda_0) = P_0. \tag{7.1}$$

Differential equations are classified into several broad categories, which are further divided into many subcategories. The broad categories include *ordinary differential equations* and *partial differential equations*. An ordinary differential equation (ODE) relates a function of a single variable to its ordinary derivatives, as in the following examples:

$$\frac{dy}{dt} = -5y + \exp(-t),$$

$$\left[1 + \left(\frac{dy}{dt}\right)^3\right]\frac{d^4y}{dt^4} - 5\frac{dy}{dt}\left(\frac{d^3y}{dt^3}\right)^2 = 0.$$

Here y is the function, and t is the independent variable.

Alternatively, a partial differential equation (PDE) deals with a function of several independent variables and its partial derivatives. The equation

$$\frac{\partial u}{\partial t} = \frac{\partial^2 u}{\partial x^2} + \frac{\partial^2 u}{\partial y^2} + \frac{\partial^2 u}{\partial z^2}$$

is an example of a partial differential equation.

Whichever the type may be, a differential equation is said to be of the nth order if it involves a derivative of the nth order, but no derivative of an order higher than this. For example, the above equations are of the first, fourth, and second order, respectively.

This chapter focuses on numerical methods for solving first-order ODEs.

7.1   Solution of a Differential Equation

Consider a first-order ODE in the form

$$y' = f(x, y). \tag{7.2}$$

A solution of Equation (7.2) is a *function* $y = y(x)$, satisfying this equation. For example, the function $y(x) = \sin x$ is a solution of differential equation $y' = \cos x$. To solve a differential equation means to find *all* functions satisfying this equation. For example, applying rules of integration to the equation $y' = \cos x$, we find that the functions $f_c(x) = \sin x + c$, where c is a real constant, represent all solutions of this equation. In this example, the solution consisting of a set of continuous functions was found *analytically*. In contrast, most *numerical* methods enable one to find only *approximate* values of y corresponding to some finite set of values of t, and the solution found this way is given by the corresponding table of values.

Consider the *initial value problem (IVP)* in the form

$$y' = f(t, y) \text{ with } y(a) = y_0, \tag{7.3}$$

where $a \equiv t_0$ and y_0 are given initial values of the independent variable t and the function y, respectively. A solution to problem (7.3) on an interval $[a, b]$ is a differentiable function $y = y(t)$ such that

$$y'(t) = f(t, y) \text{ for all } t \in [a, b] \tag{7.4}$$

and

$$y(a) = y_0. \tag{7.5}$$

Example 7.2 *The function $y(t) = \exp(2t) + t^2 + t$ is a solution of the IVP*

$$y' = 2y - 2t^2 + 1 \text{ with } y(0) = 1$$

for $t \geq 0$. Indeed,

1. *$y'(t) = 2\exp(2t) + 2t + 1 = 2y(t) - 2t^2 + 1$ for all t;*

2. *$y(0) = \exp(0) + 0^2 + 0 = 1$,*

so, both conditions (7.4) and (7.5) are satisfied.

7.2 Taylor Series and Picard's Methods

Given an IVP

$$y' = f(t, y), \quad y(t_0) = y_0, \tag{7.6}$$

a *Taylor series solution* to this IVP can be generated by recursively computing the coefficients for the Taylor series approximation

$$y_k(t) = \sum_{i=0}^{k} \frac{y^{(i)}(t_0)}{i!} (t - t_0)^i, \ k \geq 0$$

of $y(t)$ about $t = t_0$ using (7.6) as illustrated in the following example.

Example 7.3 *Consider the IVP*

$$y' = 1 + 2ty + 3y^2, \ y(0) = 0.$$

We have $y_0(t) = y(t_0) = 0$. *Using* $t = 0$ *and* $y = 0$ *in the equation we get* $y'(0) = 1$, *hence* $y_1(t) = t$. *To compute* $y''(0)$, *we differentiate the equation:*

$$y'' = (1 + 2ty + 3y^2)' = 2y + 2ty' + 6yy' \implies y''(0) = 0.$$

Thus, $y_2(t) = t$. *Similarly,*

$$y''' = (2y + 2ty' + 6yy')' = 4y' + 2ty'' + 6(y')^2 + 6yy'' \implies y'''(0) = 10,$$

and $y_3(t) = t + \frac{10}{6}t^3$, *etc.*

Note that the IVP (7.6) is equivalent to the integral equation

$$y(t) = y(t_0) + \int_{t_0}^{t} f(u, y(u)) du, \tag{7.7}$$

i.e. $y(t)$ is a solution to (7.6) $\iff$ $y(t)$ is a solution to (7.7). A solution to the integral equation (7.7) can be constructed using *Picard's method* as follows:

$$y_0(t) = y_0;$$
$$y_1(t) = y_0 + \int_{t_0}^{t} f(u, y_0(u)) du;$$
$$y_2(t) = y_0 + \int_{t_0}^{t} f(u, y_1(u)) du;$$
$$\vdots$$
$$y_{k+1}(t) = y_0 + \int_{t_0}^{t} f(u, y_k(u)) du.$$

Picard's method can be summarized as follows:

$$y_0(t) = y(t_0);$$
$$y_{k+1}(t) = y(t_0) + \int_{t_0}^{t} f(u, y_k(u))du, \ k \geq 0. \tag{7.8}$$

Example 7.4 *Consider the IVP*

$$y' = 3t^2 y^2, \ y(0) = -1.$$

Applying Picard's method twice, we obtain:

$$y_1(t) = y(0) + \int_0^t 3u^2 y(0)^2 du = -1 + \int_0^t 3u^2 du = t^3 - 1;$$

$$y_2(t) = -1 + \int_0^t 3u^2 y_1(u)^2 du = -1 + \int_0^t 3u^2(u^3 - 1)^2 du = \frac{1}{3}t^9 - t^6 + t^3 - 1.$$

Example 7.5 *We solve the following IVP using the Taylor series and Pickard's method:*

$$y' = y, \ y(0) = 1.$$

We have: $y'' = y'$, $y''' = y''$, $\ldots$ *Hence,* $1 = y(0) = y'(0) = y''(0) = y'''(0) = \ldots$, *and the Taylor series solution gives*

$$y(t) = 1 + t + \frac{t^2}{2!} + \frac{t^3}{3!} + \cdots + \frac{t^k}{k!} + \cdots = \exp(t).$$

Applying Picard's method, we obtain

$$
\begin{aligned}
y_0(t) &= 1; \\
y_1(t) &= 1 + \int_0^t y_0(u)du = 1 + t; \\
y_2(t) &= 1 + \int_0^t y_1(u)du = 1 + \int_0^t (1+u)du = 1 + t + \frac{t^2}{2}; \\
y_3(t) &= 1 + \int_0^t y_2(u)du = 1 + \int_0^t \left(1 + u + \frac{u^2}{2}\right)du = 1 + t + \frac{t^2}{2} + \frac{t^3}{3!}; \\
&\vdots \\
y_k(t) &= 1 + \int_0^t y_{k-1}(u)du = 1 + \int_0^t \left(1 + u + \frac{u^2}{2!} + \frac{u^3}{3!} + \cdots + \frac{u^{k-1}}{(k-1)!}\right)du \\
&= 1 + u + \frac{u^2}{2!} + \frac{u^3}{3!} + \cdots + \frac{u^{k-1}}{(k-1)!} + \frac{u^k}{k!}.
\end{aligned}
$$

As one can see, in this example Pickard's method also generates the Taylor series for $y(t) = \exp(t)$, *which is the exact solution of the given IVP.*

7.3   Euler's Method

Consider the IVP

$$y' = f(t, y), \quad y(t_0) = y_0. \tag{7.9}$$

Assume that $y(t) \in C^2[t_0, b]$. Applying Taylor's theorem to expand $y(t)$ about $t = t_0$, we obtain

$$y(t) = y(t_0) + y'(t_0)(t - t_0) + \frac{y''(c_t)(t - t_0)^2}{2} \tag{7.10}$$

for some $c_t \in [t_0, t]$, where $t \in [t_0, b]$. Note that $y'(t_0)$ can be replaced by $f(t_0, y(t_0))$. Then for $t = t_1$, where $t_1 = t_0 + h$, $h > 0$, we have

$$y(t_1) = y(t_0) + hf(t_0, y(t_0)) + \frac{y''(c_{t_1})h^2}{2}. \tag{7.11}$$

If h is chosen small enough, then the term involving h^2 is close to 0 and thus can be neglected, yielding the following formula:

$$y(t_1) = y(t_0) + hf(t_0, y(t_0)). \tag{7.12}$$

In the last equation we obtained the formula of a single step of the so-called *Euler's method*.

Figure 7.1 illustrates Euler's method geometrically: starting from the point (t_0, y_0), we obtain an approximation y_1 to the solution $y(t)$ in point t_1 by moving along the tangent line to $y(t)$ at t_0 until we reach the point (t_1, y_1).

Suppose that, given the IVP (7.9), we want to find a numerical solution over interval $[a, b]$, where $a \equiv t_0$, using Euler's method. First we divide $[a, b]$ into n equal subintervals, each of length $h = \frac{b-a}{n}$, thus obtaining a set of $n+1$ mesh points $t_k = a + kh, k = 0, 1, \ldots, n$. The value h is called the *step size*. Using Euler's method we find an approximate solution in the mesh points by following the scheme described in Algorithm 7.1.

Algorithm 7.1 Euler's method for solving the IVP $y' = f(t, y)$ with $y(a) = y_0$ on the interval $[a, b]$.

1: **Input:** function f, interval $[a, b]$, $y_0 = y(a)$, and the number of steps n
2: **Output:** (t_k, y_k) such that $y_k \approx y(t_k)$, $k = 1, \ldots, n$
3: $h = \frac{b-a}{n}$, $t_0 = a$
4: **for** $k = 1, \ldots, n$ **do**
5: $\quad t_k = t_{k-1} + h$
6: $\quad y_k = y_{k-1} + hf(t_{k-1}, y_{k-1})$
7: **end for**
8: **return** (t_k, y_k), $k = 1, \ldots, n$

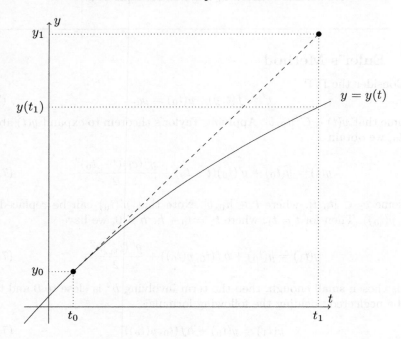

FIGURE 7.1: An illustration of the one-step Euler method.

Example 7.6 *Suppose that we have a 30-year 10% coupon bond with yield to maturity $\lambda_0 = 10\%$ and price $P_0 = 100$. We are interested to know the price of this bond when its yield changes to $\lambda_1 = 11\%$. We can use the price sensitivity formula mentioned in Example 7.1:*

$$\frac{dP}{d\lambda} = -PD_M, \quad \text{with } P(\lambda_0) = P_0.$$

Let us use Euler's method to solve the above equation. We have

$$f(\lambda, P) = -PD_M; \ \lambda_0 = 10; \ P_0 = 100; \ h = 0.11 - 0.10 = 0.01;$$

$$f(\lambda_0, P_0) = -P_0 D_M(0.10, 2, 30, 0.10) = -100 \cdot 9.47 = -947.$$

Therefore,

$$P(\lambda_1) \approx 100 + 0.01(-947) = 100 - 9.47 = 90.53.$$

Example 7.7 *Consider the equation $y' = t + y$ with $y(0) = 1$. Apply Euler's method three times with $h = 0.05$. We have:*

$$f(t, y) = t + y,$$

$$t_i = 0.05 \cdot i, \ i = 0, 1, 2, \ldots.$$

$$
\begin{aligned}
y_1 &= y_0 + hf(t_0, y_0) = y_0 + h(t_0 + y_0) = 1 + 0.05(0+1) = 1.05, \\
y_2 &= y_1 + hf(t_1, y_1) = 1.05 + 0.05(0.05 + 1.05) = 1.105, \\
y_3 &= y_2 + h(t_2, y_2) = 1.105 + 0.05(0.10 + 1.105) = 1.165.
\end{aligned}
$$

7.3.1 Discretization errors

In Euler's method we deal with truncation errors arising as a result of replacing an infinite continuous region by a finite discrete set of mesh points.

Definition 7.1 (Global and Local Discretization Errors) *Let $y = y(t)$ be the unique solution to an IVP $y' = f(t, y)$ with $y(t_0) = y_0$, and let $\{(t_k, y_k) : k = 0, \ldots n\}$ be the set of approximate values obtained using a numerical method for solving the IVP. The global discretization error of the method at step k is given by*

$$
e_k = y_k - y(t_k) \tag{7.13}
$$

for $k = 0, 1, \ldots n$. When $k = n$, the corresponding error e_n is called the final global error.

The local discretization error at step k is defined as the error of a single step from t_{k-1} to t_k:

$$
\epsilon_k = \hat{y}_k - y(t_k), \tag{7.14}
$$

where $\hat{y}_k = y(t_{k-1}) + hf(t_{k-1}, y(t_{k-1})), \ k = 1, \ldots n$.

Figure 7.2 illustrates the above definitions for $n = 3$.

7.4 Runge-Kutta Methods

The Runge-Kutta (RK) methods discussed in this section allow one to achieve a good accuracy without computing the higher-order derivatives. As before, we consider the IVP

$$
y' = f(t, y), \ y(t_0) = y_0. \tag{7.15}
$$

Assume that the value y_k of the approximate solution at point t_k has been found, and we want to find a proper value y_{k+1} of the numerical solution at the next point $t_{k+1} = t_k + h$. Then Euler's method can be generalized in the following way:

$$
y_{k+1} = y_k + hG(t_k, y_k, h), \tag{7.16}
$$

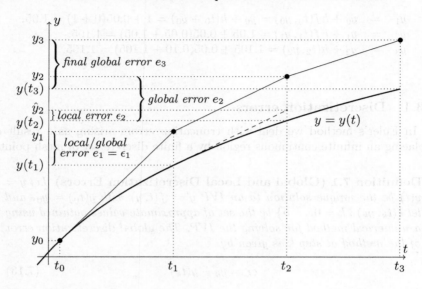

FIGURE 7.2: An illustration of discretization errors in Euler's method.

where $G(t_k, y_k, h)$ is called an *increment function* and can be written in a general form as

$$G(t_k, y_k, h) = p_1 f_1 + p_2 f_2 + \ldots + p_l f_l. \tag{7.17}$$

In this expression, l is some positive integer called the *order of the method*, p_i, $i = 1, \ldots, l$ are constants, and f_i, $i = 1, \ldots, l$ are given by

$$
\begin{aligned}
f_1 &= f(t_k, y_k), \\
f_2 &= f(t_k + \alpha_1 h, y_k + \alpha_1 h f_1), \\
f_3 &= f(t_k + \alpha_2 h, y_k + \alpha_2 h f_2), \\
&\vdots \\
f_l &= f(t_k + \alpha_{l-1} h, y_k + \alpha_{l-1} h f_{l-1}).
\end{aligned}
\tag{7.18}
$$

The coefficients $\alpha_1, \ldots, \alpha_{l-1}$, $p_1, \ldots, p_l$ are chosen such that the precision of the approximation is sufficiently good. Given y_k, one can recursively compute $f_1, \ldots, f_l$, and then y_{k+1}.

The first-order Runge-Kutta method ($l = 1$) is, in fact, the previously discussed Euler method.

7.4.1   Second-order Runge-Kutta methods

For $l = 2$ we obtain

$$y_{k+1} = y_k + h(p_1 f_1 + p_2 f_2), \tag{7.19}$$

where

$$f_1 = f(t_k, y_k),$$
$$f_2 = f(t_k + \alpha_1 h, y_k + \alpha_1 h f_1). \tag{7.20}$$

Let $y(t)$ be the solution of Equation (7.15), then for the second-order derivative of $y(t)$ we obtain

$$\frac{d^2 y}{dt^2} = \frac{d}{dt} f(t, y(t)) = \frac{\partial f}{\partial t} + \frac{\partial f}{\partial y} \frac{dy}{dt} = \frac{\partial f}{\partial t} + \frac{\partial f}{\partial y} f. \tag{7.21}$$

Therefore, substituting (7.21) into a Taylor series expansion to the second-order term,

$$y(t_k + h) = y(t_k) + h y'(t_k) + \frac{h^2}{2} y''(t_k) + O(h^3),$$

and using the fact that $y(x)$ is the solution of (7.15), we obtain:

$$\frac{y(t_k + h) - y(t_k)}{h} = \left[f + \frac{h}{2} \left(\frac{\partial f}{\partial t} + \frac{\partial f}{\partial y} f \right) \right]_{\substack{t = t_k \\ y = y(t_k)}} + O(h^2). \tag{7.22}$$

Using a first-order Taylor series expansion for f_2 in (7.20) as a function of two variables, we get

$$
\begin{aligned}
f_2 &= f(t_k + \alpha_1 h, y_k + \alpha_1 h f_1) \\
&= \left[f + \alpha_1 h \frac{\partial f}{\partial t} + \alpha_1 h \frac{\partial f}{\partial y} f \right]_{\substack{t = t_k \\ y = y(t_k)}} + O(h^2),
\end{aligned} \tag{7.23}
$$

and from (7.19),

$$
\begin{aligned}
\frac{y_{k+1} - y_k}{h} &= p_1 f_1 + p_2 f_2 \\
&= \left[(p_1 + p_2) f + p_2 \alpha_1 h \frac{\partial f}{\partial t} + p_2 \alpha_1 h \frac{\partial f}{\partial y} f \right]_{\substack{t = t_k \\ y = y(t_k)}} + O(h^2).
\end{aligned} \tag{7.24}
$$

Equating the coefficients of the corresponding terms in the right-hand sides of (7.22) and (7.24), we derive the following equations:

$$p_1 + p_2 = 1,$$

and

$$p_2 \alpha_1 = \frac{1}{2}.$$

Therefore, for a fixed α_1, substituting $p_2 = \frac{1}{2\alpha_1}$ and $p_1 = \frac{2\alpha_1 - 1}{2\alpha_1}$ in (7.19) we obtain a general scheme for the second-order RK methods:

$$y_{k+1} = y_k + h \left(\frac{2\alpha_1 - 1}{2\alpha_1} f_1 + \frac{1}{2\alpha_1} f_2 \right), \tag{7.25}$$

where
$$
\begin{aligned}
f_1 &= f(t_k, y_k), \\
f_2 &= f(t_k + \alpha_1 h, y_k + \alpha_1 h f_1).
\end{aligned}
\tag{7.26}
$$

In general, there are infinitely many values for α_1 to choose from, each yielding a different second-order RK method (although all of them give exactly the same result if the solution to the IVP is a polynomial of degree ≤ 2). We mention three of the most popular versions:

- *Heun Method with a Single Corrector*: $\alpha_1 = 1$;

- *The Improved Polygon Method*: $\alpha_1 = \frac{1}{2}$;

- *Ralston's Method*: $\alpha_1 = \frac{3}{4}$.

A general scheme of the second-order Runge-Kutta method is summarized in Algorithm 7.2.

Algorithm 7.2 Second-order Runge-Kutta method for solving the IVP $y' = f(t, y)$ with $y(a) = y_0$ on the interval $[a, b]$.

1: **Input:** f, $[a, b]$, $y_0 = y(a)$, parameter α_1, and the number of steps n
2: **Output:** (t_k, y_k) such that $y_k \approx y(t_k)$, $k = 1, \ldots, n$
3: $h = \frac{b-a}{n}$, $t_0 = a$
4: **for** $k = 1, \ldots, n$ **do**
5: $\quad t_k = t_{k-1} + h$
6: $\quad f_1 = f(t_{k-1}, y_{k-1})$
7: $\quad f_2 = f(t_{k-1} + \alpha_1 h, y_{k-1} + \alpha_1 h f_1)$
8: $\quad y_k = y_{k-1} + h\left(\frac{2\alpha_1 - 1}{2\alpha_1} f_1 + \frac{1}{2\alpha_1} f_2\right)$
9: **end for**
10: **return** (t_k, y_k), $k = 1, \ldots, n$

Example 7.8 *Consider the logistic model (also called the Verhulst-Pearl model) for population dynamics. According to this model, the population $P(t)$ of a certain species as a function of time t satisfies the following equation:*

$$
\frac{dP}{dt} = cP\left(1 - \frac{P}{M}\right),
$$

where c is a growth rate coefficient, M is a limiting size for the population, and both c and M are constants.

Assuming that $P(0) = P_0$ is given, the exact solution to this equation is

$$
P(t) = \frac{MP_0}{P_0 + (M - P_0)\exp(-ct)}.
$$

Putting $c = 0.1$, $M = 500$, and $P(0) = 300$, we obtain the following IVP:

$$
\frac{dP}{dt} = 0.1P\left(1 - \frac{P}{500}\right), \quad P(0) = 300.
$$

Next we approximate $P(1)$ using one step of the Heun method, the improved polygon method, and Ralston's method, and compare absolute errors for each method. We have $h = 1, f(t, P) = 0.1P\left(1 - \frac{P}{500}\right), t_0 = 0, P_0 = 300$, and the exact solution $P(t) = \frac{150000}{300 + 200\exp(-0.1t)}$, so $P(1) \approx 311.8713948706$.

Using the Heun method, we obtain

$$
\begin{aligned}
f_1 &= f(t_0, P_0) = 30\left(1 - \frac{300}{500}\right) = 12, \\
f_2 &= f(t_0 + h, P_0 + hf_1) = f(1, 312) = 11.7312, \\
P_1 &= P_0 + h\left(\frac{1}{2}f_1 + \frac{1}{2}f_2\right) = 300 + 0.5(12 + 11.7312) = 311.8656,
\end{aligned}
$$

and the absolute error is $P_1 - P(1) \approx 311.8656 - 311.8714 = -0.0058$.

The improved polygon method gives

$$
\begin{aligned}
f_1 &= f(t_0, P_0) = 12, \\
f_2 &= f(t_0 + 0.5h, P_0 + 0.5hf_1) = f(0.5, 306) = 11.8728, \\
P_1 &= P_0 + hf_2 = 300 + 11.8728 = 311.8728,
\end{aligned}
$$

and the absolute error is $P_1 - P(1) \approx 0.0014$.

Finally, by Ralston's method,

$$
\begin{aligned}
f_1 &= f(t_0, P_0) = 12, \\
f_2 &= f(t_0 + 0.75h, P_0 + 0.75hf_1) = f(0.75, 309) = 11.8038, \\
P_1 &= P_0 + h\left(\frac{1}{3}f_1 + \frac{2}{3}f_2\right) = 300 + 4 + 7.8692 = 311.8692,
\end{aligned}
$$

and the absolute error is $P_1 - P(1) \approx -0.0022$.

7.4.2 Fourth-order Runge-Kutta methods

Fourth-order RK methods are probably the most popular among the Runge-Kutta methods. They can be derived analogously to the second-order RK methods and are represented by an infinite number of versions as well. As an example, Algorithm 7.3 presents a scheme, which is frequently referred to as the *classical fourth-order RK method*.

Example 7.9 *We apply one step of the classical fourth-order Runge-Kutta method to the problem considered in Example 7.8. As before, we have $h = 1, f(t, P) = 0.1P\left(1 - \frac{P}{500}\right), t_0 = 0, P_0 = 300$, and the exact solution $P(1) \approx 311.8713948706$. Using the classical fourth-order Runge-Kutta method, we obtain*

$$
\begin{aligned}
f_1 &= f(t_0, P_0) = 12, \\
f_2 &= f(t_0 + 0.5h, P_0 + 0.5hf_1) = f(0.5, 306) = 11.8728, \\
f_3 &= f(t_0 + 0.5h, P_0 + 0.5hf_2) = f(0.5, 305.9364) = 11.8742, \\
f_4 &= f(t_0 + h, P_0 + hf_3) = f(1, 311.8742) = 11.7343, \\
P_1 &= P_0 + h(f_1 + 2f_2 + 2f_3 + f_4)/6 = 311.871394,
\end{aligned}
$$

and the error is $|P_1 - P(1)| < 10^{-6}$ in this case.

Algorithm 7.3 The classical fourth-order Runge-Kutta method for solving the IVP $y' = f(t, y)$ with $y(a) = y_0$ on the interval $[a, b]$.

1: **Input:** f, $[a, b]$, $y_0 = y(a)$, n
2: **Output:** (t_k, y_k) such that $y_k \approx y(t_k)$, $k = 1, \ldots, n$
3: $h = \frac{b-a}{n}$, $t_0 = a$
4: **for** $k = 1, \ldots, n$ **do**
5:     $t_k = t_{k-1} + h$
6:     $f_1 = f(t_{k-1}, y_{k-1})$
7:     $f_2 = f(t_{k-1} + 0.5h, y_{k-1} + 0.5hf_1)$
8:     $f_3 = f(t_{k-1} + 0.5h, y_{k-1} + 0.5hf_2)$
9:     $f_4 = f(t_{k-1} + h, y_{k-1} + hf_3)$
10:     $y_k = y_{k-1} + h(f_1 + 2f_2 + 2f_3 + f_4)/6$
11: **end for**
12: **return** (t_k, y_k), $k = 1, \ldots, n$

7.5   Systems of Differential Equations

Consider the *initial value problem (IVP)*:

$$\begin{array}{ll} \frac{dx}{dt} = f_1(t, x, y) \\ \frac{dy}{dt} = f_2(t, x, y) \end{array} \quad \text{with} \quad \left\{ \begin{array}{l} x(t_0) = x_0, \\ y(t_0) = y_0. \end{array} \right. \tag{7.27}$$

The solution to problem (7.27) on an interval $[t_0, b]$ is a pair of differentiable functions $x(t)$ and $y(t)$ such that for any $t \in [t_0, b]$:

$$\begin{array}{ll} x'(t) = f_1(t, x(t), y(t)) \\ y'(t) = f_2(t, x(t), y(t)) \end{array} \quad \text{with} \quad \left\{ \begin{array}{l} x(t_0) = x_0, \\ y(t_0) = y_0. \end{array} \right. \tag{7.28}$$

Example 7.10 *Verify that the pair of functions* $x(t) = \frac{1}{5}\exp(t) - \frac{1}{10}t\exp(t)$ *and* $y(t) = \frac{1}{2}\exp(t) - \frac{1}{5}t\exp(t)$ *solves the IVP*

$$\begin{array}{ll} x' = 3x - y \\ y' = 4x - y \end{array} \quad \text{with} \quad \left\{ \begin{array}{l} x(0) = 0.2, \\ y(0) = 0.5. \end{array} \right.$$

1. $x'(t) = \frac{1}{10}\exp(t) - \frac{1}{10}t\exp(t) = 3x(t) - y(t)$ *for all* t;
   $y'(t) = \frac{3}{10}\exp(t) - \frac{1}{5}t\exp(t) = 4x(t) - y(t)$.

2. $x(0) = \frac{1}{5} = 0.2$; $y(0) = \frac{1}{2} = 0.5$.

Formulas for numerical solutions of systems of differential equations are similar to those for a single equation. Algorithms 7.4 and 7.5 outline the Euler and Runge-Kutta methods for systems, respectively.

Algorithm 7.4 Euler's method for the system (7.27) on the interval $[a, b]$.

1: **Input:** f_1, f_2, $[a, b]$, $x_0 = x(a), y_0 = y(a)$, n
2: **Output:** (t_k, x_k, y_k) such that $x_k \approx x(t_k), y_k \approx y(t_k)$, $k = 1, \ldots, n$
3: $h = \frac{b-a}{n}$, $t_0 = a$
4: **for** $k = 1, \ldots, n$ **do**
5: $\quad t_k = t_{k-1} + h$
6: $\quad x_k = x_{k-1} + h f_1(t_{k-1}, x_{k-1}, y_{k-1})$
7: $\quad y_k = y_{k-1} + h f_2(t_{k-1}, x_{k-1}, y_{k-1})$
8: **end for**
9: **return** (t_k, x_k, y_k), $k = 1, \ldots, n$

Algorithm 7.5 Runge-Kutta method for the system (7.27) on $[a, b]$.

1: **Input:** f_1, f_2, $[a, b]$, $x_0 = x(a), y_0 = y(a)$, n
2: **Output:** (t_k, x_k, y_k) such that $x_k \approx x(t_k), y_k \approx y(t_k)$, $k = 1, \ldots, n$
3: $h = \frac{b-a}{n}$, $t_0 = a$
4: **for** $k = 1, \ldots, n$ **do**
5: $\quad t_k = t_{k-1} + h$
6: $\quad f_{11} = f_1(t_{k-1}, x_{k-1}, y_{k-1})$
7: $\quad f_{21} = f_2(t_{k-1}, x_{k-1}, y_{k-1})$
8: $\quad f_{12} = f_1(t_{k-1} + 0.5h, x_{k-1} + 0.5h f_{11}, y_{k-1} + 0.5h f_{21})$
9: $\quad f_{22} = f_2(t_{k-1} + 0.5h, x_{k-1} + 0.5h f_{11}, y_{k-1} + 0.5h f_{21})$
10: $\quad f_{13} = f_1(t_{k-1} + 0.5h, x_{k-1} + 0.5h f_{12}, y_{k-1} + 0.5h f_{22})$
11: $\quad f_{23} = f_2(t_{k-1} + 0.5h, x_{k-1} + 0.5h f_{12}, y_{k-1} + 0.5h f_{22})$
12: $\quad f_{14} = f_1(t_{k-1} + h, x_{k-1} + h f_{13}, y_{k-1} + h f_{23})$
13: $\quad f_{24} = f_2(t_{k-1} + h, x_{k-1} + h f_{13}, y_{k-1} + h f_{23})$
14: $\quad x_k = x_{k-1} + h(f_{11} + 2f_{12} + 2f_{13} + f_{14})/6$
15: $\quad y_k = y_{k-1} + h(f_{21} + 2f_{22} + 2f_{23} + f_{24})/6$
16: **end for**
17: **return** (t_k, x_k, y_k), $k = 1, \ldots, n$

Example 7.11 *Consider the system of differential equations:*

$$\begin{matrix} x' = y \\ y' = -ty - t^2 x \end{matrix} \quad with \quad \left\{ \begin{matrix} x(0) = 1 \\ y(0) = 1. \end{matrix} \right.$$

Use Euler's method with $h = 0.5$ to approximate $(x(1), y(1))$.
The kth iteration of Euler's method $(k = 1, 2 \ldots)$ is given by:

$$\begin{bmatrix} x_k \\ y_k \end{bmatrix} = \begin{bmatrix} x_{k-1} \\ y_{k-1} \end{bmatrix} + h \begin{bmatrix} y_{k-1} \\ -t_{k-1}y_{k-1} - t_{k-1}^2 x_{k-1} \end{bmatrix}.$$

We have $x_0 = x(0) = 1, y_0 = y(0) = 1$, and $h = 0.5$, and the first step $(k = 1)$
gives

$$\begin{bmatrix} x_1 \\ y_1 \end{bmatrix} = \begin{bmatrix} x_0 \\ y_0 \end{bmatrix} + h \begin{bmatrix} y_0 \\ -t_0 y_0 - t_0^2 x_0 \end{bmatrix} = \begin{bmatrix} 1 \\ 1 \end{bmatrix} + 0.5 \begin{bmatrix} 1 \\ 0 \end{bmatrix} = \begin{bmatrix} 1.5 \\ 1 \end{bmatrix}.$$

For the second step (k = 2) we obtain

$$\begin{bmatrix} x_2 \\ y_2 \end{bmatrix} = \begin{bmatrix} 1.5 \\ 1 \end{bmatrix} + 0.5 \begin{bmatrix} 1 \\ -0.5 \times 1 - 0.5^2 \times 1.5 \end{bmatrix} = \begin{bmatrix} 2 \\ 0.5625 \end{bmatrix}.$$

Example 7.12 *The Lotka-Volterra model describes interactions between a predator and a prey species, say, rabbits and foxes. Denote by $R(t)$ the number of rabbits and by $F(t)$ the number of foxes that are alive at time t. Then the Lotka-Volterra model consists of two differential equations describing changes in the prey and predator population, respectively:*

$$\frac{dR}{dt} = aR - bRF;$$
$$\frac{dF}{dt} = cRF - dF,$$

where a is the natural growth rate of rabbits in the absence of predation, b is the predation rate coefficient, c is the reproduction rate of foxes per 1 rabbit eaten, and d is the mortality rate of foxes.

Let time t be measured in years. Assuming $a = 0.03, b = 0.004, c = 0.0002, d = 0.1$, and the initial populations of $R(0) = 10000$ rabbits and $F(0) = 500$ foxes, we obtain the following model:

$$\frac{dR}{dt} = 0.03R - 0.004RF; \quad \text{with} \quad \begin{cases} R(0) = 10000 \\ F(0) = 500. \end{cases}$$
$$\frac{dF}{dt} = 0.0002RF - 0.1F,$$

Let us use two steps of the Runge-Kutta method to estimate the population of rabbits and foxes at time $t = 1$. We have $t_0 = 0, h = 0.5, R_0 = 10000, F_0 = 500$, $f_1(t, R, F) = 0.03R - 0.004RF$, $f_2(t, R, F) = 0.0002RF - 0.1F$.

For Step 1 we have:

$$f_{11} = f_1(0, 10000, 500) = -19700;$$
$$f_{21} = f_2(0, 10000, 500) = 950$$
$$f_{12} = f_1(0.25, 5075, 737.5) = -14819;$$
$$f_{22} = f_2(0.25, 5075, 737.5) = 674.8125$$
$$f_{13} = f_1(0.25, 6295.25, 668.7031) = -16649.7559;$$
$$f_{23} = f_2(0.25, 6295.25, 668.7031) = 775.0604$$
$$f_{14} = f_1(0.5, 1675.1221, 887.5302) = -5896.6318;$$
$$f_{24} = f_2(0.5, 1675.1221, 887.5302) = 208.5913$$
$$R_1 = R_0 + h(f_{11} + 2f_{12} + 2f_{13} + f_{14})/6 \approx 2622.1547$$
$$F_1 = F_0 + h(f_{21} + 2f_{22} + 2f_{23} + f_{24})/6 \approx 838.1947.$$

For the second step of the Runge-Kutta method, we have $t_1 = 0.5, h = 0.5, R_1 = 2622.1547, F_1 = 838.1947$, and

$$\begin{array}{ll} f_{11} = -8712.8405; & f_{21} = 355.7558; \\ f_{12} = -1633.0655; & f_{22} = -10.3942; \\ f_{13} = -7333.2500; & f_{23} = 286.4237; \\ f_{14} = 4068.8662; & f_{24} = -303.1507 \\ R_2 = 740.7709; & \\ F_2 = 888.5834. & \end{array}$$

TABLE 7.1: Runge-Kutta method for the Lotka-Volterra model in Example 7.12.

Step	t	Rabbits	Foxes
0	0	10000	500
1	0.5	2622	838
2	1	741	889
3	1.5	210	871
4	2	58	836
5	2.5	16	797
6	3	4	759
7	3.5	1	722
8	4	0	687

Thus, there will be approximately 741 rabbits and 889 foxes after 1 year.

Table 7.1 shows the result of application of 8 steps of the Runge-Kutta method to the considered problem.

7.6 Higher-Order Differential Equations

Consider a higher-order differential equation of the general form:

$$y^{(n)} = f(t, y, y', y'', \ldots, y^{(n-1)}). \tag{7.29}$$

A common way to to handle such equation (of order n) is to reduce it to an equivalent system of n first-order equations. If $n = 2$, then (7.29) becomes

$$y'' = f(t, y, y'), \tag{7.30}$$

and by substituting $z = y'$, we obtain an equivalent system

$$\begin{aligned} y' &= z \\ z' &= f(t, y, z). \end{aligned} \tag{7.31}$$

After that, the techniques for systems, such as those discussed in Section 7.5, can be applied.

Example 7.13 *Consider the differential equation*

$$\begin{aligned} y'' &= (1 - y^2)y' - y \\ y(0) &= 0.2 \\ y'(0) &= 0.2. \end{aligned}$$

Use Euler's method with $h = 0.1$ to approximate $y(0.2)$.

Denoting by $z = y'$, the given problem is reduced to the following system:

$$\begin{array}{rclcrcl} y' & = & z & & y(0) & = & 0.2 \\ z' & = & (1 - y^2)z - y & & z(0) & = & 0.2. \end{array}$$

Applying Euler's method to this system gives

$$\begin{bmatrix} y_{k+1} \\ z_{k+1} \end{bmatrix} = \begin{bmatrix} y_k \\ z_k \end{bmatrix} + h \begin{bmatrix} z_k \\ (1 - y_k^2)z_k - y_k \end{bmatrix}, k \geq 0.$$

Using $y_0 = z_0 = 0.2; h = 0.1$, we obtain $\begin{bmatrix} y_1 \\ z_1 \end{bmatrix} = \begin{bmatrix} 0.22 \\ 0.1992 \end{bmatrix}$. So, $y_2 = 0.22 + 0.1 \cdot 0.1992 = 0.23992$.

Exercises

7.1. Consider the IVP $y' = -xy, y(1) = 2$.

   (a) Verify that $y(x) = 2 \exp\left(\frac{1}{2}(1 - x^2)\right)$ is the solution.

   (b) Apply Picard's method three times.

7.2. Solve the equations

   (i) $y' = -\frac{x^2 + \exp(y)}{x(1 + \exp(y))}, \quad y(1) = 0$;

   (ii) $y' = \frac{y(1 - x^2 y^2)}{x(1 + x^2 y^2)}, \quad y(1) = 1$

on the interval $[1, 2]$ using two steps of the following methods:

   (a) Euler's method;

   (b) Heun method;

   (c) the improved polygon method;

   (d) Ralston's method;

   (e) the classical fourth-order Runge-Kutta method.

7.3. Solve the system

$$\begin{array}{rcl} x' & = & -12x + 9y \\ y' & = & 11x - 10y \end{array} \quad \text{with} \quad \left\{ \begin{array}{l} x(0) = 14 \\ y(0) = 6 \end{array} \right.$$

on the interval $[0, 0.2]$ using two steps of

(a) Euler's method for systems;

(b) Runge-Kutta method for systems.

Compute the absolute errors if the exact solution is

$$x(t) = 9\exp(-t) + 5\exp(-21t)$$
$$y(t) = 11\exp(-t) - 5\exp(-21t).$$

7.4. Consider the differential equation $y'' = y, y(0) = 1, y'(0) = 2$.

(a) Verify that $y(x) = \frac{3}{2}\exp(x) - \frac{1}{2}\exp(-x)$ is the solution.

(b) Reduce the differential equation to a first-order system. Then use Euler's method with $h = 0.5$ to approximate $y(1)$.

(c) What is the local and global discretization error for each step?

7.5. Consider the Van der Pol's equation

$$y'' - 0.1(1 - y^2)y' + y = 0$$
$$y(0) = 0$$
$$y'(0) = 1.$$

(a) Reduce it to a first-order system.

(b) Use $h = 0.5$ and Euler's method to estimate $y(1)$.

(c) Use two steps of the Runge-Kutta method to estimate $y(1)$.

7.6. Consider the differential equation

$$y''' = x^2 y'' + y^2 y',$$

$$y(1) = 0, y'(1) = -1, y''(1) = 2.$$

(a) Reduce it to a first-order system.

(b) Use $h = 0.5$ and Euler's method to estimate $y(2)$.

(c) Use two steps of the Runge-Kutta method to estimate $y(2)$.

Part III

Introduction to Optimization

Part III

Introduction to Optimization

Chapter 8

Basic Concepts

Optimization is a methodology aiming to find the best among available alternatives. The available alternatives are referred to as *feasible solutions*, and their quality is measured using some numerical function called the *objective function*. A feasible solution that yields the best (minimum or maximum) objective function value is called an *optimal solution*.

Optimization problems are of great practical interest. For example, in manufacturing, how should one cut plates of a material so that the waste is minimized? In business, how should a company allocate the available resources so that its profit is maximized? Some of the first optimization problems have been solved in ancient Greece and are regarded among the most significant discoveries of that time. In the first century A.D., the Alexandrian mathematician Heron solved the problem of finding the shortest path between two points by way of the mirror. This result, also known as Heron's theorem of the light ray, can be viewed as the origin of the theory of geometrical optics. The problem of finding extreme values gained special importance in the seventeenth century, when it served as one of the motivations in the invention of differential calculus, which is the foundation of the modern theory of mathematical optimization.

The invention of the digital computer has led to the rise of the field of numerical optimization. During World War II, optimization algorithms were used to solve military logistics and operations problems. Optimization has become one of the most important and well-studied fields in applied mathematics and engineering ever since.

This chapter provides a description of basic concepts that will be used throughout the chapters dealing with optimization.

8.1 Formulating an Optimization Problem

Consider the following example.

Example 8.1 *A company needs to design an aluminum can in the shape of a cylinder. The design is determined by the can's height h and diameter d (see Figure 8.1) and must satisfy the following requirements:*

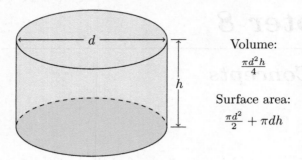

FIGURE 8.1: A cylindrical can of height h and diameter d.

- *the height of the can must be at least 50% greater than its diameter,*
- *the can's height cannot be more than twice its diameter, and*
- *the volume of the can must be at least 330 ml.*

The goal is to design a can that would require the minimum possible amount of aluminum.

We start formulating an optimization model by introducing the *decision variables*, which are the parameters whose values need to be determined in order to solve the problem. Clearly, in this example the decision variables are given by the parameters that define a can's design:

$$h \;=\; \text{the can's height, in cm}$$
$$d \;=\; \text{the can's diameter, in cm.}$$

Next we need to state our *objective function*, i.e., the quantity that we need to optimize, in mathematical terms, as a function of the decision variables. It is reasonable to assume that the aluminum sheets used have a fixed thickness. Then the amount of aluminum used is determined by the surface area of the cylinder can, which consists of two disks of diameter d (top and bottom) and a rectangular side of height h and width given by the circumference of the circle of diameter d, which is πd. Hence, the total surface area is given by:

$$f(h,d) = 2\frac{\pi d^2}{4} + \pi dh = \frac{\pi d^2}{2} + \pi dh. \tag{8.1}$$

The function $f(h,d)$ is the objective function that we want to minimize.

Finally, we need to specify the *constraints*, i.e., the conditions that the design parameters are required to satisfy. The first requirement states that the height of the can must be at least 50% greater than its diameter, which can be expressed as

$$h \geq 1.5d \quad \Leftrightarrow \quad 1.5d - h \leq 0. \tag{8.2}$$

According to the second requirement, the can's height cannot be more than twice its diameter, i.e.,

$$h \leq 2d \quad \Leftrightarrow \quad -2d + h \leq 0. \tag{8.3}$$

The third requirement is that the volume of the can must be at least 330 ml:

$$\frac{\pi d^2 h}{4} \geq 330. \tag{8.4}$$

(note that 1 ml = 1 cm^3).

It is also important to note that the can's height and diameter must always be nonnegative, i.e., $h, d \geq 0$, however, it is easy to see that this is guaranteed by the constraints (8.2) and (8.3): if we add these constraints together, we obtain $d \geq 0$, and from (8.2) we have $h \geq 1.5d \geq 0$. Thus, nonnegativity constraints are *redundant* and do not need to be included in the model.

In summary, we have the following optimization problem:

$$\begin{array}{ll} \text{minimize} & \frac{\pi d^2}{2} + \pi dh \\ \text{subject to} & 1.5d - h \ \leq \ 0 \\ & -2d + h \ \leq \ 0 \\ & \frac{\pi d^2 h}{4} \ \geq \ 330. \end{array} \tag{8.5}$$

Example 8.2 *In computational physics, one is interested in locating a set of n points $\{p^{(i)} = [x_1^{(i)}, x_2^{(i)}, x_3^{(i)}]^T : i = 1, \ldots, n\}$ on a unit sphere in $\mathbb{R}^3$ with the goal of minimizing the Lennard-Jones potential*

$$\sum_{1 \leq i \leq j \leq n} \left(\frac{1}{d_{ij}^{12}} - \frac{2}{d_{ij}^6} \right),$$

where $d_{ij} = \|p^{(i)} - p^{(j)}\|$ is the Euclidean distance between p_i and p_j, i.e.,

$$d_{ij} = \sqrt{(x_1^{(i)} - x_1^{(j)})^2 + (x_2^{(i)} - x_2^{(j)})^2 + (x_3^{(i)} - x_3^{(j)})^2}.$$

In this case, we have $3n$ decision variables describing the coordinates of n points in $\mathbb{R}^3$, and putting the sphere's center in the origin of the coordinate system in $\mathbb{R}^3$, the only constraints are

$$\|p^{(i)}\| = 1, \quad i = 1, \ldots, n.$$

Thus, the problem can be formulated as follows:

$$\begin{array}{ll} \text{minimize} & \sum_{1 \leq i \leq j \leq n} \left(\|p^{(i)} - p^{(j)}\|^{-12} - 2\|p^{(i)} - p^{(j)}\|^{-6} \right) \\ \text{subject to} & \|p^{(i)}\| = 1, \quad i = 1, \ldots, n \\ & p^{(i)} \in \mathbb{R}^3, \quad i = 1, \ldots, n. \end{array} \tag{8.6}$$

Here $p^{(i)} = [x_1^{(i)}, x_2^{(i)}, x_3^{(i)}]^T$ for $i = 1, \ldots, n$.

8.2   Mathematical Description

Let $X \subseteq \mathbb{R}^n$ be a set of n-dimensional real vectors in the form $x = [x_1, \ldots, x_n]^T$, and let $f : X \to \mathbb{R}$ be a given function. In mathematical terms, an optimization problem has the general form

$$
\begin{array}{ccc}
\text{maximize} \quad f(x) & & \text{minimize} \quad f(x) \\
\text{subject to} \quad x \in X, & \text{or} & \text{subject to} \quad x \in X
\end{array}
$$

or, equivalently,

$$
\max_{x \in X} f(x) \quad \text{or} \quad \min_{x \in X} f(x).
$$

In the above, each $x_j, j = 1, \ldots n$, is called a *decision variable*, X is called the *feasible (admissible) region*, and f is the *objective function*.

Note that a maximization problem $\max_{x \in X} f(x)$ can be easily converted into an equivalent minimization problem $\min_{x \in X}(-f(x))$ and vice versa.

We will typically use a *functional form* of an optimization problem, in which the feasible region is described by a set of *constraints* given by equalities and inequalities in the form

$$
h_i(x) = 0, \ i \in \mathcal{E},
$$

$$
g_j(x) \leq 0, \ j \in \mathcal{I},
$$

where $h_i : \mathbb{R}^n \to \mathbb{R}, i \in \mathcal{E}, g_j : \mathbb{R}^n \to \mathbb{R}, j \in \mathcal{I}$ are given functions; $\mathcal{E}$ and $\mathcal{I}$ are index sets for the equality and inequality constraints, respectively. Note that the description above covers a wide variety of possible scenarios. For example, in some problems the decision variables are required to be binary: $x_j \in \{0, 1\}$. We can easily represent this constraint in the equivalent form

$$
x_j(1 - x_j) = 0.
$$

Also, if we are given an inequality constraint in the form $g_1(x) \geq g_2(x)$, we can convert it into an equivalent inequality in the form $g_3(x) \leq 0$ by letting $g_3(x) = g_2(x) - g_1(x)$. Obviously,

$$
g_1(x) \geq g_2(x) \ \Leftrightarrow \ g_2(x) - g_1(x) \leq 0.
$$

If the feasible region X is described by a set of equality and inequality constraints,

$$
X = \{x \in \mathbb{R}^n : h_i(x) = 0, g_j(x) \leq 0, \ i \in \mathcal{E}, \ j \in \mathcal{I}\},
$$

then, obviously, a point $\bar{x}$ is feasible if and only if it satisfies each of the equality and inequality constraints defining the feasible region X. For a feasible $\bar{x} \in X$, and an inequality constraint $g_j(x) \leq 0$, we may have a strict inequality, $g_j(\bar{x}) < 0$ or an equality $g_j(\bar{x}) = 0$ satisfied.

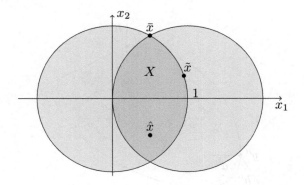

FIGURE 8.2: Illustration for Example 8.3.

> **Definition 8.1 (Binding (Active) Constraint)** *An inequality constraint* $g_j(x) \leq 0$ *is called binding or active at a feasible point* $\bar{x}$ *if* $g_j(\bar{x}) = 0$. *Otherwise, if* $g_j(\bar{x}) < 0$, *the constraint is called nonbinding or inactive at* $\bar{x}$.

Example 8.3 *Consider the feasible set* X *defined by two inequality constraints,*
$$X = \{x \in \mathbb{R}^2 : x_1^2 + x_2^2 \leq 1, (x_1 - 1)^2 + x_2^2 \leq 1\}.$$
Geometrically, this set is given by the points in the intersection of two disks corresponding to the constraints, as illustrated in Figure 8.2. In this figure, none of the constraints is active at $\hat{x}$; *only the first constraint is active at* $\tilde{x}$, *and both constraints are active at* $\bar{x}$.

Next we define the concept of a feasible direction, which characterizes a *local* property of a point in the feasible region, i.e., a property that is required to be satisfied only in an arbitrarily small neighborhood of the point.

> **Definition 8.2 (Feasible Direction)** *A vector* $d \in \mathbb{R}^n$ *is called a feasible direction for set* $X \subseteq \mathbb{R}^n$ *at* $\bar{x} \in X$ *if there exists* $\delta > 0$ *such that* $\bar{x} + \alpha d \in X$ *for any* $\alpha \leq \delta$.

Recall that $\hat{x} \in X$ is an *interior point* of X if there exists $\varepsilon > 0$ such that $\hat{x}$ is included in X together with the ε-ball centered at $\hat{x}$, i.e., $B(\hat{x}, \varepsilon) \subset X$. We call the set of all interior points of X the *interior* of X, denoted by int(X). If $\bar{x} \in X$ is not an interior point, then we call it a *boundary point* of X. Note that any direction is feasible at an interior point. Figure 8.3 shows examples

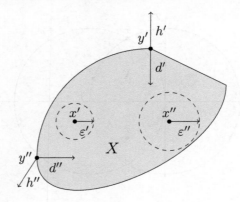

FIGURE 8.3: Examples of interior points (x' and x''), boundary points (y' and y''), feasible directions (d' and d''), and infeasible directions (h' and h'').

of interior and boundary points, as well as feasible and infeasible directions at the boundary points.

8.3   Local and Global Optimality

We consider a minimization problem in the form

$$\begin{aligned} \text{minimize} \quad & f(x) \\ \text{subject to} \quad & x \in X, \end{aligned} \tag{8.7}$$

where $f : \mathbb{R}^n \to \mathbb{R}$ is an arbitrary function of n variables $x_j, j = 1, \dots, n$.

To solve this problem, one needs to find a feasible solution x^* that minimizes f over X. Such a solution is called a *global optimal solution* or a *global minimizer* and is formally defined as follows.

Definition 8.3 (Global Minimum) *The point $x^* \in X$ is a point of global minimum (global minimizer) for the problem $\min\limits_{x \in X} f(x)$ (denoted $x^* = \arg\min\limits_{x \in X} f(x)$) if*

$$f(x^*) \le f(x) \text{ for all } x \in X.$$

A global minimizer x^* is a strict global minimizer if

$$f(x^*) < f(x) \text{ for all } x \in X \setminus \{x^*\}.$$

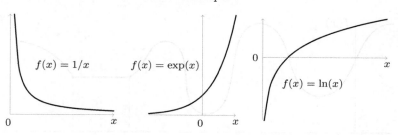

FIGURE 8.4: Examples of functions with no minimizer or maximizer.

A global minimizer does not always exist, even if the function is bounded from below. Consider, for example, the function $f(x) = \exp(x), x \in \mathbb{R}$. It is bounded from below, $f(x) \geq 0 \; \forall x \in \mathbb{R}$, however, there does not exist a point $x \in \mathbb{R}$ such that $\exp(x) = 0$. Since $\lim_{x \to -\infty} \exp(x) = 0$, $c = 0$ is the greatest lower bound for $f(x) = \exp(x)$, which is referred to as *infimum* of f.

Definition 8.4 (Infimum; Supremum) *For a function $f : X \to \mathbb{R}$, its greatest lower bound is called the infimum and is denoted by $\inf_{x \in X} f(x)$.*

Similarly, its least upper bound is called the supremum and is denoted by $\sup_{x \in X} f(x)$.

If f is unbounded from below or above, we have $\inf_{x \in X} f(x) = -\infty$ or $\sup_{x \in X} f(x) = +\infty$, respectively. Figure 8.4 shows examples of three functions that do not have a minimizer. Note that the first two of these functions are bounded from below. We have $\inf_{x \in (0, +\infty)} (1/x) = 0$, $\inf_{x \in \mathbb{R}} (\exp(x)) = 0$, and $\inf_{x \in (0, +\infty)} (\ln(x)) = -\infty$. The supremum of each of these functions over the corresponding domain is $+\infty$.

Even when a minimizer does exist, it may be extremely difficult to compute. Therefore, many of the known optimization methods focus on a less ambitious, yet still very challenging goal of finding a *local minimizer*, which is defined next.

Definition 8.5 (Local Minimum) $x^* \in X$ *is a point of local minimum (local minimizer) for the problem $\min_{x \in X} f(x)$ if there exists $\varepsilon > 0$ such that*

$$f(x) \geq f(x^*) \text{ for any } x \in X \text{ with } \|x - x^*\| \leq \varepsilon.$$

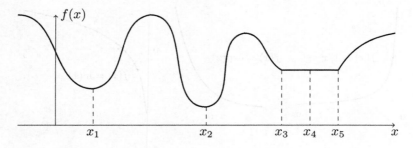

FIGURE 8.5: A function with two strict local minimizers (x_1 and x_2), one strict global minimizer (x_2), and infinitely many local minimizers that are not strict (all points in the interval $[x_3, x_5]$).

> *A local minimizer x^* is a strict local minimizer for the considered problem if there exists $\varepsilon > 0$ such that*
>
> $$f(x) > f(x^*) \text{ for any } x \in X \setminus \{x^*\} \text{ with } \|x - x^*\| \leq \varepsilon.$$

Examples of local and global minimizers are given in Figure 8.5.

8.4 Existence of an Optimal Solution

Checking whether a global minimizer exists for a given problem is an extremely difficult task in general. However, in some situations we can guarantee the existence of a global minimizer. One such case, which applies to a broad class of problems, is described by the classical Weierstrass theorem.

> **Theorem 8.1 (Weierstrass)** *The problems* $\min\limits_{x \in X} f(x)$ *and* $\max\limits_{x \in X} f(x)$, *where* $X \subset \mathbb{R}^n$ *is a compact set and* $f : X \to \mathbb{R}$ *is a continuous function, have global optimal solutions.*

In other words, a continuous function attains its minimum and maximum over a compact set.

Another case where a global minimum is guaranteed to exist is given by continuous *coercive functions*.

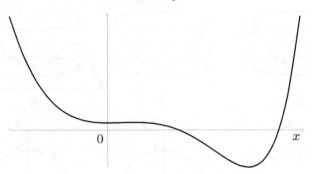

FIGURE 8.6: A coercive function, $f(x) = \frac{1}{10}(\exp(x) + \exp(-x) - x^3)$.

Definition 8.6 (Coercive Function) *A function* $f : \mathbb{R}^n \to \mathbb{R}$ *is called coercive, if*

$$\lim_{\|x\| \to \infty} f(x) = +\infty. \tag{8.8}$$

Figure 8.6 provides an example of a coercive function.

Theorem 8.2 *Any continuous coercive function* $f : \mathbb{R}^n \to \mathbb{R}$ *has a global minimizer in* $\mathbb{R}^n$.

Proof. Consider an arbitrary $\hat{x} \in \mathbb{R}^n$. Since f is coercive, there exists $C > 0$ such that for any x with $\|x\| > C$: $f(x) > f(\hat{x})$. Next, we represent $\mathbb{R}^n$ as a union of two disjoint sets, $\mathbb{R}^n = X_1 \cup X_2$, where

$$X_1 = \{x : \|x\| \le C\}, \ X_2 = \{x : \|x\| > C\}.$$

Note that $\hat{x} \in X_1$ and for any $x \in X_2$ we have $f(x) > f(\hat{x})$. Also, since X_1 is a compact set, by Weierstrass theorem there exists $x^* \in X_1$ such that $f(x^*) = \min_{x \in X_1} f(x)$. Obviously, $f(\hat{x}) \ge f(x^*)$. Thus, for any $x \in \mathbb{R}^n : f(x) \ge f(x^*)$, so x^* is a global minimizer. $\qquad\square$

8.5 Level Sets and Gradients

Definition 8.7 (Level Set) *For a function* $f : \mathbb{R}^n \to \mathbb{R}$ *and a constant* c, *the set* $\{x \in \mathbb{R}^n : f(x) = c\}$ *is called the level set of* f *at the level* c; *the set* $\{x \in \mathbb{R}^n : f(x) \le c\}$ *is called the lower level set of* f *at*

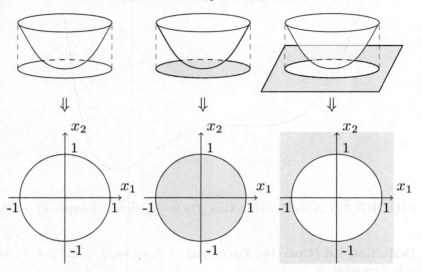

FIGURE 8.7: The level set, lower level set, and upper level set of $f(x) = x_1^2 + x_2^2$ at the level $c = 1$.

the level c; and the set $\{x \in \mathbb{R}^n : f(x) \geq c\}$ is called the upper level set of f at the level c.

Example 8.4 *For $f(x) = x_1^2 + x_2^2$, the level set, lower level set, and upper level set at the level $c = 1$ are given by $\{x \in \mathbb{R}^2 : x_1^2 + x_2^2 = 1\}$, $\{x \in \mathbb{R}^2 : x_1^2 + x_2^2 \leq 1\}$, and $\{x \in \mathbb{R}^2 : x_1^2 + x_2^2 \geq 1\}$, respectively. Geometrically, the level set is the unit radius circle centered at the origin, the lower level set consists of the circle and its interior, and the upper level set consists of the circle and its exterior. See Figure 8.7 for an illustration.*

Consider a curve γ in the level set S of f at the level c given by $\gamma = \{x(t) : t \in (a, b)\} \subset S$, where $x(t) : (a, b) \to S$ is a continuous function. Then we have

$$f(x(t)) = c, \quad \text{for } t \in (a, b). \tag{8.9}$$

Hence, the derivative of $g(t) = f(x(t))$ at any point $\bar{t} \in (a, b)$ is

$$\frac{dg(\bar{t})}{dt} = 0. \tag{8.10}$$

On the other hand, using the chain rule,

$$\frac{dg(\bar{t})}{dt} = \frac{df(x(\bar{t}))}{dt} = \nabla f(x(\bar{t}))^T x'(\bar{t}). \tag{8.11}$$

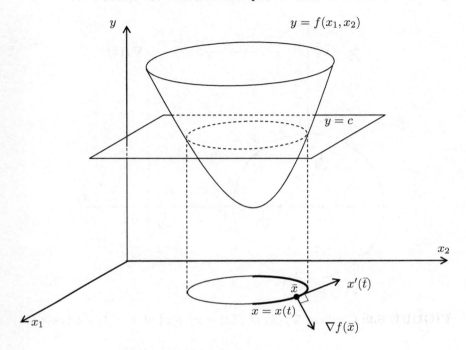

FIGURE 8.8: A level set and a gradient.

So, if we denote by $\bar{x} = x(\bar{t})$, then from the last two equations we obtain

$$\nabla f(\bar{x})^T x'(\bar{t}) = 0. \tag{8.12}$$

Geometrically, this means that vectors $\nabla f(\bar{x})$ and $x'(\bar{t})$ are orthogonal. Note that $x'(\bar{t})$ represents the tangent line to $x(t)$ at $\bar{x}$. Thus, we have the following property. Let $f(x)$ be a continuously differentiable function, and let $x(t)$ be a continuously differentiable curve passing through $\bar{x}$ in the level set of $f(x)$ at the level $c = f(\bar{x})$, where $x(\bar{t}) = \bar{x}, x'(\bar{t}) \neq 0$. Then the gradient of f at $\bar{x}$ is orthogonal to the tangent line of $x(t)$ at $\bar{x}$. This is illustrated in Figure 8.8.

Example 8.5 *For* $f(x) = x_1^2 + x_2^2$ *and* $\tilde{x} = [1/\sqrt{2}, 1/\sqrt{2}]^T$, $\nabla f(\tilde{x}) = [\sqrt{2}, \sqrt{2}]^T$. *The gradient* $\nabla f(\tilde{x})$ *is illustrated in Figure 8.9, which also shows the level set of f at the level $f(\tilde{x}) = 1$. Clearly, the gradient is orthogonal to the level set.*

Example 8.6 *Let* $f(x) = -x_1^2 - 4x_2^2$. *Then* $\nabla f(x) = [-2x_1, -8x_2]^T$. *Consider two different points,*

$$\tilde{x} = [2, 0]^T \text{ and } \bar{x} = [\sqrt{3}, 1/2]^T.$$

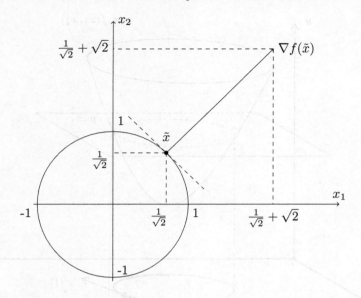

FIGURE 8.9: Gradient $\nabla f(\tilde{x})$ of $f(x) = x_1^2 + x_2^2$ at $\tilde{x} = [1/\sqrt{2}, 1/\sqrt{2}]^T$.

Then

$$\nabla f(\tilde{x}) = [-4, 0]^T \ and \ \nabla f(\bar{x}) = [-2\sqrt{3}, -4]^T.$$

Note that $f(\tilde{x}) = f(\bar{x}) = -4$. Consider the level set

$$S = \{x \in \mathbb{R}^2 : -x_1^2 - 4x_2^2 = -4\}$$

of f at the level $c = -4$ and a curve

$$\gamma = \{x(t) = [\sqrt{4 - 4t^2}, t]^T : t \in (-1, 1)\} \subset S$$

passing through $\tilde{x}$ and $\bar{x}$ in S, where $x(\tilde{t}) = \tilde{x}$ with $\tilde{t} = 0$ and $x(\bar{t}) = \bar{x}$ with $\bar{t} = 1/2$. Then

$$x'(t) = \left[-\frac{4t}{\sqrt{4 - 4t^2}}, 1 \right]^T,$$

so

$$x'(\tilde{t}) = [0, 1]^T, x'(\bar{t}) = [-2/\sqrt{3}, 1]^T,$$

and

$$\nabla f(\tilde{x})^T x'(\tilde{t}) = \nabla f(\bar{x})^T x'(\bar{t}) = 0.$$

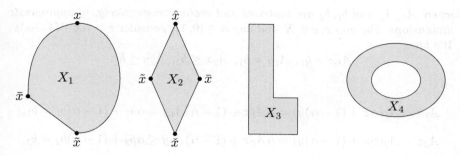

FIGURE 8.10: Examples of convex (X_1, X_2) and nonconvex (X_3, X_4) sets.

8.6 Convex Sets, Functions, and Problems

Due to their special structure, *convex problems* play a very important role in optimization. In this section, we introduce the notions of *convex sets* and *convex functions*, and provide equivalent characterizations of a convex function under conditions of continuous differentiability of the function and its gradient.

> **Definition 8.8 (Convex Combination)** *Given the points* $x_1, \ldots, x_m$ *in Euclidean n-space* $\mathbb{R}^n$ *and real numbers* $\alpha_i \geq 0$ *with* $\sum_{i=1}^{m} \alpha_i = 1$, *the point* $\sum_{i=1}^{m} \alpha_i x_i$ *is called a convex combination of these points.*

For example, geometrically, the convex combination of two points is the line segment between these two points, and the convex combination of three noncolinear points is a triangle.

> **Definition 8.9 (Convex Set)** *A set* $X \subseteq \mathbb{R}^n$ *is said to be convex if for any* $x, y \in X$ *and any* $\alpha \in (0, 1) : \alpha x + (1 - \alpha)y \in X.$

In other words, X is convex if all points located on the line segment connecting x and y are in X. See Figure 8.10 for examples.

Example 8.7 *Consider a set in* $\mathbb{R}^n$ *defined by a system of linear equations and inequalities:*

$$X = \{x \in \mathbb{R}^n : A_1 x = b_1, A_2 x \leq b_2\},$$

where A_1, A_2 and b_1, b_2 are matrices and vectors, respectively, of appropriate dimensions. For any $x, y \in X$ and any $\alpha \in (0,1)$, consider $z = \alpha x + (1 - \alpha)y$. We have

$$A_1 x = b_1, \ A_1 y = b_1, \ A_2 x \leq b_2, \ A_2 y \leq b_2,$$

so

$$A_1 z = A_1(\alpha x + (1 - \alpha)y) = \alpha A_1 x + (1 - \alpha)A_1 y = \alpha b_1 + (1 - \alpha)b_1 = b_1,$$

$$A_2 z = A_2(\alpha x + (1 - \alpha)y) = \alpha A_2 x + (1 - \alpha)A_2 y \leq \alpha b_2 + (1 - \alpha)b_2 = b_2,$$

and, therefore, $z \in X$. Thus, X is a convex set.

Definition 8.10 (Hyperplane) *A hyperplane $H \subset \mathbb{R}^n$ is a set of the form*

$$H = \{x \in \mathbb{R}^n : c^T x = b\},$$

where $c \in \mathbb{R}^n \setminus \{0\}$ and $b \in \mathbb{R}$.

A hyperplane is the intersection of the following *closed halfspaces*:

$$H_+ = \{x \in \mathbb{R}^n : c^T x \geq b\} \quad \text{and} \quad H_- = \{x \in \mathbb{R}^n : c^T x \leq b\}.$$

It is easy to see that H, H_+, H_- are all convex sets (Exercise 8.9).

Definition 8.11 (Polyhedral Set) *A set defined by linear equations and/or inequalities is called a polyhedral set or a polyhedron.*

For example, the quadrangle X_2 in Figure 8.10 is a polyhedral set that can be described by four linear inequalities, each defining a side of the quadrangle.

Definition 8.12 (Extreme Point) *Given a convex set X, a point $x \in X$ is called an extreme point of X if it cannot be represented as a convex combination of two distinct points in X, i.e., there do not exist distinct $x', x'' \in X$, $\alpha \in (0,1)$ such that $x = \alpha x' + (1 - \alpha)x''$.*

Example 8.8 *For the convex set X_1 in Figure 8.10, any point x that lies on the curve between $\bar{x}$ and $\tilde{x}$, inclusive, is an extreme point, thus X_1 has infinitely many extreme points. The other convex set in this figure, X_2, has only 4 extreme points, $\tilde{x}, \hat{x}, \bar{x}$, and $\breve{x}$.*

Extreme points of polyhedral sets are sometimes called *vertices* or *corners*. For example, $\tilde{x}, \hat{x}, \bar{x}$, and $\breve{x}$ are the vertices of X_2.

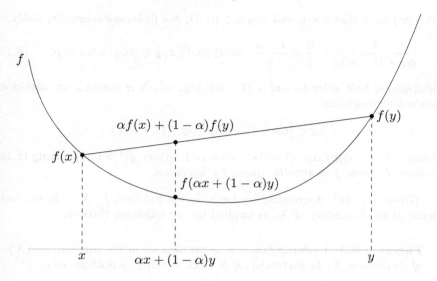

FIGURE 8.11: Geometric illustration for the definition of a convex function.

> **Definition 8.13 (Convex Function)** *Given a convex set $X \subseteq \mathbb{R}^n$, a function $f : X \to \mathbb{R}$ is convex if*
>
> $$f(\alpha x + (1-\alpha)y) \leq \alpha f(x) + (1-\alpha)f(y) \quad \forall x, y \in X, \alpha \in (0,1). \quad (8.13)$$
>
> *If the last inequality is strict whenever $x \neq y$, i.e.,*
>
> $$f(\alpha x + (1-\alpha)y) < \alpha f(x) + (1-\alpha)f(y) \,\, \forall x, y \in X, y \neq x, \alpha \in (0,1), \quad (8.14)$$
>
> *then f is called strictly convex. A function f is called concave (strictly concave) if $-f$ is convex (strictly convex).*

Geometrically, the definition means that the graph of a convex function plotted over the interval $[x, y]$ lies on or below the line segment connecting $(x, f(x))$ and $(y, f(y))$ (see Figure 8.11).

Figure 8.4 at page 167 shows plots of two convex functions ($f(x) = 1/x, f(x) = \exp(x)$) and one concave function ($f(x) = \ln(x)$). Recall that none of these functions has a minimizer.

Example 8.9 *We will demonstrate that $f(x) = 1/x$ is strictly convex on $X = (0, +\infty)$ using the definition above. We need to show that for any $x, y \in$*

$(0, +\infty)$ such that $x \neq y$ and any $\alpha \in (0, 1)$, the following inequality holds:

$$\frac{1}{\alpha x + (1-\alpha)y} < \frac{\alpha}{x} + \frac{1-\alpha}{y} \quad \forall \alpha \in (0, 1), x, y \in (0, +\infty), x \neq y. \quad (8.15)$$

Multiplying both sides by $(\alpha x + (1-\alpha)y)xy$, which is positive, we obtain an equivalent inequality

$$xy < (\alpha x + (1-\alpha)y)(\alpha y + (1-\alpha)x).$$

Since $(\alpha x + (1-\alpha)y)(\alpha y + (1-\alpha)x) - xy = \alpha(1-\alpha)(x-y)^2 > 0$, inequality (8.15) is correct, hence f is strictly convex by definition.

Given $X \subset \mathbb{R}^n$, discontinuities for a convex function $f : X \to \mathbb{R}$ can only occur at the boundary of X, as implied by the following theorem.

Theorem 8.3 *A convex function is continuous in the interior int (X) of its domain X. In particular, if $X = \mathbb{R}^n$, then f is continuous on $\mathbb{R}^n$.*

Note that a convex function is not necessarily differentiable in the interior of its domain. For example, consider $X = \mathbb{R}, f(x) = |x|$.

Definition 8.14 (Convex Problem) *A problem $\min\limits_{x \in X} f(x)$ is called a convex minimization problem (or simply a convex problem) if f is a convex function and X is a convex set.*

Example 8.10 *The problem*

$$\begin{array}{ll} minimize & 3(x_1 - 2)^2 + 2x_2^4 \\ subject\ to & x_1^2 + x_2^2 \leq 4 \end{array}$$

is convex since $f(x) = 3(x_1 - 2)^2 + 2x_2^4$ is a convex function and

$$X = \{x \in \mathbb{R}^2 : x_1^2 + x_2^2 \leq 4\}$$

is a convex set. However, the problem

$$\begin{array}{ll} minimize & 3(x_1 - 2)^2 + 2x_2^4 \\ subject\ to & x_1^2 + x_2^2 = 4 \end{array}$$

is not convex since $X = \{x \in \mathbb{R}^2 : x_1^2 + x_2^2 = 4\}$ is not a convex set.

The following theorem describes a fundamental property of convex problems concerning their optima.

Theorem 8.4 *Any local minimizer of a convex problem is its global minimizer.*

Proof. Assume that x^* is a point of local minimum of a convex problem $\min_{x \in X} f(x)$. We need to show that x^* is a global minimizer for this problem. We will use contradiction. Indeed, if we assume that $f(\hat{x}) < f(x^*)$ for some $\hat{x} \in X$, then, using the definition of convexity for f and X, for any $\alpha \in (0,1)$ we have:

$$
\begin{aligned}
f(x^* + \alpha(\hat{x} - x^*)) &= f(\alpha\hat{x} + (1-\alpha)x^*) \\
&\leq \alpha f(\hat{x}) + (1-\alpha)f(x^*) \\
&< \alpha f(x^*) + (1-\alpha)f(x^*) \\
&= f(x^*).
\end{aligned}
$$

Thus, $f(x^* + \alpha(\hat{x} - x^*)) < f(x^*)$ for any $\alpha \in (0,1)$, which contradicts the local optimality of x^*, since any ε-ball centered at x^* contains a point $x^* + \alpha(\hat{x} - x^*)$ for some $\alpha \in (0,1)$. Therefore, $\hat{x} \in X$ with $f(\hat{x}) < f(x^*)$ cannot exist, and $f(x^*) \leq f(x)$ for any $x \in X$, meaning that x^* is a global minimizer of the considered problem. $\square$

8.6.1 First-order characterization of a convex function

Theorem 8.5 *Let $X \subseteq \mathbb{R}^n$ be a convex set. If $f : X \to \mathbb{R}^n$ is differentiable on an open set containing X then it is convex on X if and only if*

$$
f(y) \geq f(x) + \nabla f(x)^T(y - x) \quad \forall x, y \in X. \tag{8.16}
$$

Proof. We first show that if f is convex, then (8.16) holds. Consider arbitrary $x, y \in X$ and the direction $d = y - x$. The directional derivative of f at x in the direction d is

$$
\begin{aligned}
\nabla f(x)^T(y - x) &= \lim_{\alpha \to 0+} \frac{f(x + \alpha(y - x)) - f(x)}{\alpha} \\
&= \lim_{\alpha \to 0+} \frac{f(\alpha y + (1-\alpha)x) - f(x)}{\alpha} \\
&\leq \lim_{\alpha \to 0+} \frac{\alpha f(y) + (1-\alpha)f(x) - f(x)}{\alpha} \\
&= f(y) - f(x).
\end{aligned}
$$

To obtain the inequality above, we used the definition of convexity for f. Thus, we have proved that if f is convex then (8.16) holds.

To prove the other direction, we assume that (8.16) holds. We need to show that this yields the convexity of f on X. For an arbitrary $\alpha \in (0,1)$ we

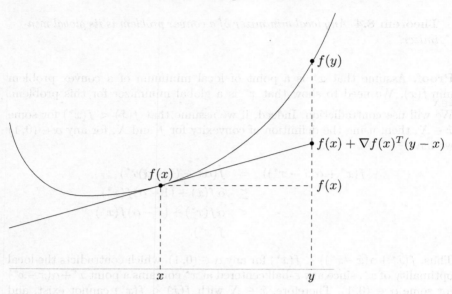

FIGURE 8.12: Geometric interpretation of the first-order characterization of a convex function.

define $z = \alpha x + (1 - \alpha)y$. Then $z \in X$ due to the convexity of X. We use inequality (8.16) two times as follows:

$$\text{for } z, x \text{ we have } f(x) \geq f(z) + \nabla f(z)^T(x - z); \qquad (8.17)$$

$$\text{for } z, y \text{ we have } f(y) \geq f(z) + \nabla f(z)^T(y - z). \qquad (8.18)$$

Multiplying (8.17) by α and (8.18) by $(1 - \alpha)$, and adding the resulting inequalities, we obtain

$$
\begin{aligned}
\alpha f(x) + (1 - \alpha)f(y) &\geq f(z) + \alpha \nabla f(z)^T(x - z) + (1 - \alpha)\nabla f(z)^T(y - z) \\
&= f(z) + \nabla f(z)^T(\alpha x + (1 - \alpha)y - z) \\
&= f(z).
\end{aligned}
$$

So, $f(\alpha x + (1 - \alpha)y) \leq \alpha f(x) + (1 - \alpha)f(y)$, and $f(x)$ is convex on X by the definition of a convex function. $\qquad \square$

Geometrically, the first-order characterization of a convex function means that the graph of f restricted to the line defined by x and y is always on or above its tangent line at x. See Figure 8.12 for an illustration.

8.6.2   Second-order characterization of a convex function

If a convex function is twice continuously differentiable, then it can be equivalently characterized using its Hessian. We discuss the corresponding second-order characterization next.

> **Theorem 8.6** *Let $X \subseteq \mathbb{R}^n$ be a convex set. If $f : X \to \mathbb{R}^n$ is twice continuously differentiable on an open set containing X then it is convex on X if and only if the Hessian $\nabla^2 f(x)$ is positive semidefinite for any $x \in X$.*

Proof. To prove the necessity, we use contradiction. Assume that $\nabla^2 f(x)$ is not positive semidefinite for some $x \in X$, then there exists $d \in \mathbb{R}^n$ with $\|d\| = 1$ such that $d^T \nabla^2 f(x) d < 0$. Let $y = x + \alpha d$ for an arbitrary $\alpha > 0$. We can express $f(y)$ using Taylor's theorem in the following form:

$$f(y) = f(x) + \nabla f(x)^T (y - x) + \frac{1}{2}(y - x)^T \nabla^2 f(x)(y - x) + o(\|y - x\|^2). \quad (8.19)$$

Taking into account that $y - x = \alpha d, \alpha > 0$, and $\|d\| = 1$, we can equivalently rewrite the last equality as

$$\frac{f(y) - f(x) - \nabla f(x)^T (y - x)}{\alpha^2} = \frac{1}{2} d^T \nabla^2 f(x) d + \frac{o(\alpha^2)}{\alpha^2}. \quad (8.20)$$

Note that if $\frac{1}{2} d^T \nabla^2 f(x) d \neq 0$, then by definition of $o(\cdot)$ there exists a sufficiently small $\alpha \neq 0$ such that $\frac{o(\alpha^2)}{\alpha^2}$ is smaller in absolute value than $\frac{1}{2} d^T \nabla^2 f(x) d$, so the sign of the right-hand side of (8.20) is the same as the sign of $d^T \nabla^2 f(x) d$. In particular, if $d^T \nabla^2 f(x) d < 0$, then there exists $\alpha > 0$ such that for $y = x + \alpha d \in X$: $f(y) - f(x) - \nabla f(x)^T (y - x) < 0$, which contradicts the first-order characterization of a convex function that must be satisfied by f.

To prove the sufficiency, assume that $\nabla^2 f(x)$ is positive semidefinite for all $x \in X$. Then we show that f satisfies the first-order characterization of a convex function and is therefore convex. Consider arbitrary $x, y \in X$. According to Taylor's theorem, there exists $\alpha \in (0, 1)$ such that

$$f(y) = f(x) + \nabla f(x)^T (y - x) + \frac{1}{2}(y - x)^T \nabla^2 f(x + \alpha(y - x))(y - x). \quad (8.21)$$

This can be equivalently written as

$$f(y) - f(x) - \nabla f(x)^T (y - x) = \frac{1}{2}(y - x)^T \nabla^2 f((1 - \alpha)x + \alpha y)(y - x). \quad (8.22)$$

Using Rayleigh's inequality, we have

$$(y - x)^T \nabla^2 f((1 - \alpha)x + \alpha y)(y - x) \geq \lambda_{\min} \|y - x\|^2, \quad (8.23)$$

where $\lambda_{\min}$ is the smallest eigenvalue of $\nabla^2 f((1 - \alpha)x + \alpha y)$. From (8.22) and (8.23) we obtain

$$f(y) - f(x) - \nabla f(x)^T (y - x) \geq \frac{1}{2}\lambda_{\min}\|y - x\|^2 \geq 0.$$

Since $x, y \in X$ were chosen arbitrarily, this inequality holds for any $x, y \in X$. Thus, by the first-order characterization of a convex function, f is convex on X. $\qquad\square$

Exercises

8.1. Consider the problem $\min\limits_{x \in \mathbb{R}^n} f(x)$, where $f : \mathbb{R}^n \to \mathbb{R}$ is an arbitrary continuous function. Prove or disprove each of the following independent statements concerning this problem.

(a) If the considered problem has a local minimizer that is not strict, then it has infinitely many local minimizers.

(b) If x^* is a strict local minimizer and a global minimizer for the considered problem, then x^* is a strict global minimizer for the same problem.

(c) If x^* is a strict global minimizer of f, then x^* is a strict local minimizer of f.

(d) If there exists a constant C such that $f(x) \geq C$ for all $x \in \mathbb{R}^n$, then f has at least one local minimizer.

(e) If f is convex, it cannot have local minimizers $\bar{x}$ and $\hat{x}$ such that $f(\hat{x}) < f(\bar{x})$.

(f) If f has a unique global minimizer and no other local minimizers, then f is convex.

(g) If f is strictly convex, then any local minimizer is its strict global minimizer.

8.2. Consider the problem $\min\limits_{x \in \{0,1\}^n} f(x)$ of minimizing a function $f : \mathbb{R}^n \to \mathbb{R}$ over the discrete set $\{0,1\}^n$. Show that if $f(x)$ is a linear function with respect to each variable (that is, fixing the values of $n-1$ variables turns $f(x)$ into a linear function of a single variable), then the considered problem is equivalent to the problem $\min\limits_{x \in [0,1]^n} f(x)$ of minimizing $f(x)$ over the n-dimensional unit hypercube $[0,1]^n$.

8.3. In graph theory, a *simple, undirected graph* is a pair $G = (V, E)$, where

V is a finite set of *vertices* and E is a set of *edges* given by unordered pairs of vertices, $E \subseteq \{\{i,j\} : i, j \in V\}$. Given a simple undirected graph $G = (V, E)$ with $n = |V|$ vertices, the *maximum independent set problem* is to find a maximum-size subset I of V such that $\{i, j\} \notin E$ for any $i, j \in I$. Let $\alpha(G)$ denote the *independence number* of G, which is the cardinality of a maximum independent set of G. Show that

$$\alpha(G) = \max_{x \in [0,1]^n} \left(e^T x - \frac{1}{2} x^T A_G x \right),$$

where $e = [1, \ldots, 1]^T \in \mathbb{R}^n$, and $A_G = [a_{ij}]_{n \times n}$ with $a_{ij} = 1$ if $\{i,j\} \in E$ and $a_{ij} = 0$ otherwise. Hint: observe that the objective function of the above problem is linear with respect to each variable.

8.4. Solve the problem $\min_{x \in X} f(x)$, where

(a) $f(x) = \frac{x^2 - x + 2}{x^2 + 2x + 2}$, $X = \mathbb{R}$;

(b) $f(x) = (x_1 + 2x_2 - 4x_3 + 1)^2 + (2x_1 - x_2 + 1)^4 + (x_1 - 3x_2 + 2)^6$, $X = \mathbb{R}^3$.

8.5. Does the function $f(x) = e^{x_1^2 + x_2^2 + x_3^2} - x_1^3 - x_2^4 - x_3^6$ have a global minimizer in $\mathbb{R}^3$? Why?

8.6. Find all local and global minimizers, if any, for the following problems graphically.

(a) minimize $\quad 2x - x^2$
    subject to $\quad 0 \le x \le 3$.

(b) minimize $\quad -(x_1^2 + x_2^2)$
    subject to $\quad x_1 \le 1$.

(c) minimize $\quad x_1 - (x_2 - 2)^3 + 1$
    subject to $\quad x_1 \ge 1$.

(d) minimize $\quad x_1^2 + x_2^2$
    subject to $\quad x_1^2 + 9x_2^2 = 9$.

(e) minimize $\quad x_1^2 + x_2^2$
    subject to $\quad x_1^2 - x_2 \le 4$
                   $x_2 - x_1 \le 2$.

(f) minimize $\quad x_2 - x_1$
    subject to $\quad x_1^2 - x_2^3 = 0$.

8.7. Let $f : \mathbb{R}^n \to \mathbb{R}$ be an arbitrary convex function. Show that for any constant c the set $X = \{x \in \mathbb{R}^n : f(x) \le c\}$ is convex.

8.8. Prove the following properties of convex sets:

(a) Let $\mathcal{F}$ be a family of convex sets. Then the intersection $\bigcap\limits_{C \in \mathcal{F}} C$ is also a convex set.

(b) Let $C \subset \mathbb{R}^n$ be a convex set and let α be a real number. Then the set $\alpha C = \{y : y = \alpha x, \ x \in C\}$ is also convex.

(c) Let $C_1, C_2 \subset \mathbb{R}^n$ be convex sets. Then the set

$$C_1 + C_2 = \{x : x = x^{(1)} + x^{(2)}, \ x^{(1)} \in C_1, \ x^{(2)} \in C_2\}$$

is also convex.

8.9. Let $c \in \mathbb{R}^n \setminus \{0\}$ and $b \in \mathbb{R}$. Show that the hyperplane

$$H = \{x \in \mathbb{R}^n : c^T x = b\}$$

and the closed halfspaces

$$H_+ = \{x \in \mathbb{R}^n : c^T x \geq b\},$$

$$H_- = \{x \in \mathbb{R}^n : c^T x \leq b\}.$$

are convex sets.

8.10. (**Jensen's inequality**) Let $f : X \to \mathbb{R}$, where $X \subseteq \mathbb{R}^n$ is a convex set. Prove that f is a convex function if and only if for any $x^{(1)}, \ldots, x^{(k)} \in X$ and coefficients $\alpha_1, \ldots, \alpha_k$ such that

$$\sum_{i=1}^{k} \alpha_i = 1, \ \alpha_i \geq 0, i = 1, \ldots, m, \tag{8.24}$$

the following inequality holds:

$$f\left(\sum_{i=1}^{k} \alpha_i x^{(i)}\right) \leq \sum_{i=1}^{k} \alpha_i f(x^{(i)}). \tag{8.25}$$

8.11. Use Jensen's inequality to show the following:

(a) If x is a convex combination of points $x^{(1)}, \ldots, x^{(k)}$, then

$$f(x) \leq \max_{1 \leq i \leq k} f(x^{(i)}).$$

(b) If we denote by

$$\Delta = \mathrm{Conv}\{x^{(1)}, \ldots, x^{(k)}\} = \left\{x = \sum_{i=1}^{k} \alpha_i x^{(i)} : \sum_{i=1}^{k} \alpha_i = 1, \ \alpha_i \geq 0\right\},$$

then

$$\max_{x \in \Delta} f(x) = \max_{1 \leq i \leq k} f(x^{(i)}).$$

8.12. Let $X \subseteq \mathbb{R}^n$ be a convex set, and let $f, g : X \to \mathbb{R}$ be convex functions. Show that the following function is convex:

$$h(x) = \max\{f(x), g(x)\}, \quad x \in X.$$

8.13. Let $X \subseteq \mathbb{R}^n$ be a convex set. Show that $f : X \to \mathbb{R}$ is convex if and only if for any $x, y \in X$ and $\beta \geq 0$ such that $y + \beta(y - x) \in X$, we have

$$f(y + \beta(y - x)) \geq f(y) + \beta(f(y) - f(x)).$$

Interpret this equivalent characterization of a convex function geometrically.

8.14. Let $X \subseteq \mathbb{R}^n$ be a convex set. For a function $f : X \to \mathbb{R}$ its *epigraph* epi(f) is the following set in $\mathbb{R}^{n+1}$:

$$\text{epi}(f) = \{(x, c) \in \mathbb{R}^{n+1} : x \in X, c \in \mathbb{R}, f(x) \leq c\}.$$

   (a) Sketch epi(f) for $f(x) = x^2$, $X = \mathbb{R}$ and for $f(x) = x_1^2 + x_2^2$, $X = \mathbb{R}^2$.

   (b) Show that f is a convex function if and only if epi(f) is a convex set.

8.15. Let $f : \mathbb{R}^m \to \mathbb{R}$ be a convex function. For an $m \times n$ matrix A and a vector $b \in \mathbb{R}^m$, show that $g : \mathbb{R}^n \to \mathbb{R}$ defined by $g(x) = f(Ax + b)$ is a convex function.

8.16. Let $x = [x_j]_{j=1}^n$. Consider the problem

$$\text{minimize} \quad \sum_{j=1}^{m_0} \alpha_{0j} \prod_{r=1}^{n} (x_r)^{\sigma_{0j}^{(r)}}$$

$$\text{subject to} \quad \sum_{j=1}^{m_i} \alpha_{ij} \prod_{r=1}^{n} (x_r)^{\sigma_{ij}^{(r)}} \leq 1, i = 1, \ldots, m$$

$$x_j > 0, j = 1, \ldots, n,$$

where for any $i = 0, \ldots, m, j = 1, \ldots, m_i$ α_{ij} are some positive coefficients and $\sigma_{ij}^{(r)}, r = 1, \ldots n$ are arbitrary real powers. Observe that this problem is not convex in general. Formulate an equivalent convex problem.

8.17. Use the second-order characterization of a convex function to show that the following function $f : \mathbb{R} \to \mathbb{R}$ is convex:

   (a) $f(x) = |x|^p$, where $p > 1$ is a given constant;

   (b) $f(x) = |x| - \ln(1 + |x|)$;

   (c) $f(x) = \frac{x^2}{1 - |x|}$.

Chapter 9

Complexity Issues

9.1 Algorithms and Complexity

By an algorithm, we usually mean a sequence of instructions carried out in order to solve some computational problem. The steps of an algorithm need to be clearly defined, so that they can be implemented and executed on a computer.

Example 9.1 *Given an integer number a and a positive integer n, a^n can be computed using the following simple algorithm.*

Algorithm 9.1 A naive algorithm for computing the nth power of an integer.

1: **Input:** a, n
2: **Output:** a^n
3: answer $= 1$
4: **for** $i = 1, \ldots, n$ **do**
5:     answer=answer$\times a$
6: **end for**
7: **return** answer

Algorithms are usually compared by their performance. The most popular performance measure of an algorithm is the time it takes to produce the final answer. However, if an algorithm is executed on a computer, the time it will take to terminate may vary significantly depending on many factors, such as the technical characteristics of a computer used to run the algorithm. Therefore, in analysis of algorithms, their time requirements are expressed in terms of the number of elementary operations needed in order to execute the algorithm. These may be arithmetic operations, assignments, comparisons, and so on. It is assumed that each such operation takes unit time.

Example 9.2 *Algorithm 9.1 consists of $n + 1$ assignments and n multiplications; therefore, its total running time is $2n + 1$.*

Given all possible inputs, the time complexity of an algorithm is defined as the number of steps that the algorithm requires in the worst case, and is usually expressed as a function of the input parameters. Thus talking about the complexity of an algorithm, one is usually interested in the performance of the algorithm when the inputs are very large. This diminishes the role of the constant multipliers in our analysis. For example, if n is very large, the relative difference between $10n^5$ and $11n^5$ can be neglected. On the other hand, if the time required by an algorithm is expressed as a sum of terms with different rates of growth, the slower-growing terms can be neglected, since they are overpowered by faster-growing terms when n is large enough. To express the worst-case complexity of an algorithm, we will use the asymptotic notations defined next.

Definition 9.1 *Let* $\{f(n), n \geq 1\}, \{g(n), n \geq 1\}$ *be sequences of positive real numbers. We say that*

- $f(n) = O(g(n))$ *if there exists a constant* $c > 0$ *and an integer* n_0 *such that* $f(n) \leq cg(n)\ \forall n \geq n_0$.

- $f(n) = \Omega(g(n))$ *if there exists a constant* $c > 0$ *and an integer* n_0 *such that* $f(n) \geq cg(n)\ \forall n \geq n_0$.

- $f(n) = \Theta(g(n))$ *if there exist constants* $c, c' > 0$ *and an integer* n_0 *such that* $cg(n) \leq f(n) \leq c'g(n)\ \forall n \geq n_0$.

Note that Θ induces an equivalence relation on the set of functions. The equivalence class of $f(n)$ is the set of all functions $g(n)$ such that $f(n) = \Theta(g(n))$. We call this class the *rate of growth* of $f(n)$. For any polynomials $f(n)$ and $g(n)$ of degree k, we have $f(n) = \Theta(g(n)) = \Theta(n^k)$. Hence, n^k is a natural representative element for the class of polynomials of degree k.

The number of steps that an algorithm takes usually depends on the input, and the run time is usually expressed in terms of the *problem size*. If the problem's input is represented by a sequence (string) of symbols, then the problem size can be defined as the length of this string, i.e., the number of bits required to store the problem's input.

Example 9.3 *Assume that the input of a problem is given by a single integer. The number of symbols required to represent an integer n in a base-β arithmetic system is* $\lceil \log_\beta n \rceil$, *where* $\beta \geq 2$. *Hence, the size of the problem is* $\Theta(\log n)$.

Example 9.4 *Consider a problem whose input is an $m \times n$ matrix A. Then the number of bits required to represent the input is*

$$\Theta(mn + \lceil \log |P| \rceil),$$

where P is the product of all nonzero entries of A. Since the number of bits allocated to numbers of the same type in modern computers can be treated as *a constant, $\Theta(mn)$ is a reasonable approximation of the problem size in this case. It is also natural to describe the running time of an algorithm for solving such a problem using a function of n and m.*

Obviously, the size of the problem input depends on the *data structures* used to represent the problem input.

Example 9.5 *In graph theory, a simple, undirected graph is a pair $G = (V, E)$, where V is a finite set of vertices and E is a set of edges, with each edge corresponding to a pair of vertices. If $G = (V, E)$ is a graph and $e = \{u, v\} \in E$, we say that u is adjacent to v and vice versa. We also say that u and v are neighbors. The neighborhood $N(v)$ of a vertex v is the set of all its neighbors in G: $N(v) = \{u : \{u, v\} \in E\}$. If $G = (V, E)$ is a graph and $e = \{u, v\} \in E$, we say that e is incident to u and v. The degree of a vertex v is the number of its incident edges.*

A graph can be represented in several ways. For example, $G = (V, E)$ can be described by its $|V| \times |V|$ adjacency matrix $A_G = [a_{ij}]_{n \times n}$, such that

$$a_{ij} = \begin{cases} 1, & \text{if } \{i, j\} \in E, \\ 0, & \text{otherwise.} \end{cases}$$

Alternatively, the same graph can be represented by its adjacency lists, where for each vertex $v \in V$ we record the set of vertices adjacent to it. In adjacency lists, there are $2|E|$ elements, each requiring $O(\log |V|)$ bits. Hence, the total space required is $O(|E| \log |V|)$ and $O(|E|)$ space is a reasonable approximation of the size of a graph.

If an algorithm runs in time $O(n^k)$, where n is a parameter describing the problem size, then we say that this is a *polynomial-time* or *efficient algorithm* with respect to n.

Most often, we will describe the running time of an algorithm using the $O(\cdot)$ notation. An algorithm that runs in $O(n)$ time is referred to as a *linear-time algorithm*; $O(n^2)$ time as a *quadratic-time algorithm*, etc.

Example 9.6 *Denote by $f(n) = 2n + 1$, $g(n) = n$. Then $f(n) = O(g(n))$, therefore Algorithm 9.1 runs in linear time.*

The next two examples analyze the running time of two classical algorithms for the sorting problem, which, given n integers asks to sort them in a non-decreasing order. In other words, given an input list of numbers $a_1, a_2, \ldots, a_n$, we need to output these numbers in a sorted list $a_{s_1}, a_{s_2}, \ldots, a_{s_n}$, where $a_{s_1} \leq a_{s_2} \leq \ldots \leq a_{s_n}$ and $\{s_1, s_2, \ldots, s_n\} = \{1, 2, \ldots, n\}$.

Example 9.7 (Quicksort) *Quicksort uses the following divide-and-conquer strategy to divide a list into two sub-lists.*

1. *Pick an element, called a pivot, from the list. We can pick, e.g., the first element in the list as the pivot.*

2. *Reorder the list so that all elements which are less than the pivot come before the pivot and all elements greater than the pivot come after it (equal values can go either way). Note that after this partitioning, the pivot assumes its final position in the sorted list.*

3. *Recursively sort the sub-list of lesser elements and the sub-list of greater elements.*

At step k of Quicksort we make at most $n - k$ comparisons (the worst case is when the list is already sorted). The worst-case running time is given by

$$T(n) = (n-1) + (n-2) + (n-3) + \ldots + 1 = \frac{n(n-1)}{2} = \Theta(n^2).$$

Thus, Quicksort is an $\Omega(n^2)$ algorithm.

Example 9.8 (Mergesort) *Mergesort proceeds as follows:*

1. *If the list is of length 0 or 1, then do nothing (list is already sorted). Otherwise, follow Steps 2–4:*

2. *Split the unsorted list into two equal parts.*

3. *Sort each sub-list recursively by re-applying Mergesort.*

4. *Merge the two sub-lists back into one sorted list.*

Note that the merge operation involves $n - 1$ comparisons. Let $T(n)$ denote the run time of Mergesort applied to a list of n numbers. Then

$$
\begin{aligned}
T(n) &\leq 2T(n/2) + n - 1 \\
&\leq 2(2T(n/4) + n/2 - 1) + n - 1 \\
&= 4T(n/4) + (n-2) + (n-1) \\
&\ldots \\
&\leq 2^k T(n/2^k) + \sum_{i=0}^{k-1} (n - 2^i) \\
&= 2^k T(n/2^k) + kn - (2^k - 1) \\
T(1) &= 0
\end{aligned}
$$

Using $k = \log_2 n$, we obtain

$$T(n) \leq n \log n - n + 1 = \Theta(n \log n).$$

In the above two examples we discussed two sorting algorithms, Quicksort running in $\Omega(n^2)$ time, and Mergesort running in $O(n \log n)$ time. Hence, Mergesort guarantees a better worst-case running time performance. However, the worst-case analysis may not be indicative of "average-case" performance.

9.2 Average Running Time

Assume that all possible inputs of a given problem of size n are equally probable. Then the *average running time* $A(n)$ of an algorithm for solving this problem is defined as the expected running time over all possible outcomes.

Example 9.9 (Average running time of Quicksort) *Let $q_1, \ldots, q_n$ be the sorted version of $a_1, \ldots, a_n$. Note that the number of comparisons used by Quicksort to sort $a_1, \ldots, a_n$ is completely defined by the relative order of $q_1, \ldots, q_n$ in the input list $a_1, \ldots, a_n$. Hence, a random input can be thought of as a random permutation of the elements of the sorted list. Assuming that all such permutations are equally probable, we need to determine the expected number of comparisons made by Quicksort.*

Denote by x_{ij} the random variable representing the number of comparisons between q_i and q_j during a Quicksort run, where $q_1 \leq q_2 \leq \ldots \leq q_n$ is the sorted version of the input $a_1, a_2, \ldots, a_n$. We have

$$x_{ij} = \begin{cases} 1, & \text{if } q_i \text{ and } q_j \text{ are compared (with probability } p_{ij}) \\ 0, & \text{if } q_i \text{ and } q_j \text{ are not compared (with probability } 1 - p_{ij}) \end{cases}$$

Then

$$\begin{aligned} A(n) &= E\left(\sum_{i=1}^{n} \sum_{j=i+1}^{n} x_{ij} \right) \\ &= \sum_{i=1}^{n} \sum_{j=i+1}^{n} E(x_{ij}) \\ &= \sum_{i=1}^{n} \sum_{j=i+1}^{n} (1 \cdot p_{ij} + 0 \cdot (1 - p_{ij})) \\ &= \sum_{i=1}^{n} \sum_{j=i+1}^{n} p_{ij}. \end{aligned}$$

Note that q_i and q_j are compared if and only if none of the numbers $q_{i+1}, \ldots, q_{j-1}$ appear before both q_i and q_j in the list $a_1, \ldots, a_n$. The probability of this happening is

$$p_{ij} = \frac{2}{j - i + 1}.$$

Hence,

$$\begin{aligned} A(n) &= \sum_{i=1}^{n} \sum_{j=i+1}^{n} p_{ij} = \sum_{i=1}^{n} \sum_{j=i+1}^{n} \frac{2}{j-i+1} \\ &= \sum_{i=1}^{n} \left(\frac{2}{2} + \frac{2}{3} + \frac{2}{4} + \ldots + \frac{2}{n-i+1} \right) \\ &< 2 \sum_{i=1}^{n} \left(\frac{1}{2} + \frac{1}{3} + \frac{1}{4} + \ldots + \frac{1}{n} \right) \\ &= 2n \sum_{i=2}^{n} \frac{1}{i} \leq 2n \ln n = O(n \log n). \end{aligned}$$

Therefore, the average running time of Quicksort is $O(n \log n)$.

9.3   Randomized Algorithms

A *randomized algorithm* is an algorithm in which some decisions depend on the outcome of a coin flip (which can be either 0 or 1). Two major types of randomized algorithms are *Las Vegas algorithms*, which always yield a correct answer, but take random running time; and *Monte Carlo algorithms*, which run in predefined time and give a correct answer with "high probability." For a problem of size n, we say that

$$p = 1 - \frac{1}{n^\alpha} = 1 - n^{-\alpha},$$

where $\alpha > 1$, is a *high probability*.

We say that a randomized algorithm takes $\tilde{O}(f(n))$ of a resource (such as space or time) if there exist numbers $C, n_0 > 0$ such that for any $n \geq n_0$ the amount of resource used is

$$\leq C\alpha f(n) \text{ with probability} \geq 1 - n^{-\alpha}, \alpha > 1.$$

Example 9.10 (Repeated element identification) *Consider the following problem. Given n numbers, $n/2$ of which have the same value and the remaining $n/2$ are all different, find the repeated number.*

We can develop a simple deterministic algorithm for this problem, in which we first sort the numbers and then scan the sorted list until two consecutive numbers are equal. The time complexity of this algorithm is $O(n \log n)$.

Next, we consider a Las Vegas algorithm for the same problem.

Algorithm 9.2 A Las Vegas algorithm for finding the repeated element.

1: **Input:** $a_1, a_2, \ldots, a_n$, where $n/2$ of the values are identical
2: **Output:** the repeated element
3: stop $= 0$
4: **while** stop$\neq 1$ **do**
5:     randomly pick two elements a_i and a_j
6:     **if** $(i \neq j)$ and $(a_i = a_j)$ **then**
7:         stop $= 1$
8:         **return** a_i
9:     **end if**
10: **end while**

Next we determine the number of iterations of the algorithm required to ensure the correct answer with high probability. The probability of success in a single try is

$$P_s = \frac{\frac{n}{2}\left(\frac{n}{2} - 1\right)}{n^2} = \frac{\frac{n^2}{4} - \frac{n}{2}}{n^2} = \frac{1}{4} - \frac{1}{2n} > \frac{1}{8} \text{ if } n > 4.$$

Hence, the probability of failure in a single try is

$$P_f = 1 - P_s < 7/8,$$

and the probability of success in k tries is $1 - (P_f)^k > 1 - (7/8)^k$. We need to find k such that the probability of success is high (i.e., at least $1 - n^{-\alpha}$).

$$1 - (7/8)^k = 1 - n^{-\alpha} \Leftrightarrow (7/8)^k = n^{-\alpha} \Leftrightarrow k = \frac{\alpha \log n}{\log(8/7)}.$$

Let $c = 1/\log(8/7)$. If $k = c\alpha \log n$, then we have a high probability of success in k tries. Thus, the running time of the algorithm is $\tilde{O}(\log n)$.

9.4 Basics of Computational Complexity Theory

Given an optimization problem Π, a natural question is:

$$\text{Is the problem } \Pi \text{ "easy" or "hard"?} \tag{9.1}$$

Methods for solving "easy" and "hard" problems have fundamental differences, therefore distinguishing between them is important in order to approach each problem properly. Questions of this type are addressed by *computational complexity theory*.

Definition 9.2 *By easy or* tractable *problems, we mean the problems that can be solved in time polynomial with respect to their size. We also call such problems* polynomially solvable *and denote the class of polynomially solvable problems by $\mathcal{P}$.*

Example 9.11   *1. The problem of sorting n integer numbers can be solved in $O(n \log n)$ time, thus it belongs to $\mathcal{P}$.*

   2. Two $n \times n$ rational matrices can be trivially multiplied in $O(n^3)$ time, thus matrix multiplication is in $\mathcal{P}$.

   3. A linear system $Ax = b$, where A is an $n \times n$ rational matrix and b is a rational vector of length n can be solved in $O(n^3)$ time, thus the problem of solving a linear system is in $\mathcal{P}$.

Now that we decided which problems will be considered easy, how do we define "hard" problems? We could say that a problem Π is hard if it is not easy, i.e., $\Pi \notin \mathcal{P}$, however some of the problems in such a class could be *too* hard, to the extent that they cannot be solved for some inputs. Such problems are referred to as *undecidable* problems.

Example 9.12 *Consider the following problem, which is referred to as the* halting problem:

Given a computer program with its input, will it ever halt?

Alan M. Turing has shown that there cannot exist an algorithm that would always solve the halting problem. The mathematical formalism he introduced in 1930s to describe a computer implementation of algorithms is called a Turing machine and is a foundational concept in computational complexity theory.

We want to define a class of hard problems that we have a chance to solve, perhaps by "gambling." For example, we could try to guess the solution, and if we are lucky and are able to easily verify that our solution is indeed correct, then we solved the problem.

Example 9.13 *Consider a lottery, where some black box contains a set W of $n/2$ distinct winning numbers randomly chosen from the set $N = \{1, 2, \dots, n\}$ of n integers. The problem is to guess all the elements of W by picking $n/2$ numbers from N. As soon as the pick is made, we can easily verify whether the set of numbers picked coincides with W. Winning such a lottery (i.e., guessing all the winning numbers correctly) is not easy, but not impossible.*

However, if we apply this logic to optimization problems, we may still encounter some major difficulties. For example, if we pick a random feasible solution $\bar{x} \in X$ of the problem $\min_{x \in X} f(x)$, even if we are lucky and $\bar{x}$ is an optimal solution of the problem, the fact of optimality may not be easy to verify, since answering this question could require solving the optimization problem. To overcome this difficulty, we can consider different versions of optimization problems, for which the correct answer may be easier to verify.

Definition 9.3 *Given a problem $\min_{x \in X} f(x)$, we consider the following three versions of this problem:*

- Optimization version: *find x from X that maximizes $f(x)$;*
  Answer: x^* maximizes $f(x)$.

- Evaluation version: *find the largest possible $f(x)$;*
  Answer: the largest possible value for $f(x)$ is f^*.

- Decision (Recognition) version: *Given f^*, does there exist an x such that $f(x) \geq f^*$?*
  Answer: "yes" or "no."

Example 9.14 *Consider the classical* maximum clique *problem. A subset C of vertices in a simple undirected graph $G = (V, E)$ is called a* clique *if $\{i, j\} \in E$ for any distinct $i, j \in C$, i.e., all pairs of vertices are adjacent in C. Given a simple undirected graph G, the three versions of the maximum clique problem according to the definition above are:*

1. *Optimization version (maximum clique): Find a clique of the largest size in G.*

2. *Evaluation version (maximum clique size): Find the size of a largest clique in G.*

3. *Recognition version (CLIQUE): Given a positive integer k, does there exist a clique of size at least k in G?*

If we randomly pick a solution (a clique) and its size is at least k (which we can verify in polynomial time), then obviously the answer to the recognition version of the problem is "yes." This clique can be viewed as a certificate proving that this is indeed a yes instance of CLIQUE.

9.4.1 Class $\mathcal{NP}$

We will focus on decision problems that we can potentially solve by providing an efficiently verifiable certificate proving that the answer is "yes" (assuming that we are dealing with a yes instance of the problem). Such a class of problems is defined next.

Definition 9.4 *If a problem Π is such that for any yes instance of Π there exists a concise (polynomial-size) certificate that can be verified in polynomial time, then we call Π a nondeterministic polynomial problem. We denote the class of nondeterministic polynomial problems by $\mathcal{NP}$.*

9.4.2 $\mathcal{P}$ vs. $\mathcal{NP}$

Note that any problem from $\mathcal{P}$ is also in $\mathcal{NP}$ (i.e., $\mathcal{P} \subseteq \mathcal{NP}$), so there are easy problems in $\mathcal{NP}$. So, our original question (9.1) of distinguishing between "easy" and "hard" problems now becomes:

$$\text{Are there "hard" problems in } \mathcal{NP}, \text{ and if there are,} \atop \text{how do we recognize them?} \qquad (9.2)$$

The first part of this question can be stated as

$$\mathcal{P} \overset{?}{=} \mathcal{NP}. \tag{9.3}$$

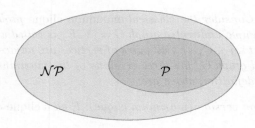

FIGURE 9.1: The conjectured relation between classes $\mathcal{P}$ and $\mathcal{NP}$.

This question is still to be answered, in fact, it is viewed as one of the most important open questions in mathematical sciences. However, it is conjectured that $\mathcal{P} \neq \mathcal{NP}$, i.e., there are problems from $\mathcal{NP}$ that are not in $\mathcal{P}$ (see Figure 9.1). Assuming that there are hard problems in $\mathcal{NP}$, it makes sense to consider a problem from $\mathcal{NP}$ hard if it is at least as hard as any other problem from $\mathcal{NP}$. In other words, we can call a problem hard if the fact that we can solve this problem would mean that we can solve any other problem in $\mathcal{NP}$ "about as easily." This idea is formalized in the notion of *polynomial time reducibility* from Π_1 to Π_2, which is an "easy" reduction, such that if we can solve Π_2 fast, then we can solve Π_1 fast.

9.4.3 Polynomial time reducibility

Definition 9.5 *We say that* Π_1 *is* polynomial time reducible *to* Π_2, *denoted by* $\Pi_1 \propto \Pi_2$, *if existence of a polynomial time algorithm for* Π_2 *would imply existence of a polynomial time algorithm for* Π_1.

In other words, $\Pi_1 \propto \Pi_2$ if there exist two polynomial time algorithms:

1. $\mathcal{A}_1$, which converts an input for Π_1 into an input for Π_2;

2. $\mathcal{A}_2$, which converts an output for Π_2 into an output for Π_1,

such that any instance π_1 of Π_1 can be solved by

(i) converting π_1 into an instance π_2 of Π_2 using $\mathcal{A}_1$,

(ii) solving π_2, and

(iii) converting the result of solving π_2 into an output for π_1 using $\mathcal{A}_2$.

An engineer and a mathematician were put in a room with an empty water boiler and a cup full of water and were told to boil the water.
Both men filled the boiler with water and turned it on to boil the water.
Next, they were put in the same room, but with the water boiler full of water and the cup emptied. Again, they were told to boil the water.
The engineer just turned on the boiler to complete the task.
The mathematician poured the water back into the cup, thus reducing it to a previously solved problem.

FIGURE 9.2: Mathematical humor touching on problem reducibility.

$\Pi_1 \propto \Pi_2$ implies that Π_2 is at least as hard as Π_1, since polynomial time solvability of Π_2 leads to polynomial time solvability of Π_1. It is easy to check that polynomial time reducibility is a *transitive* relation, i.e., if Π_1 is polynomial time reducible to Π_2 and Π_2 is polynomial time reducible to Π_3, then Π_1 is polynomial time reducible to Π_3:

$$\Pi_1 \propto \Pi_2 \text{ and } \Pi_2 \propto \Pi_3 \Rightarrow \Pi_1 \propto \Pi_3. \tag{9.4}$$

Indeed, let $\mathcal{A}_1$ and $\mathcal{A}_2$ be the two polynomial time algorithms used to reduce Π_1 to Π_2, and let $\mathcal{A}_1'$ and $\mathcal{A}_2'$ be the two polynomial time algorithms used to reduce Π_2 to Π_3. To reduce Π_1 to Π_3, we first convert the instance π_1 of Π_1 into an instance π_2 of Π_2 using $\mathcal{A}_1$, and then convert π_2 into an instance π_3 of Π_3 using $\mathcal{A}_1'$. Then we solve Π_3 and transform the answer into the answer to Π_2 using $\mathcal{A}_2'$, which we then convert into the answer to Π_1 using $\mathcal{A}_2$.

9.4.4 $\mathcal{NP}$-complete and $\mathcal{NP}$-hard problems

Definition 9.6 *A problem Π is called $\mathcal{NP}$-complete if*

1. *$\Pi \in \mathcal{NP}$, and*

2. *any problem from $\mathcal{NP}$ is polynomial time reducible to Π.*

Due to the transitivity of polynomial time reducibility, in order to show that a problem Π is $\mathcal{NP}$-complete, it is sufficient to show that

1. $\Pi \in \mathcal{NP}$, and

2. there is an $\mathcal{NP}$-complete problem Π' that can be reduced to Π in polynomial time.

To use this observation, we need to know at least one $\mathcal{NP}$-complete problem.

The first problem proved to be $\mathcal{NP}$-complete is the *satisfiability* (*SAT*) problem, which is defined next in *conjunctive normal form* representation. A Boolean variable x is a variable that can assume only the values *true* (1) and *false* (0). Boolean variables can be combined to form Boolean formulas using the following logical operations:

1. Logical AND, denoted by "$\wedge$" or "$\cdot$" (also called *conjunction*)

2. Logical OR, denoted by "$\vee$" or "$+$" (also called *disjunction*)

3. Logical NOT, denoted by $\bar{x}$.

A *clause* C is a disjunctive expression $C = \bigvee_{p=1}^{k} \ell_p$, where each *literal* ℓ_p is either x_r or $\bar{x}_r$ for some r. A *conjunctive normal form* (*CNF*) is given by

$$F = \bigwedge_{i=1}^{m} C_i,$$

where C_i is a clause. A CNF F is called *satisfiable* if there is an assignment of variables such that $F = 1$ (*TRUE*).

The Satisfiability (SAT) problem is defined as follows: Given m clauses $C_1, \ldots, C_m$ with the variables $x_1, \ldots, x_n$, is the CNF $F = \bigwedge_{i=1}^{m} C_i$ satisfiable?

Theorem 9.1 (Cook, 1971 [12]) *SAT is NP-complete.*

Cook's theorem made it possible to prove $\mathcal{NP}$-completeness of problems via polynomial time reductions from other problems that have been proved to be $\mathcal{NP}$-complete. As a result, hundreds of important problems are now known to be $\mathcal{NP}$-complete.

Example 9.15 *To demonstrate how a polynomial time reduction is used to prove $\mathcal{NP}$-completeness of a problem, consider a special case of SAT, 3-satisfiability (3-SAT) problem, in which each clause is restricted to consist of just 3 literals.*

Theorem 9.2 *3-SAT is NP-complete.*

Proof: Obviously, 3-SAT is in $\mathcal{NP}$. We will reduce SAT to 3-SAT. Given $F = \bigwedge\limits_{i=1}^{m} C_i$, we will construct a new formula F' with 3 literals per clause, such that F' is satisfiable if and only if F is satisfiable. We modify each clause C_i as follows:

1. If C_i has three literals, do nothing.

2. If C_i has $k > 3$ literals, $C_i = \bigvee\limits_{p=1}^{k} \ell_p$, we replace C_i by $k-2$ clauses

$$(\ell_1 \vee \ell_2 \vee x_1) \wedge (\bar{x}_1 \vee \ell_3 \vee x_2) \wedge (\bar{x}_2 \vee \ell_4 \vee x_3) \wedge \cdots \wedge (\bar{x}_{k-3} \vee \ell_{k-1} \vee \ell_k),$$

where $x_1, \ldots, x_{k-3}$ are new variables. The new clauses are satisfiable if and only if C_i is.

3. If $C_i = \ell$, we replace C_i by $\ell \vee y \vee z$, and if $C_i = \ell \vee \ell'$, we replace it by $\ell \vee \ell' \vee y$. We add to the formula the clauses

$$(\bar{z} \vee \alpha \vee \beta) \wedge (\bar{z} \vee \bar{\alpha} \vee \beta) \wedge (\bar{z} \vee \alpha \vee \bar{\beta}) \wedge (\bar{z} \vee \bar{\alpha} \vee \bar{\beta}) \wedge$$

$$(\bar{y} \vee \alpha \vee \beta) \wedge (\bar{y} \vee \bar{\alpha} \vee \beta) \wedge (\bar{y} \vee \alpha \vee \bar{\beta}) \wedge (\bar{y} \vee \bar{\alpha} \vee \bar{\beta}),$$

where y, z, α, β are new variables. This makes sure that z and y are false in any truth assignment satisfying F', so that the clauses $C_i = \ell$ and $C_i = \ell \vee \ell'$ are equivalent to their replacements. $\square$

Example 9.16 *Next, we will show that the recognition version of the maximum clique problem,* CLIQUE, *described in Example 9.14, is* $\mathcal{NP}$-*complete. Recall that* CLIQUE *is defined as follows. Given a simple undirected graph $G = (V, E)$ and an integer k, does there exist a clique of size $\geq k$ in G?*

Theorem 9.3 *CLIQUE is NP-complete.*

Proof: We use a polynomial time reduction from 3-SAT.

1. CLIQUE $\in NP$. Certificate: a clique of size k; can be verified in $O(k^2)$ time.

2. 3-SAT is polynomial time reducible to CLIQUE.

   Given $F = \bigwedge\limits_{i=1}^{m} C_i$, we build a graph $G = (V, E)$ as follows:

   - The vertices correspond to pairs (C_i, ℓ_{ij}), where ℓ_{ij} is a literal in $C_i (i = 1, \ldots, m; j = 1, 2, 3)$.

- Two vertices (C_i, ℓ_{ij}) and (C_p, ℓ_{pq}) are connected by an edge if and only if
$$i \neq p \text{ and } \ell_{ij} \neq \bar{\ell}_{pq}.$$

Then F is satisfiable if and only if G has a clique of size m. □

Note that many optimization problems of interest, such as the optimization version of the maximum clique problem (see Example 9.14), are unlikely to belong to class $\mathcal{NP}$. Hence, we cannot claim that the optimization version of the maximum clique problem is $\mathcal{NP}$-complete, despite the fact that it is at least as hard as CLIQUE (see Exercise 9.9). If we drop the first requirement ($\Pi \in \mathcal{NP}$), we obtain the class of problems called $\mathcal{NP}$-hard.

> **Definition 9.7** *A problem* Π *is called* $\mathcal{NP}$-hard *if any problem from* $\mathcal{NP}$ *is polynomial time reducible to* Π.

According to this definition, any $\mathcal{NP}$-complete problem is also $\mathcal{NP}$-hard, and $\mathcal{NP}$-hard problems are at least as hard as $\mathcal{NP}$-complete problems.

Example 9.17 *Since* CLIQUE *is polynomial time reducible to the optimization version of the maximum clique problem (see Exercise 9.9), the maximum clique problem is* $\mathcal{NP}$-hard.

9.5   Complexity of Local Optimization

It is generally assumed that a local optimal solution is easier to find than a global optimal solution. Next, we will show that checking local optimality for a feasible solution and checking whether a given local minimum is strict are both $\mathcal{NP}$-hard problems, even for quadratic optimization problems.

> **Theorem 9.4** *Consider the following nonconvex quadratic problem with linear constraints:*
> $$\begin{array}{rrcl} minimize & f(x) & & \\ subject\ to & Ax & \geq & b \\ & x & \geq & 0, \end{array} \qquad (9.5)$$
> *where* $f(x)$ *is an indefinite quadratic function. Then the following complexity results hold:*
>
> (a) *Given a feasible solution* x^* *of problem (9.5), checking if* x^* *is a local minimizer for this problem is* $\mathcal{NP}$-hard.

(b) *Given a local minimizer x^* of problem (9.5), checking if x^* is a strict local minimizer for this problem is $\mathcal{NP}$-hard.*

(c) *Given a global minimizer x^* of problem (9.5), checking if x^* is a strict global minimizer for this problem is $\mathcal{NP}$-hard, even if $f(x)$ is a concave quadratic function.*

Proof: We will use a polynomial time reduction from 3-SAT to problem (9.5). Let

$$F = \bigwedge_{i=1}^{m} C_i,$$

where $C_i = \ell_{i1} \vee \ell_{i2} \vee \ell_{i3}$ and each literal $\ell_{ij}, i = 1, \ldots, m, j = 1, 2, 3$ is either some variable x_k or its negation $\bar{x}_k$, $k = 1, \ldots, n$, be an arbitrary instance of the 3-SAT problem. For each such F, we construct an instance of a quadratic optimization problem with the real vector of variables $x = [x_0, x_1, \ldots, x_n]^T$ as follows. For each clause $C_i = \ell_{i1} \vee \ell_{i2} \vee \ell_{i3}$, we construct one linear inequality constraint. The left-hand side of this inequality is given by the sum of four terms, one of which is x_0, and the remaining three depend on the three literals of C_i. Namely, if $\ell_{ij} = x_k$ for some k, then we add x_k to the inequality, and if $\ell_{ij} = \bar{x}_k$, we add $(1 - x_k)$ to the inequality. The sum of these four terms is set $\geq \frac{3}{2}$ to obtain the inequality corresponding to C_i. If we denote by

$$I_i = \{k : \exists j \in \{1, 2, 3\} \text{ such that } \ell_{ij} = x_k\}$$

the set of indices of variables that are literals in C_i and by

$$\bar{I}_i = \{k : \exists j \in \{1, 2, 3\} \text{ such that } \ell_{ij} = \bar{x}_k\}$$

the set of indices of variables whose negations are literals in C_i, then we have the following correspondence:

$$\begin{array}{ccc} \text{Clause } C_i \text{ in } F & \longleftrightarrow & \text{a linear term} \\ \ell_{ij} = x_k & k \in I_i & x_k \\ \ell_{ij} = \bar{x}_k & k \in \bar{I}_i & 1 - x_k \end{array}$$

and the linear inequality representing C_i is given by

$$x_0 + \sum_{k \in I_i} x_k + \sum_{k \in \bar{I}_i} (1 - x_k) \geq \frac{3}{2}.$$

For example, for the clause $\bar{x}_1 \vee x_2 \vee x_3$ we have

$$x_0 + (1 - x_1) + x_2 + x_3 \geq \frac{3}{2}.$$

The system of m linear inequalities representing the clauses of F this way is given by

$$A'_F x \geq \frac{3}{2} e + c,$$

where A'_F is a (sparse) matrix with entries in $\{0, 1, -1\}$, $x = [x_0, \ldots, x_n]^T$, $e = [1, \ldots, 1]^T$, and c is a vector with the ith component c_i given by $c_i = -|\bar{I}_i|$.

Let A_F be the $(m + 2n) \times (n + 1)$ matrix, let b_F be an $(n + 1)$-vector of right-hand sides, and let $x = [x_0, \ldots x_n]^T$ be the $(n + 1)$-vector of variables that describe the following system of linear constraints in the form $A_F x \geq b_F$:

$$
\begin{array}{rcll}
A'_F x & \geq & \frac{3}{2} + c & \\
x_0 - x_i & \geq & -\frac{1}{2}, & i = 1, \ldots, n \\
x_0 + x_i & \geq & \frac{1}{2}, & i = 1, \ldots, n
\end{array}
\quad \equiv \quad A_F x \geq b_F. \qquad (9.6)
$$

We will use the following indefinite quadratic functions in the proof. Let

$$
f_1(x) = \sum_{i=1}^{n} (x_0 + x_i - 1/2)(x_0 - x_i + 1/2)
$$

and

$$
f_2(x) = \sum_{i=1}^{n} (x_0 + x_i - 1/2)(x_0 - x_i + 1/2) - \frac{1}{2n} \sum_{i=1}^{n} (x_i - 1/2)^2.
$$

Note that $f_1(x) = n x_0^2 - \sum_{i=1}^{n} (x_i - 1/2)^2$, i.e., it is a separable indefinite quadratic function with one convex term $n x_0^2$ and n concave terms $-(x_i - 1/2)^2, i = 1, \ldots, n$. In addition,

$$
f_2(x) = f_1(x) + q(x),
$$

where

$$
q(x) = -\frac{1}{2n} \sum_{i=1}^{n} (x_i - 1/2)^2.
$$

Also, we will use a vector $\hat{x}$ in the proof, which is defined as follows for an arbitrary x_0 and Boolean $x_1, \ldots, x_n$ satisfying F:

$$
\hat{x}_i = \begin{cases} 1/2 - x_0 & \text{if } x_i = 0 \\ 1/2 + x_0 & \text{if } x_i = 1, \end{cases} \quad i = 1, \ldots, n. \qquad (9.7)
$$

We are ready to prove statements (a)–(c) of the theorem.

(a) Given an instance F of 3-SAT, we construct the following instance of an indefinite quadratic problem as described above:

$$
\begin{array}{rl}
\text{minimize} & f_2(x) \\
\text{subject to} & A_F x \geq b_F \\
& x \geq 0.
\end{array}
\qquad (9.8)
$$

To prove (a), we show that F is satisfiable if and only if $x^* = [0, 1/2, \ldots, 1/2]$ is not a local minimum of (9.8).

First, assume that F is satisfiable. Let $x_1, \ldots, x_n$ be a variable assignment satisfying F. Given any $x_0 \geq 0$, which is arbitrarily close to 0, consider the vector $\hat{x} = [x_0, \hat{x}_1, \ldots, \hat{x}_n]^T$ defined in (9.7). Then $\hat{x}$ is feasible and

$$f_2(\hat{x}) = -\frac{x_0^2}{2} < 0 = f_2(x^*).$$

Hence, x^* is not a local minimizer.

Suppose now that x^* is not a local minimum. Then there exists a point

$$y = [y_0, \ldots, y_n]^T \text{ such that } f_2(y) < 0. \tag{9.9}$$

We show that F is satisfied with

$$x_i = \begin{cases} 0 \text{ if } y_i \leq 1/2 \\ 1 \text{ if } y_i > 1/2. \end{cases} \tag{9.10}$$

Note that x defined in (9.10) satisfies F if and only if every clause C_i of F has a literal ℓ_{ip} such that, if $\ell_{ip} = x_k$, then $y_k > 1/2$, and if $\ell_{ip} = \bar{x}_k$, then $y_k \leq 1/2$. Consider an arbitrary clause C_i of F. Without loss of generality, assume that the indices of three variables (or their negations) involved in defining its literals are 1, 2, and 3. If we define

$$y_k' = \begin{cases} y_k & \text{if } x_k \text{ is a literal in } C_i \\ (1 - y_k) & \text{if } \bar{x}_k \text{ is a literal in } C_i, \end{cases} \quad k = 1, 2, 3,$$

then having one $k \in \{1, 2, 3\}$ such that $y_k' > 1/2$ is sufficient for x in (9.10) to satisfy C_i. Next, we show that such k always exists if y satisfies (9.9). We use contradiction.

Assume, by contradiction, that $y_k' \leq 1/2, k = 1, 2, 3$ for the clause C_i above. Then the inequality corresponding to C_i in $A_F y \leq b_F$ can be written as

$$y_0 + y_1' + y_2' + y_3' \geq \frac{3}{2}.$$

For example, for the clause $\bar{x}_1 \vee x_2 \vee x_3$ we have

$$y_0 + y_1' + y_2' + y_3' = y_0 + (1 - y_1) + y_2 + y_3 \geq \frac{3}{2}.$$

For this inequality to hold, we must have $y_k' \geq \frac{1}{2} - \frac{y_0}{3}$ for at least one $k \in \{1, 2, 3\}$. Without loss of generality, consider the case $k = 1$ (other cases are established analogously). By our assumption, $y_1' \leq 1/2$, so

$$\frac{1}{2} - \frac{y_0}{3} \leq y_1' \leq \frac{1}{2} \Rightarrow -\frac{y_0}{3} \leq (1 - y_1) - \frac{1}{2} \leq 0.$$

Hence,

$$(y_1 - 1/2)^2 \leq \frac{y_0^2}{9}. \tag{9.11}$$

Also, since y is feasible for (9.8), it must satisfy the constraints in (9.6), so

$$y_0 - y_i \geq -\frac{1}{2}, y_0 + y_i \geq \frac{1}{2} \quad \Rightarrow \quad -y_0 \leq y_i - \frac{1}{2} \leq y_0, \quad i = 1, \ldots, n,$$

and thus

$$y_0^2 - (y_i - 1/2)^2 \geq 0, \quad i = 1, \ldots, n. \tag{9.12}$$

Therefore, using (9.12) and then (9.11), we obtain:

$$
\begin{aligned}
f_1(y) &= \sum_{i=1}^n (y_0 + y_i - 1/2)(y_0 - y_i + 1/2) \\
&= \sum_{i=1}^n \left(y_0^2 - (y_i - 1/2)^2 \right) \\
&\geq y_0^2 - (y_1 - 1/2)^2 \\
&\geq \frac{8}{9} y_0^2.
\end{aligned}
$$

On the other hand,

$$q(y) = -\frac{1}{2n} \sum_{i=1}^n (y_i - 1/2)^2 \geq -\frac{1}{2} y_0^2.$$

Hence

$$f_2(y) = f_1(y) + q(y) \geq \frac{8}{9} y_0^2 - \frac{1}{2} y_0^2 = \frac{7}{18} y_0^2 \geq 0,$$

a contradiction with (9.9).

(b) We associate the following indefinite quadratic problem with the given instance F of 3-SAT:

$$
\begin{aligned}
\text{minimize} \quad & f_1(x) \\
\text{subject to} \quad & A_F x \geq b_F \\
& x \geq 0.
\end{aligned}
\tag{9.13}
$$

This problem has the following properties:

i) $f_1(x) \geq 0$ for all feasible x. Therefore, the feasible solution $x^* = [0, 1/2, \ldots, 1/2]^T$ is a local and global minimum of $f(x)$ since $f(x^*) = 0$.

ii) $f_1(x) = 0$ if and only if $x_i \in \{1/2 - x_0, 1/2 + x_0\}$ for all $i = 1, \ldots, n$.

To prove (b), we show that F is satisfiable if and only if $x^* = [0, 1/2, \ldots, 1/2]^T$ is not a strict local minimum of the problem (9.13).

Let F be satisfiable, with $x_1, \ldots, x_n$ being an assignment satisfying S. For any x_0, consider the vector $\hat{x} = (x_0, \hat{x}_1, \ldots, \hat{x}_n)^T$ defined in (9.7). Then we have $f_1(\hat{x}) = 0$. Since x_0 can be chosen to be arbitrarily close to zero, x^* is not a strict local minimum.

Suppose now that $x^* = [0, 1/2, \ldots, 1/2]^T$ is not a strict local minimum, that is, there exists $y \neq x^*$ such that $f_1(y) = f(x^*) = 0$; therefore, $y_i \in \{1/2 - y_0, 1/2 + y_0\}$, $i = 1, \ldots, n$. Then the variables x_i, $i = 1, \ldots, n$ defined by

$$x_i(y) = \begin{cases} 0 \text{ if } y_i = 1/2 - y_0 \\ 1 \text{ if } y_i = 1/2 + y_0 \end{cases}$$

satisfy F.

(c) Finally, to prove (c), note that if we fix $x_0 = 1/2$ in the above indefinite quadratic problem, then the objective function $f(x)$ is concave with x^* as the global minimum. $\square$

9.6 Optimal Methods for Nonlinear Optimization

Since many of the problems of interest in nonlinear optimization are $\mathcal{NP}$-hard, it is expected that algorithms designed to solve them will run in non-polynomial time (in the worst case). Given two exponential-time algorithms for the same problem, how do we decide which one is better? Note that when we say that one method is better than another, it is important to clearly state what exactly we mean by "better." One common criterion that is used to compare two algorithms for the same problem is the worst-case running time. Now assume that we found an algorithm $\mathcal{A}$ that outperforms other known methods for the same problem Π in terms of this criterion. Then an interesting question is, can we find an algorithm with lower time complexity, or is algorithm $\mathcal{A}$ the best possible for the given problem Π? Answering this question requires establishing *lower bounds* on time complexity of algorithms that can be used to solve Π, i.e., finding a function $\mathcal{L}(n)$, where n defines the problem size, such that *any* algorithm for solving the problem will require at least $O(\mathcal{L}(n))$ time. However, if we were able to provide such a bound and show that it is exponential for an $\mathcal{NP}$-complete problem Π, this would imply that Π is not in $\mathcal{P}$, and therefore $\mathcal{P} \neq \mathcal{NP}$, which would solve the open $\mathcal{P}$ vs. $\mathcal{NP}$ problem. Hence, it is unlikely that we will be able to easily establish the lower bound that can be proved to hold for *any* algorithm solving Π, and we should set a more realistic goal of establishing a lower complexity bound that would apply to a wide class of methods for solving Π rather than all methods.

9.6.1 Classes of methods

We can define the classes of methods based on the type of information they are allowed to use about a problem. We assume that we are dealing with an optimization problem in the functional form

$$\begin{array}{ll} \text{minimize} & f(x) \\ \text{subject to} & g(x) \leq 0 \\ & h(x) = 0, \end{array} \qquad (\Pi)$$

where $f : \mathbb{R}^n \to \mathbb{R}, g : \mathbb{R}^n \to \mathbb{R}^p$, and $h : \mathbb{R}^n \to \mathbb{R}^m$.

Definition 9.8 (Zero-Order Methods) *The class of zero-order methods is defined as the class of methods that can only use zero-order local information about the functions involved in the model* (Π) *and cannot use any other information about the problem. That is, for a given point $\bar{x} \in \mathbb{R}^n$, zero-order methods can only use the values of $f(\bar{x}), g(\bar{x})$ and $h(\bar{x})$ and no other information about* (Π).

Zero-order methods are typically referred to as *direct search* or *derivative-free* methods.

Definition 9.9 (First-Order Methods) *First-order methods are the methods that can only use zero-order and first-order local information about the functions involved in the model* (Π) *and cannot use any other information about the problem. That is, for a given point $\bar{x} \in \mathbb{R}^n$, first-order methods can only use the values of $\nabla f(\bar{x}), \nabla g_j(\bar{x}), j = 1, \dots, p$, and $\nabla h_i(\bar{x}), i = 1, \dots, m$ in addition to $f(\bar{x}), g(\bar{x})$ and $h(\bar{x})$, but no other information about* (Π).

First-order methods are also known as *gradient-based* methods.

Definition 9.10 (Second-Order Methods) *Second-order methods are the methods that can only use zero-order, first-order, and second-order local information about the functions involved in the model* (Π) *and cannot use any other information about the problem. That is, for a given point $\bar{x} \in \mathbb{R}^n$, in addition to the information allowed in the first-order methods, second-order methods can only use the values of $\nabla^2 f(\bar{x}), \nabla^2 g_j(\bar{x}), j = 1, \dots, p$, and $\nabla^2 h_i(\bar{x}), i = 1, \dots, m$, but no other information about* (Π).

9.6.2 Establishing lower complexity bounds for a class of methods

Consider the problem

$$\min_{x \in B_n} f(x), \qquad (9.14)$$

where
$$B_n = [0,1]^n = \{x \in \mathbb{R}^n : 0 \le x_i \le 1, i = 1, \ldots, n\}$$
and f is *Lipschitz continuous* on B_n with respect to the infinity norm:
$$|f(x) - f(y)| \le L\|x - y\|_\infty \quad \forall x, y \in B_n,$$
with L being a Lipschitz constant. According to the Weierstrass theorem, this problem has an optimal solution. Assume that f^* is the optimal objective value. Given a small $\epsilon > 0$, we will be satisfied with an approximate solution $\tilde{x} \in B_n$ such that
$$f(\tilde{x}) - f^* \le \epsilon. \qquad (9.15)$$
Our goal in this subsection is to determine the minimum running time a zero-order method will require in order to guarantee such a solution. In other words, we want to establish the lower complexity bounds on zero-order methods for solving the considered problem.

> **Theorem 9.5** *Let $\epsilon < L/2$. Then for any zero-order method for solving problem (9.14) with an accuracy better than ϵ there exists an instance of this problem that will require at least $\left(\lfloor \frac{L}{2\epsilon} \rfloor\right)^n$ objective function evaluations.*

Proof: Let $p = \lfloor \frac{L}{2\epsilon} \rfloor$; then $p \ge 1$. Assume that there is a zero-order method $\mathcal{M}$ that needs $N < p^n$ function evaluations to solve any instance of our problem (9.14) approximately. We will use the so-called *resisting strategy* to construct a function f such that $f(x) = 0$ at any test point x used by the method $\mathcal{M}$, so that the method can only find $\tilde{x} \in B_n$ with $f(\tilde{x}) = 0$. Note that splitting $[0,1]$ into p equal segments with the mesh points $\{\frac{i}{p}, i = 0, \ldots, p\}$ for each coordinate axis defines a uniform grid partitioning of B_n into p^n equal hypercubes with the side $1/p$. Since the method $\mathcal{M}$ uses $N < p^n$ function evaluations, at least one of these hypercubes does not contain any of the points used by the method in its interior. Let $\hat{x}$ be the center of such a hypercube. We consider the function
$$\bar{f}(x) = \min\{0, L\|x - \hat{x}\|_\infty - \epsilon\}.$$

It is easy to check (Exercise 9.12) that $\bar{f}(x)$ is Lipschitz continuous with the constant L in the infinity norm, its global optimal value is $-\epsilon$, and it differs from zero only inside the box $B' = \{x : \|x - \hat{x}\|_\infty \le \epsilon/L\}$, which, since $2p \le L/\epsilon$, is a part of $B \equiv \{x : \|x - \hat{x}\|_\infty \le \frac{1}{2p}\}$. Thus, $\bar{f}(x) = 0$ at all test points of the method $\mathcal{M}$. We conclude that the accuracy of the result of our method cannot be better than ϵ if the number of the objective function evaluations is less than p^n. $\qquad \square$

Example 9.18 *If $L = 2$, $n = 10$, $\epsilon = 0.01$, in order to solve the problem (9.14) with the accuracy better than ϵ, we need at least $\left(\lfloor \frac{L}{2\epsilon} \rfloor\right)^n = 10^{20}$*

objective function evaluations. If a computer can perform 10^5 objective function evaluations per second, we will need 10^{15} seconds to solve the problem, which is approximately 31,250,000 years.

9.6.3   Defining an optimal method

Assume that the lower complexity bound for a given class of methods $\mathcal{C}$ on a given problem Π is given by $O(\mathcal{L}(n))$, and we designed a method $\mathcal{M}$ from the class $\mathcal{C}$ that solves Π in $O(\mathcal{U}(n))$ time (which can be viewed as an upper complexity bound for $\mathcal{C}$ on Π). Then it makes sense to call the method $\mathcal{M}$ an optimal class $\mathcal{C}$ method for problem Π if $\mathcal{U}(n) = O(\mathcal{L}(n))$, which is the case, e.g., when $\lim\limits_{n \to \infty} \frac{\mathcal{U}(n)}{\mathcal{L}(n)} = c$, where c is a constant.

Next we give an example of an optimal zero-order method for problem (9.14). Namely, we consider a simple method for solving problem (9.14), which we will refer to as the *uniform grid method*. We partition B_n into p^n equal hypercubes with the side $1/p$. Note that such partitioning is unique and is obtained by placing a uniform grid on B_n with the mesh points $\{\frac{i}{p}, i = 0, \ldots, p\}$ used for each coordinate axis, which split $[0,1]$ into p equal segments (see Figure 9.3 for an illustration). Let $B_{i_1 i_2 \ldots i_n}$ be the hypercube in the partition that corresponds to i_k-th such segment for the k-th coordinate axis, where $i_k \in \{1, \ldots, p\}, k = 1, \ldots, n$. Let $x(i_1, i_2, \ldots, i_n)$ be the center of the hypercube $B_{i_1 i_2 \ldots i_n}$, then the k-th coordinate of $x(i_1, i_2, \ldots, i_n)$ is $\frac{2i_k - 1}{2p}$, $k = 1, \ldots, n$. For example, $x(2,3) = [3/10, 5/10]^T$ for the illustration in Figure 9.3.

From the set

$$ X_p = \left\{ x(i_1, i_2, \ldots, i_n) = \left[\frac{2i_1 - 1}{2p}, \ldots, \frac{2i_n - 1}{2p} \right]^T : i_k \in \{1, \ldots, p\}, k = 1, \ldots, n \right\} $$

of p^n constructed points, the uniform grid algorithm simply picks one, $\bar{x}$, that minimizes $f(x)$ over X_p. Let x^* and $f^* = f(x^*)$ be a global optimum and the optimal value of problem (9.14), respectively. Then, x^* must belong to one of the hypercubes $B_{i_1 i_2 \ldots i_n}$. Let $\tilde{x}$ be the center of that hypercube. Then

$$ f(\bar{x}) - f^* \le f(\tilde{x}) - f^* \le L\|\tilde{x} - x^*\|_\infty \le \frac{L}{2p}. $$

Thus, if our goal is to find $\bar{x} \in B_n : f(\bar{x}) - f^* \le \epsilon$, then we need to choose p such that

$$ \frac{L}{2p} \le \epsilon \quad \Leftrightarrow \quad p \ge \left\lceil \frac{L}{2\epsilon} \right\rceil. $$

The resulting uniform grid method is summarized in Algorithm 9.3.

This is a *zero-order method*, since it uses only the function value information. As we have shown, the uniform grid algorithm outputs an ϵ-approximate solution by performing a total of $\lceil \frac{L}{2\epsilon} \rceil$ function evaluations.

Let the desired accuracy ϵ be given, and let $L/2 > \epsilon' > \epsilon$ be such that

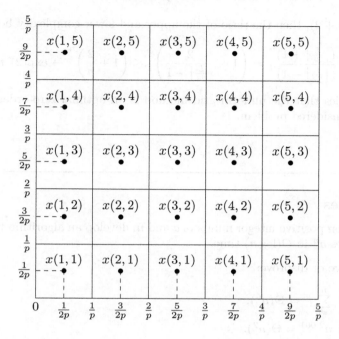

FIGURE 9.3: An illustration of the uniform grid method for $n = 2$, $p = 5$.

Algorithm 9.3 The uniform grid algorithm for minimizing f over $B_n = [0,1]^n$, where f is Lipschitz-continuous on B_n with the Lipschitz constant L in the infinity norm.

Input: $f(\cdot), \epsilon$
Output: $\bar{x} \in B_n$: $f(\bar{x}) \leq f^* + \epsilon$, where $f^* = \min\limits_{x \in B_n} f(x)$

1: $p = \lceil \frac{L}{2\epsilon} \rceil$
2: $x(i_1, i_2, \ldots, i_n) = [\frac{2i_1-1}{2p}, \frac{2i_2-1}{2p}, \ldots, \frac{2i_n-1}{2p}]^T$ $\forall i_k \in \{1, \ldots, p\}, k = 1, \ldots, n$
3: $\bar{x} = \arg\min\limits_{x \in X_p} f(x)$, where
   $$X_p = \{x(i_1, \ldots, i_n) = [\frac{2i_1-1}{2p}, \ldots, \frac{2i_n-1}{2p}]^T, i_k \in \{1, \ldots, p\}, k = 1, \ldots, n\}$$
4: **return** $\bar{x}$

$\lfloor \frac{L}{2\epsilon'} \rfloor \geq \lfloor \frac{L}{2\epsilon} \rfloor - 1$ (note that such ϵ' can always be selected). Recall from Theorem 9.5 that achieving accuracy $\epsilon < \epsilon'$ using a zero-order method will require at least $\mathcal{L}(n) = \left(\lfloor \frac{L}{2\epsilon'} \rfloor\right)^n \geq \left(\lfloor \frac{L}{2\epsilon} \rfloor - 1\right)^n$ objective function evaluations. Using the uniform grid method, we guarantee accuracy ϵ using $\mathcal{U}(n) = \left(\lceil \frac{L}{2\epsilon} \rceil\right)^n \leq \left(\lfloor \frac{L}{2\epsilon} \rfloor + 1\right)^n$ objective function evaluations. If we select

$\epsilon \leq L/(2n+2)$, then the ratio of the upper and lower complexity bounds is

$$\frac{\mathcal{U}(n)}{\mathcal{L}(n)} \leq \frac{\left(\lfloor\frac{L}{2\epsilon}\rfloor + 1\right)^n}{\left(\lfloor\frac{L}{2\epsilon}\rfloor - 1\right)^n} = \left(1 + \frac{2}{\lfloor\frac{L}{2\epsilon}\rfloor - 1}\right)^n \leq \left(1 + \frac{2}{n}\right)^n \rightarrow \exp(2), n \rightarrow \infty.$$

This implies that the uniform grid method is an optimal zero-order method for the considered problem.

Exercises

9.1. Given positive integer numbers a and n, develop an algorithm that computes a^n in $O(\log n)$ time.

9.2. Prove or disprove:

(a) $\sum_{i=0}^{n} i^5 = \Theta(n^6)$.

(b) $n^{2.001} = \Theta(n^2)$.

(c) $n^5 + 10^{12}n^4 \log n = \Theta(n^5)$.

(d) $n3^n + 2^n = O(n^3 2^n)$.

(e) If $f_1(n) = O(g_1(n))$ and $f_2(n) = O(g_2(n))$ then $\frac{f_1(n)}{f_2(n)} = O\left(\frac{g_1(n)}{g_2(n)}\right)$.

9.3. Consider an infinite array in which the first n cells contain integers in sorted order, and the rest of the cells are filled with ∞. Propose an algorithm that takes an integer k as input and finds the position of k in the array or reports that k is not in the array in $\Theta(\log n)$ time. Note that the value of n is not given.

9.4. The following *master theorem* is used to solve recurrence relations. Let $a \geq 1$ and $b > 1$ be constants. Let $f(n)$ be a function and let $T(n)$ be defined on the nonnegative integers by the recurrence of the form

$$T(n) = aT(n/b) + f(n),$$

where we interpret n/b to mean either $\lfloor n/b \rfloor$ or $\lceil n/b \rceil$. Then $T(n)$ can be bounded asymptotically as follows.

1. If $f(n) = O\left(n^{\log_b a - \epsilon}\right)$ for some $\epsilon > 0$, then $T(n) = \Theta(n^{\log_b a})$;

2. If $f(n) = \Theta\left(n^{\log_b a}\right)$, then $T(n) = \Theta(n^{\log_b a} \log n)$;

3. If $f(n) = \Omega\left(n^{\log_b a + \epsilon}\right)$ for some $\epsilon > 0$, and if $af(n/b) \leq cf(n)$ for some constant $c < 1$ and all sufficiently large n, then $T(n) = \Theta(f(n))$.

Use the master theorem to solve the following recurrence relations:

(a) $T(n) = 3T(n/3) + 3$;

(b) $T(n) = T(3n/4) + n \log n$;

(c) $T(n) = 4T(n/2) + n^2$.

9.5. Assume that we are given a Monte Carlo algorithm for solving a problem Π in T_1 time units, whose output is correct with probability $\geq \frac{1}{2}$. Also, assume that there is another algorithm that can check, in T_2 time units, whether a given answer is valid for Π. Use these two algorithms to obtain a Las Vegas algorithm for solving Π in time $\tilde{O}((T_1 + T_2) \log n)$.

9.6. Austin has ideas for n potential research papers $1, 2, \ldots, n$. To maximize his efficiency, he wants to work on one paper at a time, until it is finished. He knows with certainty that completing paper i will take exactly p_i days, $1 \leq i \leq n$. Let D be the number of days remaining until the global deadline Austin set for himself. The objective is to find the maximum number of papers that can be completed before the deadline. Propose an $O(n \log n)$ time algorithm for solving this problem. Prove that the output of your algorithm is indeed optimal.

9.7. Consider n jobs $1, 2, \ldots, n$ that must be processed by a single machine, which can process only one job at a time. Processing job i takes p_i units of time, $1 \leq i \leq n$. The objective is to minimize the average completion time, $\frac{1}{n} \sum_{i=1}^{n} c_i$, where c_i denotes the completion time for job i (for example, if job i is started at time s_i, its completion time is $s_i + p_i$. Give an algorithm that solves the problem of scheduling to minimize average completion time. Prove that your algorithm returns a correct answer. What is the running time of your algorithm?

9.8. Samyukta considers a set of n activities $1, 2, \ldots, n$ to attend during the semester. Each activity a_i has a start time s_i and a finish time f_i, where $0 \leq s_i \leq f_i < \infty$. Activity i is conducted during the half-open time interval $[s_i, f_i)$. Hence, activities i and j are compatible if the intervals $[s_i, f_i)$ and $[s_j, f_j)$ do not overlap. Since Samyukta can attend only one activity at a time, her objective is to select a maximum number of mutually compatible activities. Design an algorithm for this problem and prove that your algorithm always outputs an optimal solution. What is the running time of your algorithm?

9.9. Show that any two of the three versions of the maximum clique problem described in Example 9.14 are polynomial time reducible to each other.

9.10. Given a simple undirected graph $G = (V, E)$, a subset $C \subseteq V$ of vertices is called a *vertex cover* if each edge in E has at least one endpoint in C. Prove that the minimum vertex cover problem, which is to find a vertex cover of minimum size in G, is $\mathcal{NP}$-hard.

9.11. Given a simple undirected graph $G = (V, E)$, a subset $I \subseteq V$ of vertices is called an *independent set* if no two vertices in I are adjacent to each other. Prove that the maximum independent set problem, which is to find an independent set of maximum size in G, is $\mathcal{NP}$-hard.

9.12. Show that $\bar{f}(x) = \min\{0, L\|x - \hat{x}\|_\infty - \epsilon\}$ in the proof of Theorem 9.5 is Lipschitz continuous with the constant L in the infinity norm, its global optimal value is $-\epsilon$, and it differs from zero only inside the box $B' = \{x : \|x - \hat{x}\|_\infty \leq \epsilon/L\}$.

Chapter 10

Introduction to Linear Programming

Linear programming is a methodology for solving linear optimization problems, in which one wants to optimize a linear objective function subject to constraints on its variables expressed in terms of linear equalities and/or inequalities. Ever since the introduction of the simplex method by George Dantzig in the late 1940s, linear programming has played a major role in shaping the modern horizons of the field of optimization and its applications.

10.1 Formulating a Linear Programming Model

Example 10.1 *Consider the following problem. Heavenly Pouch, Inc. produces two types of baby carriers, non-reversible and reversible. Each non-reversible carrier sells for $23, requires 2 linear yards of a solid color fabric, and costs $8 to manufacture. Each reversible carrier sells for $35, requires 2 linear yards of a printed fabric as well as 2 linear yards of a solid color fabric, and costs $10 to manufacture. The company has 900 linear yards of solid color fabrics and 600 linear yards of printed fabrics available for its new carrier collection. It can spend up to $4,000 on manufacturing the carriers. The demand is such that all reversible carriers made are projected to sell, whereas at most 350 non-reversible carriers can be sold. Heavenly Pouch is interested in formulating a mathematical model that could be used to maximize its profit (e.g., the difference of revenues and expenses) resulting from manufacturing and selling the new carrier collection.*

10.1.1 Defining the decision variables

We start formulating a model by defining *decision variables*, which are the variables determining the outcome whose values we can control. Since Heavenly Pouch, Inc. needs to decide on how many non-reversible and reversible carriers to manufacture, it is natural to define the decision variables as follows:

$$x_1 = \text{the number of non-reversible carriers to manufacture}$$
$$x_2 = \text{the number of reversible carriers to manufacture.}$$

10.1.2   Formulating the objective function

The objective of Heavenly Pouch, Inc. is to maximize its profit, which is the difference of the revenues and manufacturing costs. Since non-reversible and reversible carriers sell for \$23 and \$35, respectively, the total revenue r resulting from selling all the carriers manufactured is given by $r = 23x_1 + 35x_2$. Similarly, the total cost c of manufacturing all the carriers is given by $c = 8x_1 + 10x_2$. Hence, the profit z can be expressed as the following linear function of the decision variables x_1 and x_2:

$$z = r - c = (23x_1 + 35x_2) - (8x_1 + 10x_2) = 15x_1 + 25x_2.$$

10.1.3   Specifying the constraints

Constraints are used to make sure that our model does not accept combinations of variable values that are physically impossible. For example, Heavenly Pouch, Inc. cannot manufacture 300 non-reversible and 300 reversible carriers simultaneously, since this would require $300 \times \$8 + 300 \times \$10 = \$5,400$ in manufacturing costs, which exceeds the given budget of \$4,000. Constraints are often associated with limited availability of various types of resources (such as materials, time, space, money) or other considerations that restrict our choice of decision variable values. From our problem's statement we know that

- the company has only 900 linear yards of solid color fabrics available,

- at most 600 linear yards of printed fabrics can be used,

- the manufacturing budget is limited to \$4,000, and

- at most 350 non-reversible carriers can be sold.

Based on the problem data and the above definition of the decision variables, these considerations can be expressed using linear inequalities as follows. Since 2 linear yards of the solid fabrics is used for each non-reversible and reversible carrier, the total amount of the solid fabrics used to manufacture x_1 non-reversible carriers and x_2 reversible carriers is $2x_1 + 2x_2$, and we have $2x_1 + 2x_2 \leq 900$, or equivalently,

$$x_1 + x_2 \leq 450 \quad \text{(solid color fabric constraint)}.$$

Since the printed fabrics are used only for reversible carriers (2 linear yards/carrier), we have $2x_2 \leq 600$, which is the same as

$$x_2 \leq 300 \quad \text{(printed fabric constraint)}.$$

Heavenly Pouch, Inc. spends \$8/carrier and \$10/carrier to manufacture a non-reversible and reversible carrier, respectively, so we have $8x_1 + 10x_2 \leq 4,000$, or

$$4x_1 + 5x_2 \leq 2,000 \quad \text{(budget constraint)}.$$

Finally, since at most 350 non-reversible carriers can be sold, we have

$$x_1 \leq 350 \quad \text{(demand constraint)}.$$

In addition, we need to specify the *nonnegativity constraints*, which are not explicitly stated in the problem description, but result from the definition of the decision variables. Namely, since the variables x_1 and x_2 represent the quantities of physical objects, their values must be nonnegative by definition:

$$x_1, x_2 \geq 0 \quad \text{(nonnegativity constraints)}.$$

Even though in some cases the nonnegativity constraints may not have any impact on the optimal solution of the problem, they play an important role in designing algorithms for linear programming, and thus must always be mentioned whenever they hold.

10.1.4 The complete linear programming formulation

In summary, we obtain the following *linear program* (LP) that models the considered problem:

$$
\begin{array}{llrcll}
\text{maximize} & 15x_1 & + & 25x_2 & & \text{(profit)} \\
\text{subject to (s.t.)} & x_1 & + & x_2 & \leq 450 & \text{(solid color fabric constraint)} \\
& & & x_2 & \leq 300 & \text{(printed fabric constraint)} \\
& 4x_1 & + & 5x_2 & \leq 2,000 & \text{(budget constraint)} \\
& x_1 & & & \leq 350 & \text{(demand constraint)} \\
& & & x_1, x_2 & \geq 0 & \text{(nonnegativity constraints)}.
\end{array}
$$

10.2 Examples of LP Models

10.2.1 A diet problem

Example 10.2 *Yiming aims to improve his diet. Based on a nutrition specialist recommendation, he wants his daily intake to contain at least 60 g of protein, 800 mg of calcium, 75 mg of vitamin C, and 2,000 calories. He would like his menu for the day to consist of five food types: almond butter, brown rice, orange juice, salmon, and wheat bread. The serving size, cost per serving, and nutrition information for each food type is provided in the table below.*

Food type	Cost ($)	Protein (g)	Calcium (mg)	Vitamin C (mg)	Calories
Almond butter (100 g)	2.90	15	270	1	600
Brown rice (200 g)	3.20	5	20	0	215
Orange juice (250 g)	0.50	2	25	106	110
Salmon (150 g)	4.50	39	23	0	280
Wheat bread (25 g)	0.30	3	35	0	66
Required ingestion	–	60	800	75	2,000

Formulate an LP that would allow Yiming to design the least expensive diet that satisfies the above requirements.

Yiming has to decide how much of each of the food types should be included in his diet. Therefore, it is natural to define a decision variable for the amount of each food type consumed daily:

$$
\begin{aligned}
x_1 &= \text{servings of almond butter eaten daily} \\
x_2 &= \text{servings of brown rice eaten daily} \\
x_3 &= \text{servings of orange juice drunk daily} \\
x_4 &= \text{servings of salmon eaten daily} \\
x_5 &= \text{servings of wheat bread eaten daily.}
\end{aligned}
$$

His objective is to minimize the cost, which can be easily written as a linear function of decision variables as follows:

$$z = 2.9x_1 + 3.2x_2 + 0.5x_3 + 4.5x_4 + 0.3x_5.$$

The constraints express the minimum daily requirements for protein:

$$15x_1 + 5x_2 + 2x_3 + 39x_4 + 3x_5 \geq 60;$$

calcium:

$$270x_1 + 20x_2 + 25x_3 + 23x_4 + 35x_5 \geq 800;$$

vitamin C:

$$x_1 + 106x_3 \geq 75;$$

and calories:

$$600x_1 + 215x_2 + 110x_3 + 280x_4 + 66x_5 \geq 2,000.$$

In addition, all of the decision variables must be nonnegative. We obtain the following LP:

$$
\begin{array}{lrcrcrcrcrl}
\text{minimize} & 2.9x_1 &+& 3.2x_2 &+& 0.5x_3 &+& 4.5x_4 &+& 0.3x_5 & \\
\text{subject to} & 15x_1 &+& 5x_2 &+& 2x_3 &+& 39x_4 &+& 3x_5 &\geq 60 \\
& 270x_1 &+& 20x_2 &+& 25x_3 &+& 23x_4 &+& 35x_5 &\geq 800 \\
& x_1 & & & +& 106x_3 & & & & &\geq 75 \\
& 600x_1 &+& 215x_2 &+& 110x_3 &+& 280x_4 &+& 66x_5 &\geq 2,000 \\
& & & & & & & & x_1, x_2, x_3, x_4, x_5 &\geq 0.
\end{array}
$$

10.2.2   A resource allocation problem

Example 10.3 *Jeff is considering 6 projects for potential investment for the upcoming year. The required investment and end-of-year payout amounts are described in the following table.*

	Project					
	1	*2*	*3*	*4*	*5*	*6*
Investment ($)	10,000	25,000	35,000	45,000	50,000	60,000
Payout ($)	12,000	30,000	41,000	55,000	65,000	77,000

Partial investment (i.e., financing only a fraction of the project instead of the whole project) is allowed for each project, with the payout proportional to the investment amount. For example, if Jeff decides to invest $5,000 in project 2, the corresponding payout will be $30,000× ($5,000/$25,000)=$6,000. Jeff has $100,000 available for investment. Formulate an LP to maximize the end-of-year payout resulting from the investment.

Let
$$x_i = \text{fraction of project } i \text{ financed}, i = 1, \ldots, 6.$$

Then we have the following LP formulation:

maximize $\quad 1000(12x_1 + 30x_2 + 41x_3 + 55x_4 + 65x_5 + 77x_6)$

subject to $\quad 10x_1 + 25x_2 + 35x_3 + 45x_4 + 50x_5 + 60x_6 \leq 100$

$$x_1 \leq 1$$
$$x_2 \leq 1$$
$$x_3 \leq 1$$
$$x_4 \leq 1$$
$$x_5 \leq 1$$
$$x_6 \leq 1$$
$$x_1, x_2, x_3, x_4, x_5, x_6 \geq 0.$$

10.2.3 A scheduling problem

Example 10.4 *St. Tatiana Hospital uses a 12-hour shift schedule for its nurses, with each nurse working either day shifts (7:00 am–7:00 pm) or night shifts (7:00 pm–7:00 am). Each nurse works 3 consecutive day shifts or 3 consecutive night shifts and then has 4 days off. The hospital is aiming to design a schedule for day-shift nurses that minimizes the total number of nurses employed. The minimum number of nurses required for each day shift during a week is given in the following table:*

Day of week/shift	Nurses required
Monday (Mo)	16
Tuesday (Tu)	12
Wednesday (We)	18
Thursday (Th)	13
Friday (Fr)	15
Saturday (Sa)	9
Sunday (Su)	7

In addition, it is required that at least half of the day-shift nurses have week-ends (Saturday and Sunday) off. Formulate this problem as an LP.

Note that a nurse's schedule can be defined by the first day of the three-day working cycle. Thus, we can define the decision variables as follows:

$$
\begin{aligned}
x_1 &= \text{ the number of nurses working Mo-Tu-We schedule}\\
x_2 &= \text{ the number of nurses working Tu-We-Th schedule}\\
x_3 &= \text{ the number of nurses working We-Th-Fr schedule}\\
x_4 &= \text{ the number of nurses working Th-Fr-Sa schedule}\\
x_5 &= \text{ the number of nurses working Fr-Sa-Su schedule}\\
x_6 &= \text{ the number of nurses working Sa-Su-Mo schedule}\\
x_7 &= \text{ the number of nurses working Su-Mo-Tu schedule.}
\end{aligned}
$$

Then our objective is to minimize

$$z = x_1 + x_2 + x_3 + x_4 + x_5 + x_6 + x_7.$$

To ensure the required number of nurses for Monday, the total number of nurses that have Monday on their working schedule should be at least 16:

$$x_1 + x_6 + x_7 \geq 16.$$

The demand constraints for the remaining 6 days of the week are formulated in the same fashion:

$$
\begin{aligned}
x_1 + x_2 + x_7 &\geq 12 \quad \text{(Tuesday)}\\
x_1 + x_2 + x_3 &\geq 18 \quad \text{(Wednesday)}\\
x_2 + x_3 + x_4 &\geq 13 \quad \text{(Thursday)}\\
x_3 + x_4 + x_5 &\geq 15 \quad \text{(Friday)}\\
x_4 + x_5 + x_6 &\geq 9 \quad \text{(Saturday)}\\
x_5 + x_6 + x_7 &\geq 7 \quad \text{(Sunday).}
\end{aligned}
$$

Note that only the first three schedules do not involve working on weekends. Therefore, the requirement that at least half of the nurses have weekends off can be expressed as

$$\frac{x_1 + x_2 + x_3}{x_1 + x_2 + x_3 + x_4 + x_5 + x_6 + x_7} \geq \frac{1}{2}.$$

Multiplying both sides of this inequality by $2(x_1 + x_2 + x_3 + x_4 + x_5 + x_6 + x_7)$, we obtain the following equivalent linear inequality:

$$x_1 + x_2 + x_3 - x_4 - x_5 - x_6 - x_7 \geq 0.$$

In summary, we obtain the following LP:

$$
\begin{array}{ll}
\text{minimize} & x_1 + x_2 + x_3 + x_4 + x_5 + x_6 + x_7 \\
\text{subject to} & x_1 \quad\quad\quad\quad\quad\quad + x_6 + x_7 \geq 16 \\
& x_1 + x_2 \quad\quad\quad\quad\quad + x_7 \geq 12 \\
& x_1 + x_2 + x_3 \quad\quad\quad\quad \geq 18 \\
& \quad\quad x_2 + x_3 + x_4 \quad\quad\quad \geq 13 \\
& \quad\quad\quad x_3 + x_4 + x_5 \quad\quad \geq 15 \\
& \quad\quad\quad\quad x_4 + x_5 + x_6 \quad \geq 9 \\
& \quad\quad\quad\quad\quad x_5 + x_6 + x_7 \geq 7 \\
& x_1 + x_2 + x_3 - x_4 - x_5 - x_6 - x_7 \geq 0 \\
& x_1, x_2, x_3, x_4, x_5, x_6, x_7 \geq 0.
\end{array}
$$

This problem has multiple optimal solutions with $z^* = 31$. One of them is given by

$$x_1^* = 11, x_2^* = 0, x_3^* = 10, x_4^* = 3, x_5^* = 2, x_6^* = 4, x_7^* = 1.$$

According to this schedule, only 10 out of 31 nurses will be scheduled to work on weekends.

10.2.4 A mixing problem

Example 10.5 *Painter Joe needs to complete a job that requires 50 gallons of brown paint and 50 gallons of gray paint. The required shades of brown and gray can be obtained my mixing the primary colors (red, yellow, and blue) in the proportions given in the following table.*

Color	Red	Yellow	Blue
Brown	40%	30%	30%
Gray	30%	30%	40%

The same shades can be obtained by mixing secondary colors (orange, green, and purple), each of which is based on mixing two out of three primary colors in equal proportions (red/yellow for orange, yellow/blue for green, and red/blue for purple). Joe currently has 20 gallons each of red, yellow, and blue paint, and 10 gallons each of orange, green, and purple paint. If needed, he can purchase any of the primary color paints for $20 per gallon, however he would like to save by utilizing the existing paint supplies as much as possible. Formulate an LP helping Joe to minimize his costs.

We will use index $i \in \{1, \ldots, 6\}$ for red, yellow, blue, orange, green, and purple colors, respectively, and index $j \in \{1, 2\}$ for brown and gray colors, respectively. Then our decision variables can be defined as

$$x_{ij} = \text{gallons of paint of color } i \text{ used to obtain color } j \text{ paint}$$

for $i \in \{1, \dots, 6\}, j \in \{1, 2\}$,

$$x_i = \text{gallons of paint } i \text{ purchased}, i = 1, 2, 3.$$

Then our objective is to minimize

$$z = 20x_1 + 20x_2 + 20x_3.$$

Next we specify the constraints. The total amount of brown and gray paint made must be at least 50 gallons each:

$$x_{11} + x_{21} + x_{31} + x_{41} + x_{51} + x_{61} \geq 50,$$
$$x_{12} + x_{22} + x_{32} + x_{42} + x_{52} + x_{62} \geq 50.$$

The amount of each paint used for mixing must not exceed its availability:

$$
\begin{aligned}
x_{11} + x_{12} &\leq 20 + x_1 \\
x_{21} + x_{22} &\leq 20 + x_2 \\
x_{31} + x_{32} &\leq 20 + x_3 \\
x_{41} + x_{42} &\leq 10 \\
x_{51} + x_{52} &\leq 10 \\
x_{61} + x_{62} &\leq 10.
\end{aligned}
$$

To express the constraints ensuring that the mixing yields the right shade of brown, note that only three out of six colors used for mixing contain red, and the total amount of red paint (including that coming from orange and purple paints) used in the brown mix is

$$x_{11} + 0.5x_{41} + 0.5x_{61}.$$

Hence, a constraint for the proportion of red color in the brown mix can be written as follows:

$$\frac{x_{11} + 0.5x_{41} + 0.5x_{61}}{x_{11} + x_{21} + x_{31} + x_{41} + x_{51} + x_{61}} = 0.4.$$

This equation can be easily expressed as a linear equality constraint:

$$0.6x_{11} - 0.4x_{21} - 0.4x_{31} + 0.1x_{41} - 0.4x_{51} + 0.1x_{61} = 0.$$

Similarly, the proportion of yellow and blue colors in the brown mix is given by:

$$\frac{x_{21} + 0.5x_{41} + 0.5x_{51}}{x_{11} + x_{21} + x_{31} + x_{41} + x_{51} + x_{61}} = 0.3$$

and

$$\frac{x_{31} + 0.5x_{51} + 0.5x_{61}}{x_{11} + x_{21} + x_{31} + x_{41} + x_{51} + x_{61}} = 0.3,$$

which can be equivalently written as

$$-0.3x_{11} + 0.7x_{21} - 0.3x_{31} + 0.2x_{41} + 0.2x_{51} - 0.3x_{61} = 0$$

and

$$-0.3x_{11} - 0.3x_{21} + 0.7x_{31} - 0.3x_{41} + 0.2x_{51} + 0.2x_{61} = 0,$$

respectively. The constraints describing the proportion of each of the primary colors in the gray paint mix can be derived analogously:

$$0.7x_{12} - 0.3x_{22} - 0.3x_{32} + 0.2x_{42} - 0.3x_{52} + 0.2x_{62} = 0 \quad \text{(red)}$$
$$-0.3x_{12} + 0.7x_{22} - 0.3x_{32} + 0.2x_{42} + 0.2x_{52} - 0.3x_{62} = 0 \quad \text{(yellow)}$$
$$-0.4x_{12} - 0.4x_{22} + 0.6x_{32} - 0.4x_{42} + 0.1x_{52} + 0.1x_{62} = 0 \quad \text{(blue)}.$$

Finally, note that the fact that $20(x_1 + x_2 + x_3)$ is minimized will force each of the variables x_1, x_2, and x_3 to be 0 unless additional red, yellow, or blue paint is required. The resulting formulation is given by

$$\begin{aligned}
\text{minimize} \quad & 20x_1 + 20x_2 + 20x_3 \\
\text{subject to} \quad & 0.6x_{11} - 0.4x_{21} - 0.4x_{31} + 0.1x_{41} - 0.4x_{51} + 0.1x_{61} = 0 \\
& -0.3x_{11} + 0.7x_{21} - 0.3x_{31} + 0.2x_{41} + 0.2x_{51} - 0.3x_{61} = 0 \\
& -0.3x_{11} - 0.3x_{21} + 0.7x_{31} - 0.3x_{41} + 0.2x_{51} + 0.2x_{61} = 0 \\
& 0.7x_{12} - 0.3x_{22} - 0.3x_{32} + 0.2x_{42} - 0.3x_{52} + 0.2x_{62} = 0 \\
& -0.3x_{12} + 0.7x_{22} - 0.3x_{32} + 0.2x_{42} + 0.2x_{52} - 0.3x_{62} = 0 \\
& -0.4x_{12} - 0.4x_{22} + 0.6x_{32} - 0.4x_{42} + 0.1x_{52} + 0.1x_{62} = 0 \\
& x_{11} + x_{21} + x_{31} + x_{41} + x_{51} + x_{61} \geq 50 \\
& x_{12} + x_{22} + x_{32} + x_{42} + x_{52} + x_{62} \geq 50 \\
& x_{11} + x_{12} - x_1 \leq 20 \\
& x_{21} + x_{22} - x_2 \leq 20 \\
& x_{31} + x_{32} - x_3 \leq 20 \\
& x_{41} + x_{42} \leq 10 \\
& x_{51} + x_{52} \leq 10 \\
& x_{61} + x_{62} \leq 10 \\
& x_{11}, x_{12}, x_{21}, x_{22}, x_{31}, x_{32}, x_{41}, x_{42}, x_{51}, x_{52}, x_{61}, x_{62}, x_1, x_2, x_3 \geq 0.
\end{aligned}$$

10.2.5 A transportation problem

Example 10.6 *A wholesale company specializing in one product has m warehouses $W_i, i = 1, \ldots, m$ serving n retail locations $R_j, j = 1, \ldots, n$. Transporting one unit of the product from W_i to R_j costs c_{ij} dollars, $i = 1, \ldots, m, j = 1, \ldots, n$. The company has s_i units of product available to ship from $W_i, i = 1, \ldots, m$. To satisfy the demand, at least d_j units of the product must be delivered to R_j. Formulate an LP to decide how many units of the product should be shipped from each warehouse to each retail location so that the company's overall transportation costs are minimized.*

The decision variables are

x_{ij} = the product quantity shipped from W_i to $R_j, i = 1, \ldots, m; j = 1, \ldots, n$.

All variables must be nonnegative. The objective is to minimize the total cost of transportation:

$$z = \sum_{i=1}^{m} \sum_{j=1}^{n} c_{ij} x_{ij}.$$

We need to make sure that the number of units shipped out of W_i does not exceed s_i:

$$\sum_{j=1}^{n} x_{ij} \leq s_i, \quad i = 1, \ldots, m.$$

Also, to satisfy the demand at R_j we must have

$$\sum_{i=1}^{m} x_{ij} \geq d_j, \quad j = 1, \ldots, n.$$

In summary, we obtain the following LP:

$$
\begin{aligned}
\text{minimize} \quad & \sum_{i=1}^{m} \sum_{j=1}^{n} c_{ij} x_{ij} \\
\text{subject to} \quad & \sum_{j=1}^{n} x_{ij} \leq s_i, \quad i = 1, \ldots, m \\
& \sum_{i=1}^{m} x_{ij} \geq d_j, \quad j = 1, \ldots, n \\
& x_{ij} \geq 0, \quad i = 1, \ldots, m, j = 1, \ldots, n.
\end{aligned}
\tag{10.1}
$$

10.2.6   A production planning problem

Example 10.7 *MIA Corporation manufactures n products $P_1, \ldots, P_n$ using m different types of resources (such as raw material, labor, etc.), $R_1, \ldots, R_m$. There are b_i units of resource i available per week. Manufacturing one unit of product P_j requires a_{ij} units of resource R_i. It is known that each unit of product P_j will sell for c_j dollars. MIA Corporation needs to decide how many units of each product to manufacture in order to maximize its weekly profit.*

We will use index i for the ith resource, $i = 1, \ldots, m$; and index j for the jth product, $j = 1, \ldots, n$. We define the decision variables as

$$x_j = \text{the number of units of product } j \text{ to manufacture}, \quad j = 1, \ldots, n.$$

Then the objective function is to maximize

$$c_1 x_1 + \ldots + c_n x_n.$$

The resource constraints, which make sure that the corporation does not exceed the availability of each resource, are given by

$$a_{i1} x_1 + \ldots + a_{in} x_n \leq b_i, \ i = 1, 2, \ldots, m.$$

Including the nonnegativity constraints, we obtain the following formulation:

$$
\begin{array}{rccccc}
\text{maximize} & c_1 x_1 & + & \ldots & + & c_n x_n \\
\text{subject to} & a_{11} x_1 & + & \ldots & + & a_{1n} x_n & \leq & b_1 \\
& \vdots & & \ddots & & \vdots & & \vdots \\
& a_{m1} x_1 & + & \ldots & + & a_{mn} x_n & \leq & b_m \\
& & & & & x_1, \ldots, x_n & \geq & 0,
\end{array}
$$

or, equivalently,

$$
\begin{array}{rl}
\text{maximize} & \sum_{j=1}^{n} c_j x_j \\
\text{subject to} & \sum_{j=1}^{n} a_{ij} x_j \leq b_i, \quad i = 1, 2, \ldots, m \\
& x_j \geq 0, \quad j = 1, 2, \ldots, n.
\end{array}
$$

Denoting by

$$
A = \begin{bmatrix} a_{11} & \cdots & a_{1n} \\ \vdots & \ddots & \vdots \\ a_{m1} & \cdots & a_{mn} \end{bmatrix}, \quad b = \begin{bmatrix} b_1 \\ \vdots \\ b_n \end{bmatrix}, \quad c = \begin{bmatrix} c_1 \\ \vdots \\ c_n \end{bmatrix},
$$

we represent the LP in a matrix form:

$$
\begin{array}{rl}
\text{maximize} & c^T x \\
\text{subject to} & Ax \leq b \\
& x \geq 0.
\end{array}
$$

10.3 Practical Implications of Using LP Models

Just like any other methodology used to solve practical problems, linear programming has its strengths and limitations. Its main advantages are in the relative simplicity of LP models and scalability of state-of-the-art algorithms, which are implemented in software packages capable of solving LP models with millions of variables and constraints. These features are sufficiently appealing to make LP by far the most frequently solved type of optimization problems in practice. However, once LP is selected as the modeling tool for a particular practical problem, it is important to recognize the assumptions that such a choice implicitly entails, as well as practical implications of making such assumptions.

Of course, the first major assumption we make by representing a problem as an LP is the *linearity* and *certainty* of dependencies involved in formulating the model's objective function and constraints. On the one hand, this is a rather strong assumption, especially given that most of the real-life processes we attempt to model are nonlinear in nature and are typically influenced by some uncertainties. On the other hand, any mathematical model is only an approximation of reality, and in many situations a linear approximation is sufficiently reasonable to serve the purpose. Recall that a linear function $f(x)$ in $\mathbb{R}^n$ is given by

$$f(x) = c_1 x_1 + \ldots + c_n x_n,$$

where $c_1, \ldots, c_n$ are constant real coefficients. This implies the properties of *additivity*,

$$f(x + y) = f(x) + f(y) \text{ for any } x, y \in \mathbb{R}^n,$$

and *proportionality*,

$$f(\alpha x) = \alpha f(x) \text{ for any } x \in \mathbb{R}^n, \alpha \in \mathbb{R}.$$

In particular, contributions of each variable to the objective function value is independent of contributions of the other variables, and if we change the value of one of the variables, say x_j, by Δ, while keeping the remaining variables unchanged, then the function value will change by $c_j \Delta$, i.e., the change in the function value is proportional to the change in a variable value.

In addition to additivity, proportionality, and certainty, another important assumption that is made in LP models is *divisibility*, meaning that fractional values of decision variables are acceptable. In reality, it may be essential that some of the decision variables are integer, however, introducing integrality constraints would turn the LP into a (mixed) integer linear programming problem, which is much harder to solve in general.

10.4   Solving Two-Variable LPs Graphically

If a linear program involves only two variables, it can be solved geometrically, by plotting the lines representing the constraints and level sets of the objective function. We will illustrate this approach by graphically solving the linear program formulated in Section 10.1:

$$
\begin{array}{llrcll}
\text{maximize} & 15x_1 & + & 25x_2 & & \text{(profit)} \\
\text{subject to} & x_1 & + & x_2 & \leq 450 & \text{(solid color fabric constraint)} \\
& & & x_2 & \leq 300 & \text{(printed fabric constraint)} \\
& 4x_1 & + & 5x_2 & \leq 2{,}000 & \text{(budget constraint)} \\
& x_1 & & & \leq 350 & \text{(demand constraint)} \\
& & & x_1, x_2 & \geq 0 & \text{(nonnegativity constraints)}.
\end{array}
$$

First, consider the line representing the points where the solid color fabric constraint is satisfied with equality, $x_1 + x_2 = 450$. This line passes through

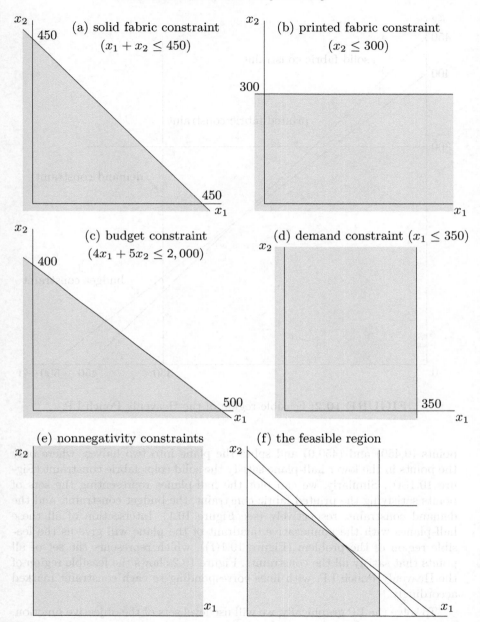

FIGURE 10.1: Drawing the feasible region in the Heavenly Pouch, Inc. example.

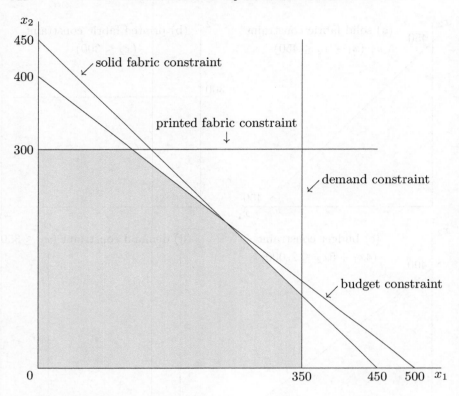

FIGURE 10.2: Feasible region of the Heavenly Pouch LP.

points (0,450) and (450,0) and splits the plane into two halves, where only the points in the lower half-plane satisfy the solid color fabric constraint (Figure 10.1(a)). Similarly, we can plot the half-planes representing the sets of points satisfying the printed fabric constraint, the budget constraint, and the demand constraint, respectively (see Figure 10.1). Intersection of all these half-planes with the nonnegative quadrant of the plane will give us the feasible region of the problem (Figure 10.1(f)), which represents the set of all points that satisfy all the constraints. Figure 10.2 shows the feasible region of the Heavenly Pouch LP, with lines corresponding to each constraint marked accordingly.

To solve the LP graphically, we will use level sets of the objective function, which in case of a maximization LP are sometimes referred to as *iso-profit lines*. Given a target objective function value (profit) $\bar{z}$, the iso-profit line is the set of points on the plane where $z = \bar{z}$, i.e., it is just the level set of the objective function z at the level $\bar{z}$. The iso-profit lines corresponding to different profit values $\bar{z}$ may or may not overlap with the feasible region. We typically start by plotting the iso-profit line for a reasonably low value of $\bar{z}$ to

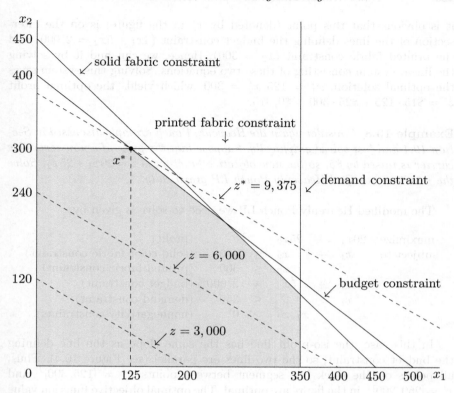

FIGURE 10.3: Solving the Heavenly Pouch LP graphically.

ensure that it contains feasible points and hence can be conveniently shown on the same plot as the feasible region. For the Heavenly Pouch LP, it appears reasonable to first plot the iso-profit line for $z = 15x_1 + 25x_2 = 3,000$, which passes through the points $(200, 0)$ and $(0, 120)$ (see Figure 10.3). We see that this profit level is feasible, so we can try a higher value, say $z = 6,000$. We see from the illustration that as we increased the target profit value, the new iso-profit line is parallel to the previous one (since the slope remained the same), and can be thought of as the result of movement of the previous iso-profit line up or to the right. If we keep increasing the target value of z, the corresponding iso-profit line will keep moving toward the upper right corner of the figure. It is clear that if we select the profit value that is too optimistic (say $z = 10,000$), the iso-profit line will have no common points with the feasible region.

However, we do not need to keep guessing which values of z would work. Instead, we observe that the optimal solution in our example corresponds to the last point that the iso-profit line will have in common with the feasible region as we move the line toward the upper right corner. From the figure, it is obvious that this point (denoted by x^* in the figure) is on the intersection of the lines defining the budget constraint $(4x_1 + 5x_2 = 2,000)$ and the printed fabric constraint $(x_2 = 300)$. Hence, we can find it by solving the linear system consisting of these two equations. Solving this system gives the optimal solution, $x_1^* = 125, x_2^* = 300$, which yields the optimal profit $z^* = \$15 \cdot 125 + \$25 \cdot 300 = \$9,375$.

Example 10.8 *Consider again the Heavenly Pouch example discussed in Section 10.1 and just solved graphically. Suppose that the price of a non-reversible carrier is raised by \$5, so the new objective function is $z = 20x_1 + 25x_2$. Solve the resulting modified Heavenly Pouch LP graphically.*

The modified Heavenly Pouch LP we need to solve is given by

$$
\begin{array}{llrcll}
\text{maximize} & 20x_1 & + & 25x_2 & & \text{(profit)} \\
\text{subject to} & x_1 & + & x_2 & \leq 450 & \text{(solid color fabric constraint)} \\
& & & x_2 & \leq 300 & \text{(printed fabric constraint)} \\
& 4x_1 & + & 5x_2 & \leq 2,000 & \text{(budget constraint)} \\
& x_1 & & & \leq 350 & \text{(demand constraint)} \\
& & & x_1, x_2 & \geq 0 & \text{(nonnegativity constraints)}.
\end{array}
$$

In this case, the iso-profit line has the same slope as the line defining the budget constraint, so the two lines are parallel (see Figure 10.4). Thus, all points on the thick line segment between points $x^* = [125, 300]^T$ and $x' = [250, 200]^T$ in the figure are optimal. The optimal objective function value is $z^* = 10,000$. Thus, there are infinitely many solutions, all of which belong to the convex combination of two extreme points (also known as vertices or corners) of the feasible region. As we will see later, any LP that has an optimal solution must have at least one corner optimum.

Example 10.9 *A retail store is planning an advertising campaign aiming to increase the number of customers visiting its physical location, as well as its online store. The store manager would like to advertise through a local magazine and through an online social network. She estimates that each 1,000 dollars invested in magazine ads will attract 100 new customers to the store, as well as 500 new website visitors. In addition, each 1,000 dollars invested in online advertising will attract 50 new local store customers, as well as 1,000 new website visitors. Her target for this campaign is to bring at least 500 new guests to the physical store and at least 5,000 new visitors to the online store. Formulate an LP to help the store minimize the cost of its advertising campaign. Solve the LP graphically.*

The decision variables are

$$
\begin{array}{rll}
x_1 & = & \text{budget for magazine advertising (in thousands of dollars)} \\
x_2 & = & \text{budget for online advertising (in thousands of dollars)},
\end{array}
$$

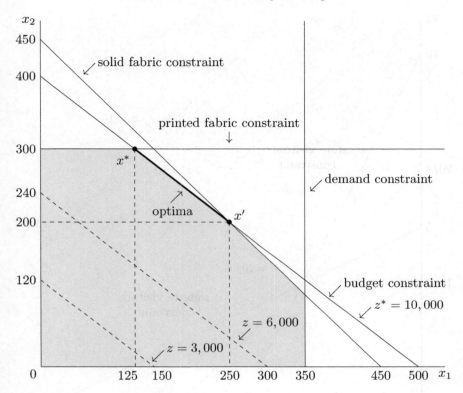

FIGURE 10.4: Graphical solution of the modified Heavenly Pouch LP.

and the problem can be formulated as the following LP:

$$
\begin{array}{rrcrcll}
\text{minimize} & x_1 & + & x_2 & & & \\
\text{subject to} & 100x_1 & + & 50x_2 & \geq & 500 & \text{(store visitors)} \\
& 500x_1 & + & 1,000x_2 & \geq & 5,000 & \text{(website visitors)} \\
& & & x_1, x_2 & \geq & 0. & \text{(nonnegativity)}
\end{array}
$$

We start solving the problem graphically by plotting the lines describing the constraints and drawing the feasible region (Figure 10.5). Then we plot two level sets for the objective function, which in case of minimization problems are called *iso-cost lines*, for $z = 15$ and $z = 10$. We observe that as the value of z decreases, the iso-cost line moves down, toward the origin. If we keep decreasing z, the iso-cost line will keep moving down, and at some point, will contain no feasible points. It is clear from the figure that the last feasible point the iso-cost line will pass through as we keep decreasing the value of z will be the point of intersection of the lines defining the store visitors constraint and

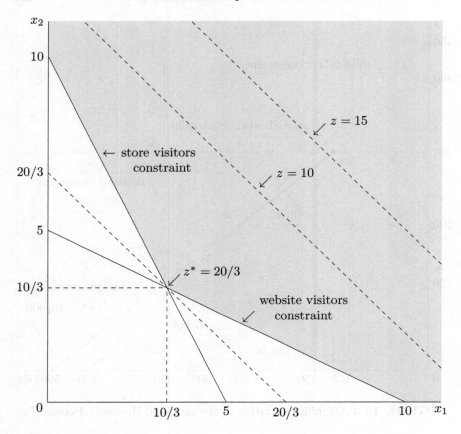

FIGURE 10.5: Solving Example 10.9 LP graphically.

the website visitor constraint. Hence, solving the system

$$\begin{aligned} 100x_1 &+ 50x_2 &= 500 \\ 500x_1 &+ 1,000x_2 &= 5,000, \end{aligned}$$

we find the optimal solution $x_1^* = x_2^* = 10/3 \approx 3.333$, $z^* = 20/3 \approx 6.666$. Thus, the store should spend \$6,666 on advertising and split this budget evenly between the magazine and online advertising to reach its goals.

Note that even though the LP we just solved has an unbounded feasible region, it still has an optimal solution. However, if the objective function was improving along one of the directions in which the feasible region is unbounded (called direction of unboundedness), an optimal solution would not exist.

Example 10.10 *Consider a problem that has the same objective function and feasible region as the LP in the previous example, but change the objective to* *maximization:*

$$\begin{array}{rrcl}
maximize & x_1 + & x_2 & \\
subject\ to & 100x_1 + & 50x_2 & \geq 500 \\
& 500x_1 + & 1{,}000x_2 & \geq 5{,}000 \\
& & x_1, x_2 & \geq 0.
\end{array}$$

Clearly the objective function value tends to infinity if one of the variables is increased toward infinity; thus this LP has no optimal solution.

Encountering an LP with a feasible region containing points with any desirable objective value is highly unlikely in practice. Next we consider an example of an LP that may arise in decision-making situations with limited resources. Namely, having insufficient resources may lead to a model that not only has no optimal solution, but does not even have a feasible solution.

Example 10.11 *Assume that the retail store in Example 10.9 has an advertising budget limited to \$5,000. This condition is reflected in the budget constraint in the corresponding LP model:*

$$\begin{array}{rrcll}
minimize & x_1 + & x_2 & & \\
subject\ to & 100x_1 + & 50x_2 & \geq 500 & (store\ visitors) \\
& 500x_1 + & 1{,}000x_2 & \geq 5{,}000 & (website\ visitors) \\
& x_1 + & x_2 & \leq 5 & (budget) \\
& & x_1, x_2 & \geq 0. & (nonnegativity)
\end{array}$$

From the illustration in Figure 10.5 it is clear that the set of points such that $x_1 + x_2 \leq 5$ does not overlap with the set of points satisfying each of the remaining constraints of the LP. Thus, no feasible point exists for this LP. Indeed, we previously determined that an advertising campaign that will yield the target results will cost at least \$6,666.

10.5 Classification of LPs

LPs can be classified in terms of their feasibility and optimality properties. We saw in the previous section that some LPs have one or more optimal solutions, whereas others may have no optimal or even feasible solutions.

Definition 10.1 *An LP is called*

- **feasible** *if it has at least one feasible solution and* **infeasible,** *otherwise;*
- **optimal** *if it has an optimal solution;*

- **unbounded** *if it is feasible and its objective function is not bounded (from above for a maximization problem and from below for a minimization problem) in the feasible region.*

For example, the Heavenly Pouch LP, as well as LPs in Examples 10.8 and 10.9 are optimal (which also implies that they are feasible LPs). In particular, the LP in Example 10.9 is optimal despite its feasible region being unbounded. The LP considered in Example 10.10 is unbounded, and the LP in Example 10.11 is infeasible.

Later (Theorem 11.5 at page 266) we will establish the following fact.

If an LP is not optimal, it is either unbounded or infeasible.

If an optimal LP has more than one optimal solution, then it is easy to show that it has infinitely many optimal solutions. Indeed, consider a maximization LP
$$\max_{x \in X} c^T x,$$
where X is a polyhedral set, and assume that the LP has two alternative optimal solutions x^* and x' with the optimal objective value $z^* = c^T x^* = c^T x'$. Then, for an arbitrary $\alpha \in (0,1)$ consider a convex combination of x^* and x', $y = \alpha x^* + (1-\alpha)x' \in X$. We have

$$c^T y = c^T(\alpha x^* + (1-\alpha)x') = \alpha c^T x^* + (1-\alpha)c^T x' = \alpha z^* + (1-\alpha)z^* = z^*,$$

so, y is also an optimal solution of the LP. Thus, we established that the following property holds.

If an LP is optimal, it either has a unique optimal solution or infinitely many optimal solutions. Moreover, the set of all optimal solutions to an LP is a convex set.

Exercises

10.1. Romeo Winery produces two types of wines, Bordeaux and Romerlot, by blending Merlot and Cabernet Sauvignon grapes. Making one barrel of Bordeaux blend requires 250 pounds of Merlot and 250 pounds of Cabernet Sauvignon, whereas making one barrel of Romerlot requires 450 pounds of Merlot and 50 pounds of Cabernet Sauvignon. The profit received from selling Bordeaux is $800 per barrel, and from selling

Romerlot, $600 per barrel. Romeo Winery has 9,000 pounds of Merlot and 5,000 pounds of Cabernet Sauvignon available. Formulate an LP model aiming to maximize the winery's profit. Solve the LP graphically.

10.2. O&M Painters produce orange and maroon paints by mixing the so-called primary red, yellow, and blue paint colors. The proportions of red, yellow, and blue paints used to get the required shades are 50%, 40%, and 10%, respectively for orange, and 60%, 10%, and 30%, respectively for maroon. What is the maximum combined amount of orange and maroon paints that O&M Painters can produce, given that they have 6 gallons of red paint, 4 gallons of yellow paint, and 1.8 gallons of blue paint available for mixing? Formulate this problem as an LP and solve it graphically.

10.3. The Concrete Guys make two types of concrete by mixing cement, sand, and stone. The regular mix contains 30% of cement, 15% of sand, and 55% of stone (by weight), and sells for 5 cents/lb. The extra-strong mix must contain at least 50% of cement, at least 5% of sand, and at least 20% of stone, and sells for 8 cents/lb. The Concrete Guys have 100,000 lb of cement, 50,000 lb of sand, and 100,000 lb of stone in their warehouse. Formulate an LP to determine the amount of each mix the Concrete Guys should make in order to maximize their profit.

10.4. A football coach is deciding on an offensive plan for a game. He estimates that a passing play will yield 7 yards on average, whereas a rushing play will average 5 yards. To balance the offense, he wants neither passing nor rushing to constitute more than 2/3 of all offensive plays. Propose an LP formulation to determine the distribution of passing and rushing plays that maximizes an average gain per play. Solve the formulated LP graphically.

10.5. A plant uses two alternative processes to manufacture three different products, X, Y, and Z, of equal value. An hour of the first process costs $3 and yields 3 units of X, 2 units of Y, and 1 unit of Z, whereas an hour of the second process costs $2 and yields 2 units of each product. The daily demand for X, Y, and Z is 25, 20, and 10 units, respectively. Each machine can be used for at most 10 hours a day. Formulate an LP to minimize the daily manufacturing costs while satisfying the demand and time constraints. Solve the LP graphically.

10.6. SuperCleats Inc. is a small business manufacturing 4 models of soccer shoes, Dynamo, Spartacus, Torpedo, and Kanga. Manufacturing one pair of each model is associated with material, labor, and overhead costs given in the table below, which also lists the company's annual budget for each kind of resource.

	Dynamo	Spartacus	Torpedo	Kanga	Annual Budget
Price	100	80	60	50	
Material Cost	15	10	8	5	45,000
Labor Cost	35	22	18	10	100,000
Overhead Cost	5	5	4	4	20,000

In addition, the annual demand for Dynamo and Spartacus shoes is limited to 1,000 pairs each. Formulate an LP to help SuperCleats Inc. decide on the quantity of each shoe to manufacture so that its overall profit is maximized.

10.7.   A company is looking to hire retail assistants for its new store. The number of assistants required on different days of the week is as follows: Monday – 4, Tuesday – 5, Wednesday – 5, Thursday – 6, Friday – 7, Saturday – 8, and Sunday – 8. Each assistant is expected to work four consecutive days and then have three days off. Formulate an LP aiming to meet the requirements while minimizing the number of hires.

10.8.   A financial company considers five different investment options for the next two years, as described in the following table:

	Inv.1	Inv.2	Inv.3	Inv.4	Inv.5
Year 0 cash required	$11	$8	$7	$10	$0
Year 1 cash received	$2	$0	$0	$12	$0
Year 1 cash required	$0	$8	$0	$0	$10
Year 2 cash received	$12	$20	$11	$0	$12

Here, year 0 represents the present time, year 1 represents one year from now, and year 2 represents two years from now. For example, investment 1 requires an $11 million cash investment now and yields $2 million and $12 million cash in one and two years from now, respectively; investment 2 requires two $8 million cash deposits (now and one year from now), with a $20-million payday 2 years from now, etc. Any fraction of each investment alternative can be purchased. For example, the company could purchase 0.25 of investment 2, which would require two $2-million cash investments (now and in one year), yielding $5 million in two years. The company expects to have $20 million to invest now, plus $10 million to invest at year 1 (in addition to the cash received from the original investment). Formulate an LP to determine an investment strategy that would maximize cash on hand after 2 years.

10.9.   A farmer maintains a pasture for his 30 cows via watering and fertilizing. The grass grows uniformly and at a rate that is constant for the given level of watering and fertilizing. Also, a cow grazing reduces the amount of grass uniformly and at a constant rate. From the past experience, the farmer knows that under regular conditions (no watering or fertilizing) the grass will run out in 20 days with 40 cows grazing, and in 30 days with 30 cows using the pasture. Spending \$1 per week on watering increases the rate of grass growth by 1%, and spending \$1 per week on fertilizer increases the grass growth rate by 2%. For the fertilizer to be effective, the grass must be watered properly, meaning that for each dollar spent on fertilizing, the farmer must spend at least 50 cents on watering. Other than that, it can be assumed that the grass growth rate increase due to watering is independent of that due to fertilizing. Formulate an LP to minimize the cost of maintaining the pasture while making sure that the grass never runs out. Solve the LP graphically.

10.10. Assume that in the diet problem (Section 10.2.1, p. 213) Yiming wants calories from fat not to exceed 30% of the total calories he consumes. Calories from fat for each considered food type are as follows: almond butter – 480, brown rice – 15, salmon – 110, orange juice – 6, and wheat bread – 8. Modify the corresponding LP formulation to include a constraint limiting the fraction of fat calories.

10.11. Solve the following LPs graphically:

(a)
$$
\begin{aligned}
\text{maximize}\quad & 3x_1 + 5x_2 \\
\text{subject to}\quad & x_1 + x_2 \le 5 \\
& x_2 \le 3 \\
& 3x_1 + 5x_2 \le 18 \\
& x_1 \le 4 \\
& x_1, x_2 \ge 0
\end{aligned}
$$

(b)
$$
\begin{aligned}
\text{minimize}\quad & x_1 + 5x_2 \\
\text{subject to}\quad & x_1 + 2x_2 \ge 3 \\
& x_1 + x_2 \ge 2 \\
& 5x_1 + x_2 \ge 5 \\
& x_1 \le 4 \\
& x_1, x_2 \ge 0.
\end{aligned}
$$

10.12. Design an algorithm to solve the resource allocation problem in Section 10.2.2.

Chapter 11

The Simplex Method for Linear Programming

In this chapter, we discuss one of the first and most popular methods for solving LPs, the *simplex method* originally proposed by George Dantzig in 1940s for solving problems arising in military operations. In order to apply this method, an LP is first converted to its *standard form*, as discussed in the following section.

11.1 The Standard Form of LP

LP in the standard form has only equality and nonnegativity constraints. An inequality constraint can easily be converted into an equality constraint by introducing a new variable in the left-hand side as follows. If the ith constraint is of "$\leq$" type, we add a nonnegative *slack variable* s_i in the left-hand side of the inequality to turn it into an equality constraint. Say, if our i^{th} constraint was

$$x_1 + 2x_2 + 3x_3 \leq 4,$$

we would replace it with the equality constraint

$$x_1 + 2x_2 + 3x_3 + s_i = 4,$$

where

$$s_i = 4 - x_1 - 2x_2 - 3x_3 \geq 0.$$

Similarly, if the ith constraint is of "$\geq$" type, we introduce a nonnegative *excess variable* e_j and subtract it from the left-hand side of the constraint to obtain the corresponding equality constraint. For example, for

$$x_1 + 2x_2 + 3x_3 \geq 4,$$

we have

$$x_1 + 2x_2 + 3x_3 - e_i = 4,$$

where

$$e_i = -4 + x_1 + 2x_2 + 3x_3 \geq 0.$$

Example 11.1 *The standard form of the LP*

$$
\begin{array}{rrrrrrl}
maximize & 3x_1 & - & 5x_2 & + & 7x_3 & \\
subject\ to & 2x_1 & + & 4x_2 & - & x_3 & \geq -3 \\
& 4x_1 & - & 2x_2 & + & 8x_3 & \leq 7 \\
& 9x_1 & + & x_2 & + & 3x_3 & = 11 \\
& & & & x_1, x_2, x_3 & \geq 0
\end{array}
$$

is given by

$$
\begin{array}{rrrrrrrrrl}
maximize & 3x_1 & - & 5x_2 & + & 7x_3 & & & & \\
subject\ to & 2x_1 & + & 4x_2 & - & x_3 & - & e_1 & & = -3 \\
& 4x_1 & - & 2x_2 & + & 8x_3 & & & + s_2 & = 7 \\
& 9x_1 & + & x_2 & + & 3x_3 & & & & = 11 \\
& & & & & x_1, x_2, x_3, e_1, s_2 & \geq & 0.
\end{array}
$$

We require that all variables are nonnegative in the standard form representation, however, in general, some of the LP variables may be *unrestricted in sign* or *free*. We write $x_j \in \mathbb{R}$ to denote a free variable x_j. If a free variable x_j is present in our LP, we can represent it as the difference of two nonnegative variables as follows:

$$
x_j = x'_j - x''_j, \quad \text{where } x'_j, x''_j \geq 0.
$$

Example 11.2 *The standard form of the LP*

$$
\begin{array}{rrrrrrl}
maximize & 3x_1 & - & 5x_2 & + & 7x_3 & \\
subject\ to & 2x_1 & + & 4x_2 & - & x_3 & \geq -3 \\
& 4x_1 & - & 2x_2 & + & 8x_3 & \leq 7 \\
& 9x_1 & + & x_2 & + & 3x_3 & = 11 \\
& & x_1 \in \mathbb{R}, & & x_2, x_3 & \geq & 0
\end{array}
$$

is given by

$$
\begin{array}{rrrrrrrrrrrl}
maximize & 3x'_1 & - & 3x''_1 & - & 5x_2 & + & 7x_3 & & & & \\
subject\ to & 2x'_1 & - & 2x''_1 & + & 4x_2 & - & x_3 & - & e_1 & & = -3 \\
& 4x'_1 & - & 4x''_1 & - & 2x_2 & + & 8x_3 & & & + s_2 & = 7 \\
& 9x'_1 & - & 9x''_1 & + & x_2 & + & 3x_3 & & & & = 11 \\
& & & & & x'_1, x''_1, x_2, x_3, e_1, s_2 & \geq & 0.
\end{array}
$$

Recall that we can always convert a "$\geq$"-type constraints into a "$\leq$"-type constraint by multiplying both sides of the "$\geq$"-type constraint by -1, thus it is reasonable to assume that all inequality constraints are given in "$\leq$" form. Consider a general LP

$$
\begin{array}{rrrl}
maximize & c^T x & & \\
subject\ to & A'x & \leq & b' \\
& A''x & = & b'' \\
& x & \geq & 0,
\end{array}
\tag{11.1}
$$

where A' and b' are an $m' \times n$ matrix and m'-vector describing m' inequality constraints, and A'' and b'' are an $m'' \times n$ matrix and m''-vector describing m'' equality constraints, respectively. Let $m = m' + m''$ be the total number of constraints in this LP other than the nonnegativity constraints. Then the standard form of LP (11.1) is given by

$$\begin{aligned} \text{maximize} \quad & c^T x \\ \text{subject to} \quad Ax \ &= \ b \\ x \ &\geq \ 0, \end{aligned} \qquad (11.2)$$

where the vector x of $m' + n$ variables includes m' slack variables $x_{n+1}, \ldots, x_{n+m'}$ in addition to the n original variables $x_1, \ldots, x_n$; A is an $m \times (n+m)$ matrix,

$$A = \left[\begin{array}{cc} A' & I_{m'} \\ A'' & O \end{array} \right],$$

where $I_{m'}$ is the $m' \times m'$ identity matrix and O is the $m'' \times m'$ zero matrix; c is a $(n+m')$-vector of objective function coefficients (with $c_{n+1} = \ldots = c_{n+m'} = 0$); and b is an m-vector of right-hand sides.

11.2 The Simplex Method

The simplex method, originally developed by George Dantzig in the 1940s, is still one of the most commonly used approaches to solving LP problems in practice. To start with, we will restrict our discussion of the method to LPs in the form

$$\begin{aligned} \text{maximize} \quad & c^T x \\ \text{subject to} \quad Ax \ &\leq \ b \quad \text{where } b \geq 0. \\ x \ &\geq \ 0 \end{aligned} \qquad (11.3)$$

The reason for this is that finding an initial feasible solution $x^{(0)}$ for an LP in the form (11.3) is very easy: we can just use $x^{(0)} = 0$, which obviously satisfies all the constraints. This is important, since the simplex method needs a starting feasible solution $x^{(0)}$, which it will use to generate a finite sequence of feasible points $x^{(0)}, x^{(1)}, x^{(2)}, \ldots, x^{(N)}$, such that each next point in the sequence has an objective value at least as good as that of the previous point, i.e., $z(x^{(k+1)}) \geq z(x^{(k)})$, $k = 0, \ldots, N-1$, and $x^* = x^{(N)}$ is an optimal solution of the LP. If we do not require $b \geq 0$ in the LP above, finding a starting point $x^{(0)}$ is more challenging, and we will address this case after discussing how to solve the LPs with $b \geq 0$.

We will introduce the basic idea of the method using the Heavenly Pouch LP formulated in Section 10.1 and solved graphically in Section 10.4. Consider the following LP:

maximize $\quad 15x_1 \;+\; 25x_2$

subject to $\quad\quad x_1 \;+\;\quad x_2 \;\leq\; 450 \quad$ (solid color fabric constraint)

$\quad\quad\quad\quad\quad\quad\quad x_2 \;\leq\; 300 \quad$ (printed fabric constraint)

$\quad\quad\quad 4x_1 \;+\; 5x_2 \;\leq\; 2,000 \quad$ (budget constraint)

$\quad\quad\quad x_1 \;\quad\quad\quad\;\leq\; 350 \quad$ (demand constraint)

$\quad\quad\quad\quad\quad\quad x_1, x_2 \;\geq\; 0 \quad$ (nonnegativity constraints).

First, we convert this LP to the standard form by introducing a slack variable s_i for each constraint $i, i = 1, \ldots, 4$:

maximize $\quad 15x_1 \;+\; 25x_2$

$$
\begin{aligned}
\text{subject to} \quad x_1 \;+\;\quad x_2 \;+\; s_1 \quad\quad\quad\quad\quad\quad &= 450 \\
x_2 \quad\quad\; + s_2 \quad\quad\quad\quad &= 300 \\
4x_1 \;+\; 5x_2 \quad\quad\quad\quad + s_3 \quad &= 2,000 \\
x_1 \quad\quad\quad\quad\quad\quad\quad\; + s_4 &= 350 \\
x_1, x_2, s_1, s_2, s_3, s_4 \;&\geq\; 0.
\end{aligned}
\tag{11.4}
$$

We can write it in the equivalent *dictionary format*, where the slack variables s_1, s_2, s_3, s_4 are expressed through the remaining variables as follows:

$$
\begin{array}{rcrcrcr}
z &=& & & 15x_1 &+& 25x_2 \\
\hline
s_1 &=& 450 &-& x_1 &-& x_2 \\
s_2 &=& 300 & & &-& x_2 \\
s_3 &=& 2,000 &-& 4x_1 &-& 5x_2 \\
s_4 &=& 350 &-& x_1 & &
\end{array}
\tag{11.5}
$$

In this representation, we will call the variables kept in the left-hand side the *basic variables*, and the remaining variables *nonbasic variables*. Obviously, the number of basic variables is the same as the number of constraints in our LP, and the number of nonbasic variables equals the number of variables in the original LP, before the slack variables were introduced. The sets of basic and nonbasic variables will be updated step by step, using an operation called *pivot*, as we proceed with the iterations of the simplex method. We will denote the sets of all the basic and nonbasic variables at step k of the simplex method by BV_k and NV_k, respectively. We assume that the initial dictionary corresponds to step $k = 0$. Thus, in our example

$$
BV_0 = \{s_1, s_2, s_3, s_4\}, \quad NV_0 = \{x_1, x_2\}.
$$

Note that to get a feasible solution to the considered LP, we can set all the nonbasic variables to 0, and this will uniquely determine the corresponding values of the basic variables and the objective function. We have:

$$
x_1 = x_2 = 0 \;\Rightarrow\; s_1 = 450, s_2 = 300, s_3 = 2,000, s_4 = 350; \; z = 0.
$$

We call this solution the *basic solution* corresponding to the basis BV_0. If all variables have nonnegative values in a basic solution, then the solution is called a *basic feasible solution* (*bfs*) and the corresponding dictionary is called feasible. Note that the basic solution with the basis BV_0 in our example is, in fact, a basic feasible solution.

Our LP can also be conveniently represented in the *tableau format*:

z	x_1	x_2	s_1	s_2	s_3	s_4	rhs	$Basis$
1	−15	−25	0	0	0	0	0	z
0	1	1	1	0	0	0	450	s_1
0	0	1	0	1	0	0	300	s_2
0	4	5	0	0	1	0	2,000	s_3
0	1	0	0	0	0	1	350	s_4

(11.6)

Here rhs stands for right-hand side. The entries in the tableau are just the coefficients of LP in the standard form (11.4), where the z-row is modified by moving all variables to the left-hand side, so instead of $z = 15x_1 + 25x_2$ we write

$$z - 15x_1 - 25x_2 = 0.$$

In this format, z is treated as a variable that is always basic. Since the dictionary format is helpful for visual explanation of the ideas behind the method and the tableau format is more handy for performing the computations, we will use both representations as we describe the steps of the simplex method below. In both dictionary and tableau formats, we number the rows starting with 0, so the top row is referred to as row 0 or the z-row, and row i corresponds to the ith constraint. The basic feasible solution at step 0 is given in the following table.

Step 0 basic feasible solution
BV_0 : $\quad s_1, s_2, s_3, s_4$
NV_0 : $\quad x_1, x_2$
bfs : $\quad x_1 = x_2 = 0$
$\qquad\quad s_1 = 450, s_2 = 300, s_3 = 2{,}000, s_4 = 350$
$\qquad\quad z = 0$

We are ready to perform the first iteration of the simplex method.

11.2.1 Step 1

Let us analyze our problem written in the dictionary form as in (11.5) above, taking into account that the current basic feasible solution has all the nonbasic variables at 0, $x_1 = x_2 = 0$. We have:

$$
\begin{aligned}
z   &=   & 15x_1 &+& 25x_2 \\
s_1 &= \phantom{2,{}}450 &-\ x_1 &-& x_2 \\
s_2 &= \phantom{2,{}}300 &   &-& x_2 \\
s_3 &= 2{,}000 &-\ 4x_1 &-& 5x_2 \\
s_4 &= \phantom{2,{}}350 &-\ x_1 & &
\end{aligned}
$$

Since the objective function is expressed in terms of nonbasic variables only, the only way for us to change the value of z is by changing at least one of the nonbasic variables from 0 to some positive value (recall that all variables must be nonnegative). To increase the value of z, we can increase the value of a nonbasic variable that has a positive coefficient in the objective. Due to the linearity of the objective function, increasing a variable by 1 unit will change the objective by value equal to the coefficient of that variable in the z-row of the dictionary. Thus, to get the highest possible increase in the objective per unit of increase in the variable value, it makes sense to try to increase the value of the variable with the highest coefficient. We have

$$z = 15x_1 + 25x_2,$$

we pick variable x_2 as the one whose value will be increased. We call this variable the *pivot variable* and the corresponding column in the dictionary is called the *pivot column*. We want to increase the value of x_2 as much as possible while keeping the other nonbasic variable equal to 0. The amount by which we can increase x_2 is restricted by the nonnegativity constraints for the basic variables, which must be satisfied to ensure feasibility:

$$
\begin{aligned}
s_1 &= 450 - x_2 \geq 0 \\
s_2 &= 300 - x_2 \geq 0 \\
s_3 &= 2,000 - 5x_2 \geq 0 \\
s_4 &= 350 - 0x_2 \geq 0
\end{aligned}
$$

(Note that the column of the dictionary corresponding to variable x_1 is ignored since $x_1 = 0$ in the current basic feasible solution and we do not change its value). For all of these inequalities to be satisfied, we must have $x_2 \leq 300$. Thus, the largest feasible increase for x_2 is equal to 300. Note that the largest possible increase corresponds to the smallest ratio of the free coefficient to the absolute value of the coefficient for x_2 in the same row, assuming that the coefficient for x_2 is negative. The rows where the coefficient for x_2 is nonnegative can be ignored, since the corresponding inequalities are redundant. For example, if we had an inequality $500 + 5x_2 \geq 0$, it is always satisfied due to nonnegativity of x_2. We say that the row in which the smallest ratio is achieved wins the *ratio test*. This row is called the *pivot row*. In our example, the second row, which has s_2 as the basic variable, is the pivot row.

To carry out the pivot, we express the nonbasic variable in the pivot column through the basic variable in the pivot row:

$$x_2 = 300 - s_2.$$

Then we substitute this expression for x_2 in the remaining rows of the dictionary:

$$z = 15x_1 + 25x_2 = 15x_1 + 25(300 - s_2) = 7,500 + 15x_1 - 25s_2,$$

$$s_1 = 450 - x_1 - x_2 = 450 - x_1 - (300 - s_2) = 150 - x_1 + s_2,$$

$$s_3 = 2,000 - 4x_1 - 5x_2 = 2,000 - 4x_1 - 5(300 - s_2) = 500 - 4x_1 + 5s_2,$$

$$s_4 = 350 - x_1 - 0x_2 = 350 - x_1.$$

We obtain the step 1 dictionary:

$$
\begin{array}{rrrrr}
z & = & 7,500 & + & 15x_1 & - & 25s_2 \\
\hline
s_1 & = & 150 & - & x_1 & + & s_2 \\
x_2 & = & 300 & & & - & s_2 \\
s_3 & = & 500 & - & 4x_1 & + & 5s_2 \\
s_4 & = & 350 & - & x_1 &
\end{array}
\qquad (11.7)
$$

To complete the same step using the tableau format, we consider the tableau (11.6). We find the most negative coefficient in the z-row; it corresponds to x_2, thus the corresponding column is the pivot column. We perform the ratio test by dividing the entries in the *rhs* column by the corresponding entries in the pivot column that are positive. The minimum such ratio, 300, corresponds to the second row, which is the pivot row. The coefficient on the intersection of the pivot row and pivot column in the table is the *pivot element*.

z	x_1	x_2	s_1	s_2	s_3	s_4	*rhs*	*Basis*	*Ratio*
1	-15	-25	0	0	0	0	0	z	
0	1	1	1	0	0	0	450	s_1	450
0	0	$\boxed{1}$	0	1	0	0	300	s_2	300 $\leftarrow$
0	4	5	0	0	1	0	2,000	s_3	400
0	1	0	0	0	0	1	350	s_4	—

To perform the pivot, we use elementary row operations involving the pivot row with the goal of turning all pivot column entries in the non-pivot rows into 0s and the pivot element into 1. In particular, we multiply the pivot row by $25, -1$, and -5, add the result to rows 0, 1, and 3, respectively, and update the corresponding rows. Since the pivot element is already 1, the pivot row is kept unchanged, but the corresponding basic variable is now x_2 instead of s_2. As a result, we obtain the following step 1 tableau:

z	x_1	x_2	s_1	s_2	s_3	s_4	*rhs*	*Basis*	
1	-15	0	0	25	0	0	7,500	z	
0	1	0	1	-1	0	0	150	s_1	
0	0	1	0	1	0	0	300	x_2	(11.8)
0	4	0	0	-5	1	0	500	s_3	
0	1	0	0	0	0	1	350	s_4	

Compare this tableau to the corresponding dictionary (11.7). Clearly, the dictionary and the tableau describe the same system. The basic feasible solution after step 1 is summarized in the following table.

Step 1 basic feasible solution
BV_1 : $\quad s_1, x_2, s_3, s_4$
NV_1 : $\quad x_1, s_2$
bfs : $\quad x_1 = 0, x_2 = 300$
$\quad\quad\quad s_1 = 150, s_2 = 0, s_3 = 500, s_4 = 350$
$\quad\quad\quad z = 7,500$

We saw that, as a result of the pivot operation, one of the previously nonbasic variables, x_2, has become basic, whereas s_2, which was basic, has become nonbasic. We will call the variable that is entering the basis during the current iteration the *entering variable* and the variable that is leaving the basis the *leaving variable*.

11.2.2 Step 2

The next step is performed analogously to the first step. Again, we analyze the current dictionary (11.7) and try to increase the objective function value by updating the current basic feasible solution.

$$
\begin{array}{rcrcrcr}
z & = & 7,500 & + & 15x_1 & - & 25s_2 \\
\hline
s_1 & = & 150 & - & x_1 & + & s_2 \\
x_2 & = & 300 & & & - & s_2 \\
s_3 & = & 500 & - & 4x_1 & + & 5s_2 \\
s_4 & = & 350 & - & x_1 & &
\end{array}
$$

Notice that, as before, the basic feasible solution can be obtained by setting all nonbasic variables equal to 0, so the same considerations as in step 1 apply when we decide which nonbasic variable should be increased in value and hence enter the basis. Since only one variable, x_1, has a positive coefficient in the objective, it is the only entering variable candidate. Row 3 wins the ratio test, so s_3 is the leaving variable and we have

$$x_1 = 125 + \frac{5}{4}s_2 - \frac{1}{4}s_3,$$

$$z = 7,500 + 15x_1 - 25s_2 = 9,375 - \frac{25}{4}s_2 - \frac{15}{4}s_3,$$

$$s_1 = 150 - x_1 + s_2 = 25 - \frac{1}{4}s_2 + \frac{1}{4}s_3,$$

$$s_4 = 350 - x_1 = 225 - \frac{5}{4}s_2 + \frac{1}{4}s_3.$$

The resulting step 2 dictionary is given by:

$$z = 9,375 - \frac{25}{4}s_2 - \frac{15}{4}s_3$$

$$\begin{aligned} s_1 &= 25 - \tfrac{1}{4}s_2 + \tfrac{1}{4}s_3 \\ x_2 &= 300 - s_2 \\ x_1 &= 125 + \tfrac{5}{4}s_2 - \tfrac{1}{4}s_3 \\ s_4 &= 225 - \tfrac{5}{4}s_2 + \tfrac{1}{4}s_3 \end{aligned}$$

(11.9)

Next, we carry out the computations for step 2 in the tableau format. In the step 1 tableau (11.8), we find the most negative coefficient in the z-row, which leads to selecting x_1 as the entering variable. Row 3 wins the ratio test, so s_3 is the leaving variable.

z	x_1	x_2	s_1	s_2	s_3	s_4	rhs	Basis	Ratio
1	-15	0	0	25	0	0	7,500	z	
0	1	0	1	-1	0	0	150	s_1	150
0	0	1	0	1	0	0	300	x_2	–
0	$\boxed{4}$	0	0	-5	1	0	500	s_3	125 ←
0	1	0	0	0	0	1	350	s_4	350

We first divide row 3 by the pivot element value, which is 4, and then use it to eliminate the remaining nonzero coefficients in the pivot column. We obtain the following step 2 tableau:

z	x_1	x_2	s_1	s_2	s_3	s_4	rhs	Basis
1	0	0	0	25/4	15/4	0	9,375	z
0	0	0	1	1/4	$-1/4$	0	25	s_1
0	0	1	0	1	0	0	300	x_2
0	1	0	0	$-5/4$	1/4	0	125	x_1
0	0	0	0	5/4	$-1/4$	1	225	s_4

(11.10)

Again, this tableau is equivalent to the corresponding dictionary (11.9). We summarize the basic feasible solution after step 2 below.

Step 2 basic feasible solution
BV_2: x_1, x_2, s_1, s_4
NV_2: s_2, s_3
bfs: $x_1 = 125, x_2 = 300$
$s_1 = 25, s_2 = 0, s_3 = 0, s_4 = 225$
$z = 9,375$

11.2.3　Recognizing optimality

Aiming to improve the current solution, we analyze the step 2 dictionary (11.9):

$$z \;=\; 9,375 \;-\; \tfrac{25}{4}s_2 \;-\; \tfrac{15}{4}s_3$$

$$
\begin{aligned}
s_1 &= 25 - \tfrac{1}{4}s_2 + \tfrac{1}{4}s_3 \\
x_2 &= 300 - s_2 \\
x_1 &= 125 + \tfrac{5}{4}s_2 - \tfrac{1}{4}s_3 \\
s_4 &= 225 - \tfrac{5}{4}s_2 + \tfrac{1}{4}s_3.
\end{aligned}
$$

The objective function is given by

$$z = 9,375 - \frac{25}{4}s_2 - \frac{15}{4}s_3,$$

where both s_2 and s_3 are nonnegative. Since the coefficients for s_2 and s_3 are negative, it is clear that the highest possible value z^* of z is obtained by putting $s_2 = s_3 = 0$. Thus, the current basic feasible solution is optimal. When reporting the optimal solution, we can ignore the slack variables as they were not a part of the original LP we were solving. Thus, the optimal solution is given by

$$x_1^* = 125, \; x_2^* = 300, \; z^* = 9,375.$$

Recognize that this is the same solution as the one we obtained graphically in Section 10.4. Since a negative nonbasic variable coefficient in the dictionary format is positive in the tableau format and vice versa, a tableau is deemed optimal if all nonbasic variables have nonnegative coefficients in row 0.

> If in a feasible dictionary, all nonbasic variables have nonpositive coefficients in the z-row, then the corresponding basic feasible solution is an optimal solution of the LP.
>
> If we use the tableau format, then the basic feasible solution is optimal if all nonbasic variables have nonnegative coefficients in row 0 of the corresponding tableau.

11.2.4　Recognizing unbounded LPs

A step of the simplex method consists of selecting an entering variable, leaving variable, and updating the dictionary by performing a pivot. Any nonbasic variable with a positive coefficient in the z-row of the dictionary (or a negative coefficient in row 0 of the corresponding tableau) can be selected as the entering variable. If there is no such variable, it means that the current basic feasible solution is optimal, and the LP is solved. The leaving variable is the basic variable representing a row that wins the ratio test. However, if all coefficients in the pivot column of the dictionary are positive, the ratio test produces no result. For example, consider the following dictionary:

$$
\begin{array}{rrrrr}
z & = & 90 & - & 25x_1 & + & 4x_2 \\
s_1 & = & 25 & - & 14x_1 & + & x_2 \\
s_2 & = & 30 & - & x_1 \\
s_3 & = & 12 & + & 5x_1 & + & 14x_2 \\
s_4 & = & 22 & - & 4x_1 & + & 7x_2.
\end{array}
$$

The only entering variable candidate is x_2. However, when we try to do the ratio test, none of the rows participates. Because the coefficient for x_2 is nonnegative in each row, we can increase x_2 to $+\infty$ without violating any of the constraints. When x_2 increases to $+\infty$, so does z, thus the problem is unbounded. The tableau corresponding to the dictionary above is given by

z	x_1	x_2	s_1	s_2	s_3	s_4	rhs	$Basis$
1	25	-4	0	0	0	0	90	z
0	14	-1	1	0	0	0	25	s_1
0	1	0	0	1	0	0	30	s_2
0	-5	-14	0	0	1	0	12	s_3
0	4	-7	0	0	0	1	22	s_4

Thus, if we use the tableau format, we conclude that the problem is unbounded if at some point we obtain a tableau that has a column such that none of its entries are positive.

If during the execution of the simplex method we encounter a variable that has all nonnegative coefficients in the dictionary format, then the LP is unbounded.

In tableau format, an LP is proved to be unbounded as soon as a column with no positive entries is detected.

11.2.5 Degeneracy and cycling

When applied to the Heavenly Pouch LP, the simplex method produced a sequence of basic feasible solutions, such that each next solution had a strictly higher objective value than the previous one. However, this may not always be the case, as illustrated by the next example. Consider the following LP:

$$\text{maximize} \quad 5x_1 + 4x_2 - 20x_3 - 2x_4$$
$$\text{subject to} \quad \tfrac{1}{4}x_1 - \tfrac{1}{8}x_2 + 12x_3 + 10x_4 \le 0$$
$$\tfrac{1}{10}x_1 + \tfrac{1}{20}x_2 + \tfrac{1}{20}x_3 + \tfrac{1}{5}x_4 \le 0 \qquad (11.11)$$
$$x_1, x_2, x_3, x_4 \ge 0.$$

We apply the simplex method to this problem. We will use the tableau format. Let x_5 and x_6 be the slack variables for the first and second constraints, respectively. We always select the nonbasic variable with the most negative coefficient in row 0 as the entering variable. In case there are multiple ratio test winners, we select the basic variable with the lowest index as the leaving variable. The step 0 tableau is given by

z	x_1	x_2	x_3	x_4	x_5	x_6	rhs	$Basis$
1	-5	-4	20	2	0	0	0	z
0	$\boxed{\tfrac{1}{4}}$	$-\tfrac{1}{8}$	12	10	1	0	0	x_5
0	$\tfrac{1}{10}$	$\tfrac{1}{20}$	$\tfrac{1}{20}$	$\tfrac{1}{5}$	0	1	0	x_6

(11.12)

Note that both basic variables are equal to 0 in the starting basic feasible solution.

> **Definition 11.1** *Basic solutions with one or more basic variables equal to 0 are called degenerate.*

We carry out the first step. The entering variable is x_1, since it is the variable with the most negative coefficient in the z-row, and the leaving variable is x_5, which is the basic variable in the row winning the ratio test. The step 1 tableau is given by:

z	x_1	x_2	x_3	x_4	x_5	x_6	rhs	$Basis$
1	0	$-\tfrac{13}{2}$	260	202	20	0	0	z
0	1	$-\tfrac{1}{2}$	48	40	4	0	0	x_1
0	0	$\boxed{\tfrac{1}{10}}$	$-\tfrac{19}{4}$	$-\tfrac{19}{5}$	$-\tfrac{2}{5}$	1	0	x_6

(11.13)

The basic feasible solution is the same as at step 0, even though the basis has changed.

> **Definition 11.2** *An iteration of the simplex method, which results in a new basis with the basic feasible solution that is identical to the previous*

basic feasible solution is called a degenerate iteration and the corresponding phenomenon is referred to as degeneracy.

Continuing with the computations, we obtain the step 2 tableau:

z	x_1	x_2	x_3	x_4	x_5	x_6	rhs	$Basis$
1	0	0	$-\frac{195}{4}$	-45	-6	65	0	z
0	1	0	$\boxed{\frac{97}{4}}$	21	2	5	0	x_1
0	0	1	$-\frac{95}{2}$	-38	-4	10	0	x_2

(11.14)

step 3 tableau:

z	x_1	x_2	x_3	x_4	x_5	x_6	rhs	$Basis$
1	$\frac{195}{97}$	0	0	$-\frac{270}{97}$	$-\frac{192}{97}$	$\frac{7280}{97}$	0	z
0	$\frac{190}{97}$	1	0	$\boxed{\frac{304}{97}}$	$-\frac{8}{97}$	$\frac{1920}{97}$	0	x_2
0	$\frac{4}{97}$	0	1	$\frac{84}{97}$	$\frac{8}{97}$	$\frac{20}{97}$	0	x_3

(11.15)

step 4 tableau:

z	x_1	x_2	x_3	x_4	x_5	x_6	rhs	$Basis$
1	$\frac{15}{4}$	$\frac{135}{152}$	0	0	$-\frac{39}{19}$	$\frac{1760}{19}$	0	z
0	$-\frac{1}{2}$	$-\frac{21}{76}$	1	0	$\boxed{\frac{2}{19}}$	$-\frac{100}{19}$	0	x_3
0	$\frac{5}{8}$	$\frac{97}{304}$	0	1	$-\frac{1}{38}$	$\frac{120}{19}$	0	x_4

(11.16)

step 5 tableau:

z	x_1	x_2	x_3	x_4	x_5	x_6	rhs	$Basis$
1	-6	$-\frac{9}{2}$	$\frac{39}{2}$	0	0	-10	0	z
0	$\frac{1}{2}$	$\frac{1}{4}$	$\frac{1}{4}$	1	0	$\boxed{5}$	0	x_4
0	$-\frac{19}{4}$	$-\frac{21}{8}$	$\frac{19}{2}$	0	1	-50	0	x_5

(11.17)

step 6 tableau:

z	x_1	x_2	x_3	x_4	x_5	x_6	rhs	$Basis$
1	-5	-4	20	2	0	0	0	z
0	$\boxed{\frac{1}{4}}$	$-\frac{1}{8}$	12	10	1	0	0	x_5
0	$\frac{1}{10}$	$\frac{1}{20}$	$\frac{1}{20}$	$\frac{1}{5}$	0	1	0	x_6

(11.18)

The last tableau is exactly the same as the step 0 tableau (11.12). Thus, if we continue with the execution of the simplex method, we will keep repeating the calculations performed in steps 1–6 and will never be able to leave the same solution.

> **Definition 11.3** *A situation when the simplex method goes through a series of degenerate steps, and as a result, revisits a basis it encountered previously is called* cycling.

Several methods are available that guarantee that cycling is avoided. One of them is Bland's rule, which is discussed next. According to this rule, the variables are ordered in a certain way, for example, in the increasing order of their indices, i.e., $x_1, x_2, \ldots, x_{n+m}$. Then, whenever there are multiple candidates for the entering or the leaving variable, the preference is given to the variable that appears earlier in the ordering. All nonbasic variables with a positive coefficient in the z-row of the dictionary (or a negative coefficient in row 0 of the tableau) are candidates for the entering variable, and all the basic variables representing the rows that win the ratio test are candidates for the leaving variable.

> **Theorem 11.1** *If Bland's rule is used to select the entering and leaving variables in the simplex method, then cycling never occurs.*

To illustrate Bland's rule, we will apply it to the cycling example above. The first 5 steps are identical to the computations above, and we arrive at the following tableau:

z	x_1	x_2	x_3	x_4	x_5	x_6	rhs	$Basis$
1	-6	$-\frac{9}{2}$	$\frac{39}{2}$	0	0	-10	0	z
0	$\boxed{\frac{1}{2}}$	$\frac{1}{4}$	$\frac{1}{4}$	1	0	5	0	x_4
0	$-\frac{19}{4}$	$-\frac{21}{8}$	$\frac{19}{2}$	0	1	-50	0	x_5

(11.19)

The candidates for entering the basis are x_1, x_2, and x_6, thus, according to Bland's rule, x_1 is chosen as the entering variable.

z	x_1	x_2	x_3	x_4	x_5	x_6	rhs	$Basis$
1	0	$-\frac{3}{2}$	$\frac{45}{2}$	12	0	50	0	z
0	1	$\boxed{\frac{1}{2}}$	$\frac{1}{2}$	2	0	10	0	x_1
0	0	$-\frac{1}{4}$	$\frac{95}{8}$	$\frac{19}{2}$	1	$-\frac{5}{2}$	0	x_5

(11.20)

After one more step, we obtain an optimal tableau:

z	x_1	x_2	x_3	x_4	x_5	x_6	rhs	$Basis$
1	3	0	24	18	0	80	0	z
0	2	1	1	4	0	20	0	x_2
0	$\frac{1}{2}$	0	$\frac{97}{8}$	$\frac{21}{2}$	1	$\frac{5}{2}$	0	x_5

$$(11.21)$$

In fact, the solution has not changed compared to the basic feasible solution we had at step 0, however, the last tableau proves its optimality.

11.2.6 Properties of LP dictionaries and the simplex method

Note that the dictionary format we used to execute the simplex steps above is just a linear system in which m basic variables and z are expressed through n nonbasic variables, where n and m are the number of variables and constraints, respectively, in the original LP ($n = 2$ and $m = 4$ for the Heavenly Pouch LP). Namely, for the problem

$$\text{maximize} \quad \sum_{j=1}^{n} c_j x_j$$
$$\text{subject to} \quad \sum_{j=1}^{n} a_{ij} x_j \ \leq \ b_i, \quad i = 1, \ldots, m \qquad (11.22)$$
$$x_j \ \geq \ 0, \quad j = 1, \ldots, n,$$

where $b \geq 0$, we constructed the initial feasible dictionary as follows:

$$z = \sum_{j=1}^{n} c_j x_j$$
$$\rule{6cm}{0.4pt}$$
$$s_i = b_i - \sum_{j=1}^{n} a_{ij} x_j, \quad i = 1, \ldots, m. \qquad (11.23)$$

We first show that any two dictionaries with the same basis must be identical. Indeed, consider two dictionaries corresponding to the same basis. Let $\mathcal{B}$ be the set of indices of the basic variables and let $\mathcal{N}$ be the set of indices of nonbasic variables.

$$z = \bar{z} + \sum_{j \in \mathcal{N}} \bar{c}_j x_j \qquad\qquad z = \tilde{z} + \sum_{j \in \mathcal{N}} \tilde{c}_j x_j$$
$$x_i = \bar{b}_i - \sum_{j \in \mathcal{N}} \bar{a}_{ij} x_j, \quad i \in \mathcal{B} \qquad x_i = \tilde{b}_i - \sum_{j \in \mathcal{N}} \tilde{a}_{ij} x_j, \quad i \in \mathcal{B}.$$

For a nonbasic variable x_k, set $x_k = t > 0$, and for $j \in \mathcal{N}, j \neq k$, set $x_j = 0$. Then we have

$$\bar{b}_i - \bar{a}_{ik} t = \tilde{b}_i - \tilde{a}_{ik} t \text{ for all } i \in \mathcal{B}, \text{ and } \bar{z} + \bar{c}_k t = \tilde{z} + \tilde{c}_k t.$$

Since these identities hold for any t, we have

$$\bar{b}_i = \tilde{b}_i, \bar{a}_{ik} = \tilde{a}_{ik}, \bar{z} = \tilde{z}, \bar{c}_k = \tilde{c}_k \text{ for all } i \in \mathcal{B}, k \in \mathcal{N}.$$

Therefore, the following property holds.

> Any two dictionaries of the same LP with the same basis are identical.

The initial dictionary (11.23) is, essentially, a linear system that represents the original LP written in the standard form. The only transformations we apply to this linear system at each subsequent iteration of the simplex method are the elementary row operations used to express the new set of basic variables through the remaining variables. Since applying an elementary row operation to a linear system results in an equivalent linear system, we have the following property.

> Every solution of the set of equations comprising the dictionary obtained at any step of the simplex method is also a solution of the step 0 dictionary, and vice versa.

The ratio test we use to determine the leaving variable at each step is designed to ensure that the constant term in the right-hand side of each equation is nonnegative, so that setting all the nonbasic variables to 0 yields nonnegative values for all the basic variables and thus the corresponding basic solution is feasible. Thus, if we start with a feasible dictionary, feasibility is preserved throughout execution of the simplex method.

> If step 0 dictionary is feasible, then each consecutive dictionary generated using the simplex method is feasible.

The entering variable on each step of the simplex method is chosen so that each next basic feasible solution obtained during the simplex method execution is at least as good as the previous solution. However, we saw that the simplex method may go through some consecutive degenerate iterations with no change in the objective function value, in which case cycling can occur. It appears that this is the only case where the method may not terminate.

> **Theorem 11.2** *If the simplex method avoids cycling, it must terminate by either finding an optimal solution or by detecting that the LP is unbounded.*

Proof. There are only $\binom{n+m}{m}$ different ways of choosing a set of m basic variables from the set of $n+m$ variables. Since any two dictionaries corresponding to the same basis are identical, there can only be a finite number of the simplex method steps that are different. Therefore, if the simplex method does not terminate, it must eventually revisit a previously visited basis, meaning that cycling occurs. $\square$

We can use Bland's rule, discussed in Section 11.2.5, or other methods to make sure that cycling does not occur, and thus the simplex method terminates.

11.3 Geometry of the Simplex Method

We illustrate the simplex steps geometrically using the Heavenly Pouch LP formulated in Section 10.1, solved graphically in Section 10.4, and solved using the simplex method in Section 11.2. The Heavenly Pouch LP is given by:

$$
\begin{array}{lrcrcll}
\text{maximize} & 15x_1 & + & 25x_2 & & & \\
\text{subject to} & x_1 & + & x_2 & \leq & 450 & \text{(solid color fabric constraint)} \\
& & & x_2 & \leq & 300 & \text{(printed fabric constraint)} \\
& 4x_1 & + & 5x_2 & \leq & 2{,}000 & \text{(budget constraint)} \\
& x_1 & & & \leq & 350 & \text{(demand constraint)} \\
& & x_1, x_2 & \geq & 0 & & \text{(nonnegativity constraints)},
\end{array}
$$

and has the following representation in the standard form:

$$
\begin{array}{lrcrcrcll}
\text{maximize} & 15x_1 & + & 25x_2 & & & & & \\
\text{subject to} & x_1 & + & x_2 & + & s_1 & & = & 450 \\
& & & x_2 & + & s_2 & & = & 300 \\
& 4x_1 & + & 5x_2 & + & s_3 & & = & 2{,}000 \\
& x_1 & & & + & s_4 & & = & 350 \\
& & & x_1, x_2, s_1, s_2, s_3, s_4 & \geq & 0. & & &
\end{array}
\tag{11.24}
$$

The LP is solved graphically in Figure 11.1. Also, solving this LP with the simplex method produced the following basic feasible solutions:

Step	Basic variables	Basic feasible solution
0	s_1, s_2, s_3, s_4	$x_1 = x_2 = 0, s_1 = 450, s_2 = 300, s_3 = 2{,}000, s_4 = 350$
1	s_1, x_2, s_3, s_4	$x_1 = 0, x_2 = 300, s_1 = 150, s_2 = 0, s_3 = 500, s_4 = 350$
2	x_1, x_2, s_1, s_4	$x_1 = 125, x_2 = 300, s_1 = 25, s_2 = 0, s_3 = 0, s_4 = 225$

These basic feasible solutions are represented by extreme points A, B, and C of the feasible region (see Figure 11.1). Note that each pair of consecutive

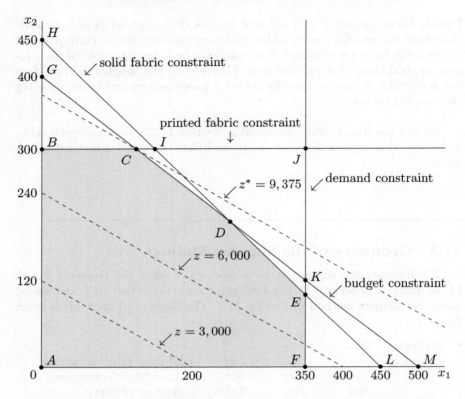

FIGURE 11.1: Graphical solution of the Heavenly Pouch LP.

basic feasible solutions generated by the simplex method, which have all but one basic variables in common, represent vertices of the polyhedron that are connected by an edge of the feasible region.

> **Definition 11.4** *We will call two basic feasible solutions* adjacent *if their sets of basic variables differ by just one element.*

In other words, if the LP has m constraints, then adjacent basic feasible solutions have $m - 1$ variables in common. Using this terminology, at any given step the simplex method moves from the current basic feasible solution to an adjacent basic feasible solution. Next, we will show that any basic feasible solution represents an extreme point (vertex) of the feasible region, which is exactly what we observed in our example. We consider the feasible region $X = \{x : Ax = b, x \geq 0\}$ of an LP in the standard form.

Theorem 11.3 *A point $\bar{x} \in \mathbb{R}^n$ is an extreme point of $X = \{x : Ax = b, x \geq 0\}$ if and only if it can be represented as a basic feasible solution of an LP with the feasible region given by X.*

Proof. Let $\bar{x}$ be an extreme point of X. Let $\mathcal{J}^+ = \{j : \bar{x}_j > 0\}$ be the set of indices that represent positive components of $\bar{x}$. We construct a basic feasible solution for an arbitrary LP with the feasible region X as follows. We need to select a set of m basic variables $\{x_j : j \in \mathcal{B}\}$, where $\mathcal{B}$ is an m-element subset of $\{1, \ldots, m + n\}$. If $\bar{x}$ has no more than m positive components, we select an arbitrary $\mathcal{B}$ such that $\mathcal{J}^+ \subseteq \mathcal{B}$. Then, since $\bar{x} \in X$, $\bar{x}$ is a basic feasible solution for the corresponding dictionary. If, on the other hand, $\bar{x}$ has more than m positive components, we select $\mathcal{B} \subset \mathcal{J}^+$, and consider the dictionary that has $\{x_j : j \in \mathcal{B}\}$ as the basic variables. Then $\bar{x}$ satisfies the system given by the dictionary, where at least one of the nonbasic variables, $\bar{x}_k (k \notin \mathcal{B})$, is positive. If we reduce or increase the value of $\bar{x}_k$ by $\epsilon > 0$ and keep the other nonbasic variable values unchanged, we can always select ϵ small enough for the basic variables to stay positive after this modification. Let $\bar{x}^-$ and $\bar{x}^+$ be the feasible solutions obtained from $\bar{x}$ by setting $\bar{x}_k^- = \bar{x}_k - \epsilon, \bar{x}_k^+ = \bar{x}_k + \epsilon$ for sufficiently small ϵ, while keeping the remaining $\bar{x}_j, j \notin \mathcal{B}$ unchanged. Then, obviously $\bar{x}^-, \bar{x}^+ \in X, \bar{x}^- \neq \bar{x}^+$ and $\bar{x} = 0.5\bar{x}^- + 0.5\bar{x}^+$, which contradicts the assumption that $\bar{x}$ is an extreme point of X.

Now, assume that $\bar{x}$ is a basic feasible solution of an arbitrary LP with the feasible region X that is not an extreme point of X. Then there exist $\tilde{x}, \hat{x} \in X, \tilde{x} \neq \hat{x}$, and $\alpha \in (0, 1)$ such that

$$\bar{x} = \alpha\tilde{x} + (1 - \alpha)\hat{x}.$$

Let $\{x_j : j \in \mathcal{B}\}$ be the set of basic variables, then for $\tilde{x}$ to be different from $\bar{x}$, we must have $\tilde{x}_k > 0$ for at least one $k \notin \mathcal{B}$ (otherwise, if all nonbasic variables are set to 0, the remaining variables must have the same values as they have in $\bar{x}$, meaning that $\tilde{x} = \bar{x}$). From the equation above we have $\bar{x}_k = \alpha\tilde{x}_k + (1 - \alpha)\hat{x}_k$, where $\bar{x}_k = 0$ and $\tilde{x}_k > 0$. Then $\hat{x}_k$ must be negative, contradicting the feasibility of $\hat{x}$. □

Returning to our example, observe that the feasible region in Figure 11.1 has 6 extreme points, A, B, C, D, E, and F. The correspondence between these extreme points and basic feasible solutions of the LP is shown in Table 11.1.

Recall that any basic (not necessarily feasible) solution is obtained by setting the nonbasic variables to 0. In our two-dimensional case, we have two nonbasic variables. If one of the original variables is 0, then the corresponding basic solution lies on the line defining the corresponding coordinate axis, and if a slack variable is 0 then the constraint it represents is binding for the corresponding basic solution. Thus, basic solutions correspond to pairs of lines defining the feasible region (including the two coordinate axes). The total number of basic solutions that we may potentially have in our example

TABLE 11.1: The correspondence between extreme points of the feasible region, as shown in Figure 11.1, and basic feasible solutions of the LP.

| Basic | Basic feasible solution | | | | | | Extreme |
variables	x_1	x_2	s_1	s_2	s_3	s_4	point
s_1, s_2, s_3, s_4	0	0	450	300	2,000	350	A
x_2, s_1, s_3, s_4	0	300	150	0	500	350	B
x_1, x_2, s_1, s_4	125	300	25	0	0	225	C
x_1, x_2, s_2, s_4	250	200	0	100	0	100	D
x_1, x_2, s_2, s_3	350	100	0	200	100	0	E
x_1, s_1, s_2, s_3	350	0	100	300	600	0	F

is $\binom{2+4}{4} = 15$, however, not every set of 4 variables may form a basis. For example, the basis consisting of variables x_1, s_1, s_3, and s_4 is not possible, since this would imply that $x_2 = 0$ and $s_2 = 300 - x_2 = 0$ at the same time, which is impossible. Geometrically, this corresponds to parallel lines defining a pair of constraints (in our case, the line for the printed fabric constraint, $x_2 = 300$, is parallel to the line $x_2 = 0$), meaning that both constraints cannot be binding at the same time. In our example, x_1 and s_4 cannot be nonbasic simultaneously as well, since the demand constraint line is parallel to the x_2-axis. Excluding these two cases, there are $15 - 2 = 13$ potential basic solutions. As we already discussed, six of them (A, B, C, D, E, F) are basic feasible solutions, and we can see that the remaining basic solutions, which lie on pairwise intersections of lines defining the constraints (including nonnegativity), are infeasible (points G, H, I, J, K, L, and M in Figure 11.1). Establishing the correspondence between these points and basic solutions is left as an exercise (Exercise 11.4 at page 278).

In summary, from a geometric viewpoint, the simplex method starts from one of the vertices of the feasible region and proceeds by moving to a better (or, at least, no worse) adjacent vertex, whenever one exists. The method terminates at a vertex that has no adjacent vertex with a better objective.

11.4   The Simplex Method for a General LP

Assume that there is an additional minimum manufacturing order constraint in the Heavenly Pouch problem formulated in Section 10.1, which requires that at least 100 carriers are made. After adding the constraint, which can be expressed as $x_1 + x_2 \geq 100$, we obtain the following LP:

$$\begin{array}{llll}
\text{maximize} & 15x_1 + 25x_2 & & \\
\text{subject to} & x_1 + x_2 & \leq 450 & \text{(solid color fabric constraint)} \\
& x_2 & \leq 300 & \text{(printed fabric constraint)} \\
& 4x_1 + 5x_2 & \leq 2,000 & \text{(budget constraint)} \\
& x_1 & \leq 350 & \text{(demand constraint)} \\
& x_1 + x_2 & \geq 100 & \text{(manufacturing constraint)} \\
& x_1, x_2 & \geq 0 & \text{(nonnegativity constraints)}.
\end{array}$$

As before, we first convert the LP to the standard form by introducing the slack variables s_i, $i = 1, \ldots, 4$ for the first four constraints, and an excess variable e_5 for the fifth constraint.

$$\begin{array}{llr}
\text{maximize} & 15x_1 + 25x_2 & \\
\text{subject to} & x_1 + x_2 + s_1 & = 450 \\
& x_2 + s_2 & = 300 \\
& 4x_1 + 5x_2 + s_3 & = 2,000 \quad (11.25) \\
& x_1 + s_4 & = 350 \\
& x_1 + x_2 - e_5 & = 100 \\
& x_1, x_2, s_1, s_2, s_3, s_4, e_5 & \geq 0.
\end{array}$$

The corresponding dictionary,

$$\begin{array}{rrlll}
z = & & 15x_1 & + & 25x_2 \\
\hline
s_1 = & 450 & - x_1 & - & x_2 \\
s_2 = & 300 & & - & x_2 \\
s_3 = & 2,000 & - 4x_1 & - & 5x_2 \\
s_4 = & 350 & - x_1 & & \\
e_5 = & -100 & + x_1 & + & x_2
\end{array} \quad (11.26)$$

is clearly infeasible, since setting $x_1 = x_2 = 0$ results in $e_5 = -100 < 0$, and the excess variable e_5 cannot be used as the basic variable for the fifth constraint. Therefore we cannot initialize the simplex method with this dictionary.

Next, we will discuss two variants of the simplex method designed to overcome this obstacle, the *big-M method* and the *two-phase simplex method*. Both methods are based on a similar idea of introducing an *artificial variable* for each constraint where the starting basic variable (such as the slack or excess variable) is not readily available. For example, the fifth constraint in the last LP above,

$$x_1 + x_2 - e_5 = 100,$$

could be modified to include an artificial variable $a_5 \geq 0$ as follows:

$$x_1 + x_2 - e_5 + a_5 = 100.$$

With this modification, we can use a_5 as the starting basic variable for this constraint, thus obtaining a basic feasible solution for the resulting LP. However, to get a feasible solution for the original problem (if there is one), we need to make sure that in the end the artificial variable $a_5 = 0$. The methods we are about to discuss utilize alternative ways of driving the artificial variables out of the basis, thus ensuring that they all eventually vanish whenever the LP is feasible. We first discuss the general setup for both methods and then proceed to describe each of them in more detail.

Consider a general LP

$$\begin{aligned}
\text{maximize} \quad & \sum_{j=1}^{n} c_j x_j \\
\text{subject to} \quad & \sum_{j=1}^{n} a'_{ij} x_j \leq b'_i, \quad i = 1, \ldots, m' \\
& \sum_{j=1}^{n} a''_{ij} x_j = b''_i, \quad i = 1, \ldots, m'' \\
& x_j \geq 0, \quad j = 1, \ldots, n.
\end{aligned} \quad \text{(P)}$$

After introducing m' slack variables $x_{n+1}, \ldots, x_{n+m'}$ and updating the problem coefficients accordingly, we can write this LP in the standard form as follows:

$$\begin{aligned}
\text{maximize} \quad & \sum_{j=1}^{n+m'} c_j x_j \\
\text{subject to} \quad & \sum_{j=1}^{n+m'} a_{ij} x_j = b_i, \quad i = 1, \ldots, m \\
& x_j \geq 0, \quad j = 1, \ldots, n+m',
\end{aligned} \quad \text{(PS)}$$

where $m = m' + m''$, indices $i = m' + 1, \ldots, m$ correspond to the original equality constraints, and $b_i \geq 0$ for all $i = m' + 1, \ldots, m$ (this can be easily guaranteed by multiplying both sides of the original equality constraints by -1 whenever $b''_i < 0$, $i = 1, \ldots, m''$).

Let $\mathcal{I}^- = \{i \in \{1, \ldots, m'\} : b_i < 0\}$ be the set of indices that correspond to the original inequality constraints that have a negative right-hand side in (PS). Each of these constraints will require introducing an artificial variable in order to obtain a starting basic feasible solution. In addition, we will add an artificial variable to each of the last m'' constraints (i.e., the equations corresponding to the original equality constraints in (P)), which will serve as the basic variables for these constraints in the initial dictionary. For convenience, we denote by

$$\mathcal{I}^a = \mathcal{I}^- \cup \{m' + 1, \ldots, m\}$$

the set of indices of all constraints, both inequality and equality, that require an artificial variable to initialize the simplex method.

We will associate the following two problems, which share the same feasible region, with LP (P) (and (PS)):

minimize $\quad \sum_{i \in \mathcal{I}^a} a_i$

subject to $\quad \sum_{j=1}^{n+m'} a_{ij} x_j \qquad = \quad b_i, \quad i \in \{1, \ldots, m'\} \setminus \mathcal{I}^a$

$$\sum_{j=1}^{n+m'} a_{ij} x_j \ + \ a_i \ = \ b_i, \quad i \in \mathcal{I}^a \tag{A}$$

$$x_j, a_i \ \geq \ 0, \quad j = 1, \ldots, n + m', i \in \mathcal{I}^a,$$

maximize $\quad \sum_{j=1}^{n+m'} c_j x_j - \sum_{i \in \mathcal{I}^a} M a_i$

subject to $\quad \sum_{j=1}^{n+m'} a_{ij} x_j \qquad = \quad b_i, \quad i \in \{1, \ldots, m'\} \setminus \mathcal{I}^a$

$$\sum_{j=1}^{n+m'} a_{ij} x_j \ + \ a_i \ = \ b_i, \quad i \in \mathcal{I}^a \tag{B}$$

$$x_j, a_i \ \geq \ 0, \quad j = 1, \ldots, n + m', i \in \mathcal{I}^a,$$

where M is some sufficiently large positive constant referred to as the "*big M.*" Problem (A) is called the *auxiliary problem* and problem (B) is the *big-M problem* associated with (P).

Example 11.3 *Consider the following LP:*

$$\begin{aligned}
maximize \quad & x_1 - 2x_2 + 3x_3 \\
subject\ to \quad & -2x_1 + 3x_2 + 4x_3 \ \geq \ 12 \\
& 3x_1 + 2x_2 + x_3 \ \geq \ 6 \\
& x_1 + x_2 + x_3 \ \leq \ 9 \\
& x_1, x_2, x_3 \ \geq \ 0.
\end{aligned} \tag{11.27}$$

This LP in the standard form is given by:

$$\begin{aligned}
maximize \quad & x_1 - 2x_2 + 3x_3 \\
subject\ to \quad & -2x_1 + 3x_2 + 4x_3 - x_4 \qquad\qquad\quad = 12 \\
& 3x_1 + 2x_2 + x_3 \qquad - x_5 \qquad\quad = 6 \\
& x_1 + x_2 + x_3 \qquad\qquad + x_6 = 9 \\
& x_1, x_2, x_3, x_4, x_5, x_6 \ \geq \ 0.
\end{aligned} \tag{11.28}$$

Clearly, the basic solution with the basis consisting of $x_4, x_5,$ and x_6 is infeasible since it has negative values for x_4 and x_5 ($x_4 = -12$ and $x_5 = -6$). Hence, we introduce artificial variables for the first two constraints. Then the corresponding auxiliary problem can be written as a maximization problem as follows:

$$\begin{aligned}
maximize \quad & \qquad\qquad\qquad\qquad\qquad\qquad - a_1 - a_2 \\
subject\ to \quad & -2x_1 + 3x_2 + 4x_3 - x_4 \qquad + a_1 \qquad\; = 12 \\
& 3x_1 + 2x_2 + x_3 \qquad - x_5 \qquad + a_2 = 6 \\
& x_1 + x_2 + x_3 \qquad\qquad + x_6 \qquad\qquad = 9 \\
& x_1, x_2, x_3, x_4, x_5, x_6, a_1, a_2 \ \geq \ 0,
\end{aligned} \tag{11.29}$$

and the big-M problem associated with LP (11.27) *is given by:*

$$
\begin{array}{lrcrcrcrcrcrcl}
\text{maximize} & x_1 & - & 2x_2 & + & 3x_3 & & & & & - & Ma_1 & - & Ma_2 & & \\
\text{subject to} & -2x_1 & + & 3x_2 & + & 4x_3 & - & x_4 & + & & & a_1 & & & = & 12 \\
& 3x_1 & + & 2x_2 & + & x_3 & & & - & x_5 & & & + & a_2 & = & 6 \\
& x_1 & + & x_2 & + & x_3 & & & & & + & x_6 & & & = & 9 \\
& & & & & & & x_1, x_2, x_3, x_4, x_5, x_6, a_1, a_2 & & & & & & \geq & 0. &
\end{array}
\qquad (11.30)
$$

Note that we can easily obtain a basic feasible solution for both (A) and (B) by selecting the basis consisting of the slack variables $x_{n+i}, i \in \{1, \ldots, m'\} \setminus \mathcal{I}^a$ for rows where artificial variables were not needed, and the artificial variables $a_i, i \in \mathcal{I}^a$ for the remaining rows. Also, the objective function of the auxiliary problem (A) is always nonnegative and thus any feasible solution of this problem with $a_i = 0, i \in \mathcal{I}^a$ is optimal.

In addition, observe that if a feasible solution of (B) with $a_i > 0$ for at least one $i \in \mathcal{I}^a$ exists, then the objective function value can be made arbitrarily poor (i.e., very large, negative) by selecting a sufficiently large constant $M > 0$.

Theorem 11.4 *The following properties hold for LP* (P) *and the associated auxiliary problem* (A) *and big-M problem* (B):

1. *LP* (P) *is feasible if and only if the optimal objective function value of the associated auxiliary problem* (A) *is 0.*

2. *LP* (P) *is optimal and a vector x^* is an optimal solution to its standard form LP* (PS) *if and only if there exists a positive real constant M such that $x = x^*$, $a_i = 0, i \in \mathcal{I}^a$ is an optimal solution of the associated big-M problem* (B).

3. *LP* (P) *is unbounded if and only if the associated big-M problem* (B) *is unbounded for any $M > 0$.*

Proof: 1. If LP (P) is feasible, then any feasible solution x to (P) and $a_i = 0, i \in \mathcal{I}^a$ provide an optimal solution for (A). Also, if 0 is not the optimal objective function value for (A), then (A) does not have a feasible solution with $a_i = 0, i \in \mathcal{I}^a$, implying that LP (P) is infeasible.

2. If LP (P) is optimal and a vector x^* is an optimal solution to its standard form LP (PS), then $x = x^*$, $a_i = 0, i \in \mathcal{I}^a$ give an optimal solution to (B). Indeed, if one of the artificial variables is positive then taking $M \to +\infty$, the objective function of (B) will tend to $-\infty$. If, on the other hand, there exists M such that $x = x^*$, $a_i = 0, i \in \mathcal{I}^a$ is an optimal solution of (B), then x^* is an optimal solution for (P), since if there was a better solution $\tilde{x}$ for (P), taking $x = \tilde{x}$, $a_i = 0, i \in \mathcal{I}^a$ would give a better than optimal solution for (B), which is impossible.

3. If (P) is unbounded, then setting all the artificial variables to 0 shows that (B) is also unbounded for any $M > 0$. If (B) is unbounded for any

$M > 0$, then taking $M \to 0$ shows that (P) is unbounded. □

The last theorem provides foundations for the two-phase simplex and the big-M methods discussed next.

11.4.1 The two-phase simplex method

The two-phase simplex method consists of the following two phases:

- Phase I: Solve the auxiliary problem (A), and, as a result, either obtain a feasible tableau for the original problem (P) (if the optimal objective value is 0), or conclude that the problem is infeasible (if the optimal objective value is positive).

- Phase II: If the problem was not judged infeasible in Phase I, solve the original problem using the optimal tableau of the auxiliary problem (A) to get the starting tableau for the original LP (P).

To get a feasible tableau for (P) from an optimal tableau for (A) in Phase II, we need to get rid of the artificial variables. If all artificial variables are non-basic in the optimal tableau for (A), we just drop the corresponding columns and express the objective of (P) through nonbasic variables to obtain a feasible tableau for (P). If some of the auxiliary variables are basic in the obtained optimal solution for (P), the solution must be degenerate since the basic artificial variables are equal to 0, and we attempt to drive them out of the basis by performing additional degenerate pivots. This is always possible, unless there is a basic artificial variable in the optimal tableau such that its corresponding row does not have nonzero coefficients in columns other than artificial variable columns, in which case we can just remove the corresponding row. This can be the case only when the original LP in the standard form has linearly dependent constraints, as will be illustrated in Example 11.6. We consider other examples first.

Example 11.4 *Use the two-phase simplex method to solve LP (11.27) from Example 11.3:*

$$
\begin{array}{rrrrrr}
maximize & x_1 & - & 2x_2 & + & 3x_3 \\
subject\ to & -2x_1 & + & 3x_2 & + & 4x_3 & \geq & 12 \\
& 3x_1 & + & 2x_2 & + & x_3 & \geq & 6 \\
& x_1 & + & x_2 & + & x_3 & \leq & 9 \\
& & & & & x_1, x_2, x_3 & \geq & 0.
\end{array}
$$

The corresponding auxiliary problem was presented in (11.29) and is given by (here we use e_1, e_2, and s_3 instead of x_4, x_5, and x_6 for convenience):

$$
\begin{array}{llll}
\text{maximize} & & -a_1 - a_2 \\
\text{subject to} & -2x_1 + 3x_2 + 4x_3 - e_1 \quad\quad + a_1 \quad\quad = 12 \\
& 3x_1 + 2x_2 + x_3 \quad\quad - e_2 \quad\quad + a_2 = 6 \\
& x_1 + x_2 + x_3 \quad\quad\quad\quad + s_3 \quad\quad\quad = 9 \\
& x_1, x_2, x_3, e_1, e_2, s_3, a_1, a_2 \geq 0.
\end{array}
$$

We solve the Phase I LP using the simplex method in the tableau format. Expressing the objective through nonbasic variables,

$$
\begin{aligned}
z &= -a_1 - a_2 \\
&= -(12 + 2x_1 - 3x_2 - 4x_3 + e_1) - (6 - 3x_1 - 2x_2 - x_3 + e_2) \\
&= -18 + x_1 + 5x_2 + 5x_3 - e_1 - e_2,
\end{aligned}
$$

we obtain the following step 0 tableau:

z	x_1	x_2	x_3	e_1	e_2	s_3	a_1	a_2	rhs	$Basis$
1	-1	-5	-5	1	1	0	0	0	-18	z
0	-2	3	4	-1	0	0	1	0	12	a_1
0	3	$\boxed{2}$	1	0	-1	0	0	1	6	a_2
0	1	1	1	0	0	1	0	0	9	s_3

Step 1 tableau:

z	x_1	x_2	x_3	e_1	e_2	s_3	a_1	a_2	rhs	$Basis$
1	$\frac{13}{2}$	0	$-\frac{5}{2}$	1	$-\frac{3}{2}$	0	0	$\frac{5}{2}$	-3	z
0	$-\frac{13}{2}$	0	$\boxed{\frac{5}{2}}$	-1	$\frac{3}{2}$	0	1	$-\frac{3}{2}$	3	a_1
0	$\frac{3}{2}$	1	$\frac{1}{2}$	0	$-\frac{1}{2}$	0	0	$\frac{1}{2}$	3	x_2
0	$-\frac{1}{2}$	0	$\frac{1}{2}$	0	$\frac{1}{2}$	1	0	$-\frac{1}{2}$	6	s_3

Step 2 tableau:

z	x_1	x_2	x_3	e_1	e_2	s_3	a_1	a_2	rhs	$Basis$
1	0	0	0	0	0	0	1	1	0	z
0	$-\frac{13}{5}$	0	1	$-\frac{2}{5}$	$\frac{3}{5}$	0	$\frac{2}{5}$	$-\frac{3}{5}$	$\frac{6}{5}$	x_3
0	$\frac{14}{5}$	1	0	$\frac{1}{5}$	$-\frac{4}{5}$	0	$-\frac{1}{5}$	$\frac{4}{5}$	$\frac{12}{5}$	x_2
0	$\frac{4}{5}$	0	0	$\frac{1}{5}$	$\frac{1}{5}$	1	$-\frac{1}{5}$	$-\frac{1}{5}$	$\frac{27}{5}$	s_3

This tableau is optimal. Since the optimal objective value is 0, the original LP is feasible. To obtain a feasible tableau for the original LP, we drop the columns for a_1 and a_2 in the tableau above and replace the basic variables x_2 and x_3 in the objective function of the original LP,

$$z = x_1 - 2x_2 + 3x_3,$$

with their expressions through the nonbasic variables from the optimal Phase I tableau,

$$x_2 = \tfrac{12}{5} - \tfrac{14}{5}x_1 - \tfrac{1}{5}e_1 + \tfrac{4}{5}e_2;$$
$$x_3 = \tfrac{6}{5} + \tfrac{13}{5}x_1 + \tfrac{2}{5}e_1 - \tfrac{3}{5}e_2.$$

We obtain

$$z = x_1 - 2x_2 + 3x_3 = -\tfrac{6}{5} + \tfrac{72}{5}x_1 + \tfrac{8}{5}e_1 - \tfrac{17}{5}e_2.$$

Hence, the initial phase II tableau is given by

z	x_1	x_2	x_3	e_1	e_2	s_3	rhs	Basis
1	$-\tfrac{72}{5}$	0	0	$-\tfrac{8}{5}$	$\tfrac{17}{5}$	0	$-\tfrac{6}{5}$	z
0	$-\tfrac{13}{5}$	0	1	$-\tfrac{2}{5}$	$\tfrac{3}{5}$	0	$\tfrac{6}{5}$	x_3
0	$\boxed{\tfrac{14}{5}}$	1	0	$\tfrac{1}{5}$	$-\tfrac{4}{5}$	0	$\tfrac{12}{5}$	x_2
0	$\tfrac{4}{5}$	0	0	$\tfrac{1}{5}$	$\tfrac{1}{5}$	1	$\tfrac{27}{5}$	s_3

Step 1 (phase II) tableau:

z	x_1	x_2	x_3	e_1	e_2	s_3	rhs	Basis
1	0	$\tfrac{36}{7}$	0	$-\tfrac{4}{7}$	$-\tfrac{5}{7}$	0	$\tfrac{78}{7}$	z
0	0	$\tfrac{13}{14}$	1	$-\tfrac{3}{14}$	$-\tfrac{1}{7}$	0	$\tfrac{24}{7}$	x_3
0	1	$\tfrac{5}{14}$	0	$\tfrac{1}{14}$	$-\tfrac{2}{7}$	0	$\tfrac{6}{7}$	x_1
0	0	$-\tfrac{2}{7}$	0	$\tfrac{1}{7}$	$\boxed{\tfrac{3}{7}}$	1	$\tfrac{33}{7}$	s_3

Step 2 (phase II) tableau:

z	x_1	x_2	x_3	e_1	e_2	s_3	rhs	Basis
1	0	$\tfrac{14}{3}$	0	$-\tfrac{1}{3}$	0	$\tfrac{5}{3}$	19	z
0	0	$\tfrac{5}{6}$	1	$-\tfrac{1}{6}$	0	$\tfrac{1}{3}$	5	x_3
0	1	$\tfrac{1}{6}$	0	$\boxed{\tfrac{1}{6}}$	0	$\tfrac{2}{3}$	4	x_1
0	0	$-\tfrac{2}{3}$	0	$\tfrac{1}{3}$	1	$\tfrac{7}{3}$	11	e_2

Step 3 (phase II) tableau:

z	x_1	x_2	x_3	e_1	e_2	s_3	rhs	$Basis$
1	2	5	0	0	0	3	27	z
0	1	1	1	0	0	1	9	x_3
0	6	1	0	1	0	4	24	e_1
0	−2	−1	0	0	1	1	3	e_2

This tableau is optimal, so the optimal solution to the original LP is

$$x_1^* = x_2^* = 0, x_3^* = 9, \ z^* = 27.$$

Example 11.5 *Consider the following LP:*

$$
\begin{aligned}
maximize \quad & x_1 + x_2 + x_3 \\
subject\ to \quad & 2x_1 + 2x_2 + 3x_3 = 6 \\
& x_1 + 3x_2 + 6x_3 = 12.
\end{aligned}
$$

Use the two-phase simplex method to solve this LP.

We introduce artificial variables a_1 and a_2 and write down the phase I starting tableau:

z	x_1	x_2	x_3	a_1	a_2	rhs	$Basis$
1	−3	−5	−9	0	0	−18	z
0	2	2	$\boxed{3}$	1	0	6	a_1
0	1	3	6	0	1	12	a_2

After one pivot, we obtain an optimal tableau:

z	x_1	x_2	x_3	a_1	a_2	rhs	$Basis$
1	3	1	0	3	0	0	z
0	$\frac{2}{3}$	$\frac{2}{3}$	1	$\frac{1}{3}$	0	2	x_3
0	−3	−1	0	−2	1	0	a_2

The corresponding basic feasible solution is degenerate, and the optimality will not be altered if we make an additional pivot with x_2 as the entering variable and a_2 as the leaving variable:

z	x_1	x_2	x_3	a_1	a_2	rhs	$Basis$
1	0	0	0	1	1	0	z
0	$-\frac{4}{3}$	0	1	−1	$\frac{2}{3}$	2	x_3
0	3	1	0	2	−1	0	x_2

Removing the artificial variables and expressing the objective in terms of the nonbasic variable x_1, we obtain the following phase II initial tableau:

z	x_1	x_2	x_3	rhs	$Basis$
1	$\frac{2}{3}$	0	0	2	z
0	$-\frac{4}{3}$	0	1	2	x_3
0	3	1	0	0	x_2

This tableau happens to be optimal, with the optimal solution given by

$$x_1^* = x_2^* = 0, x_3^* = 2, \ z^* = 2.$$

Example 11.6 *Consider the following LP with linearly dependent constraints:*

$$\begin{aligned} \text{maximize} \quad & x_1 + x_2 + 2x_3 \\ \text{subject to} \quad & x_1 + 2x_2 + 3x_3 = 6 \\ & 2x_1 + 4x_2 + 6x_3 = 12. \end{aligned}$$

Solve the problem using the two-phase simplex method.

We introduce artificial variables a_1 and a_2 and write down the phase I starting tableau:

z	x_1	x_2	x_3	a_1	a_2	rhs	$Basis$
1	-3	-6	-9	0	0	-18	z
0	1	2	$\boxed{3}$	1	0	6	a_1
0	2	4	6	0	1	12	a_2

After one pivot, we obtain an optimal phase I tableau:

z	x_1	x_2	x_3	a_1	a_2	rhs	$Basis$
1	0	0	0	3	0	0	z
0	$\frac{1}{3}$	$\frac{2}{3}$	1	$\frac{1}{3}$	0	2	x_3
0	0	0	0	-2	1	0	a_2

Again, the corresponding basic optimal solution is degenerate, but this time we cannot make both a_1 and a_2 nonbasic at the same time like we did in the previous example. However, we can see that the second row does not involve the original variables. Hence, if we remove this row along with the column for a_1 and express the objective function through the nonbasic variables,

$$z = x_1 + x_2 + 2x_3 = x_1 + x_2 + 4 - \frac{2}{3}x_1 - \frac{4}{3}x_2 = 4 + \frac{1}{3}x_1 - \frac{1}{3}x_2,$$

we obtain a feasible tableau for the original problem:

z	x_1	x_2	x_3	rhs	$Basis$
1	$-\frac{1}{3}$	$\frac{1}{3}$	0	4	z
0	$\boxed{\frac{1}{3}}$	$\frac{2}{3}$	1	2	x_3

Carrying out a pivot, we obtain:

z	x_1	x_2	x_3	rhs	Basis
1	0	1	1	6	z
0	1	2	3	6	x_1

This tableau is optimal, with the corresponding basic optimal solution given by

$$x_1^* = 6, x_2^* = x_3^* = 0, \ z^* = 6.$$

11.4.2 The big-M method

Unlike the two-phase simplex method, the big-M method consists of just one phase that solves the big-M problem (B) in order to solve (P). Note that there is something unusual about the big-M problem (B)–it involves a non-numerical parameter M as the objective function coefficient for the artificial variables. When performing a pivot, the parameter M is treated as $M \to +\infty$. Thus, when we compare two nonbasic variable coefficients involving M, the coefficient with a greater multiplier for M is considered greater. For example, $3M - 200$ is greater than $2M + 500$ for a sufficiently large M.

After solving the big-M LP (B), we check if all artificial variables are equal to 0 in the optimal solution. If they are, the corresponding solution for x is optimal for (P). If there is a positive artificial variable in the optimal solution for (B), then the original LP (P) is infeasible.

Example 11.7 *Use the big-M method to solve LP* (11.27) *from Example 11.3:*

$$
\begin{array}{rrrrcl}
maximize & x_1 & - & 2x_2 & + & 3x_3 \\
subject\ to & -2x_1 & + & 3x_2 & + & 4x_3 & \geq & 12 \\
& 3x_1 & + & 2x_2 & + & x_3 & \geq & 6 \\
& x_1 & + & x_2 & + & x_3 & \leq & 9 \\
& & & & & x_1, x_2, x_3 & \geq & 0.
\end{array}
$$

The corresponding big-M problem is given by:

$$
\begin{array}{rl}
maximize & x_1 - 2x_2 + 3x_3 \qquad\qquad\qquad - Ma_1 - Ma_2 \\
subject\ to & -2x_1 + 3x_2 + 4x_3 - e_1 \qquad\quad + a_1 \qquad\qquad = 12 \\
& 3x_1 + 2x_2 + x_3 \qquad - e_2 \qquad\qquad + a_2 = 6 \\
& x_1 + x_2 + x_3 \qquad\qquad + s_3 \qquad\qquad = 9 \\
& x_1, x_2, x_3, e_1, e_2, s_3, a_1, a_2 \geq 0.
\end{array}
$$

Expressing the artificial variables in row 0 through the nonbasic variables, we obtain the following step 0 tableau:

z	x_1	x_2	x_3	e_1	e_2	s_3	a_1	a_2	rhs	$Basis$
1	$-M-1$	$-5M+2$	$-5M-3$	M	M	0	0	0	$-18M$	z
0	-2	3	$\boxed{4}$	-1	0	0	1	0	12	a_1
0	3	2	1	0	-1	0	0	1	6	a_2
0	1	1	1	0	0	1	0	0	9	s_3

Step 1 tableau:

z	x_1	x_2	x_3	e_1	e_2	s_3	a_1	a_2	rhs	$Basis$
1	$-\frac{7M+5}{2}$	$-\frac{5M-17}{4}$	0	$-\frac{M+3}{4}$	M	0	$\frac{5M+3}{4}$	0	$-3M+9$	z
0	$-\frac{1}{2}$	$\frac{3}{4}$	1	$-\frac{1}{4}$	0	0	$\frac{1}{4}$	0	3	x_3
0	$\boxed{\frac{7}{2}}$	$\frac{5}{4}$	0	$\frac{1}{4}$	-1	0	$-\frac{1}{4}$	1	3	a_2
0	$\frac{3}{2}$	$\frac{1}{4}$	0	$\frac{1}{4}$	0	1	$-\frac{1}{4}$	0	6	s_3

Step 2 tableau:

z	x_1	x_2	x_3	e_1	e_2	s_3	a_1	a_2	rhs	$Basis$
1	0	$\frac{36}{7}$	0	$-\frac{4}{7}$	$-\frac{5}{7}$	0	$\frac{7M+4}{7}$	$\frac{7M+5}{7}$	$\frac{78}{7}$	z
0	0	$\frac{13}{14}$	1	$-\frac{3}{14}$	$-\frac{1}{7}$	0	$\frac{3}{14}$	$\frac{1}{7}$	$\frac{24}{7}$	x_3
0	1	$\frac{5}{14}$	0	$\frac{1}{14}$	$-\frac{2}{7}$	0	$-\frac{1}{14}$	$\frac{2}{7}$	$\frac{6}{7}$	x_1
0	0	$-\frac{2}{7}$	0	$\frac{1}{7}$	$\boxed{\frac{3}{7}}$	1	$-\frac{1}{7}$	$-\frac{3}{7}$	$\frac{33}{7}$	s_3

Step 3 tableau:

z	x_1	x_2	x_3	e_1	e_2	s_3	a_1	a_2	rhs	$Basis$
1	0	$\frac{14}{3}$	0	$-\frac{1}{3}$	0	$\frac{5}{3}$	$\frac{3M+1}{3}$	M	19	z
0	0	$\frac{5}{6}$	1	$-\frac{1}{6}$	0	$\frac{1}{3}$	$\frac{1}{6}$	0	5	x_3
0	1	$\frac{1}{6}$	0	$\boxed{\frac{1}{6}}$	0	$\frac{2}{3}$	$-\frac{1}{6}$	0	4	x_1
0	0	$-\frac{2}{3}$	0	$\frac{1}{3}$	1	$\frac{7}{3}$	$-\frac{1}{3}$	-1	11	e_2

Step 4 tableau:

z	x_1	x_2	x_3	e_1	e_2	s_3	a_1	a_2	rhs	$Basis$
1	2	5	0	0	0	3	M	M	27	z
0	1	1	1	0	0	1	0	0	9	x_3
0	6	1	0	1	0	4	-1	0	24	e_1
0	-2	-1	0	0	1	1	0	-1	3	e_2

This tableau is optimal. The optimal solution is

$$x_1^* = x_2^* = 0, x_3^* = 9, z^* = 27.$$

11.5 The Fundamental Theorem of LP

We are now ready to state the fundamental theorem of linear programming.

Theorem 11.5 (The fundamental theorem of LP) *Every LP has the following properties:*

1. *If it has no optimal solution, then it is either infeasible or unbounded.*

2. *If it has a feasible solution, then it has a standard-form representation in which it has a basic feasible solution.*

3. *If it has an optimal solution, then it has a standard-form representation in which it has a basic optimal solution.*

Proof. The proof follows from the analysis of the two-phase simplex method above. If an LP has a feasible solution, then Phase I of the two-phase simplex method will find a basic feasible solution. If the LP is optimal, then Phase II of the two-phase simplex method will find a basic optimal solution. If the LP has no optimal solution and was not proved infeasible at Phase I of the two-phase simplex method, then we start Phase II. Phase II will always terminate if, e.g., we use Bland's rule to avoid cycling. If the LP is not optimal, Phase II will prove that the problem is unbounded, since this is the only remaining possibility. $\qquad\square$

11.6 The Revised Simplex Method

In this section, we will discuss how the simplex method steps can be performed more efficiently using matrix operations. We will consider a problem in the form

$$\text{maximize} \quad \sum_{j=1}^{n} c_j x_j$$
$$\text{subject to} \quad \sum_{j=1}^{n} a_{ij} x_j \leq b_i, \quad i = 1, \ldots, m \tag{11.31}$$
$$x_j \geq 0, \quad j = 1, \ldots, n.$$

Before solving the problem using the simplex method, we introduce the slack variables $x_{n+1}, x_{n+2}, \ldots, x_{n+m}$ and write this problem in the standard form,

$$\text{maximize} \quad c^T x$$
$$\text{subject to} \quad Ax = b \tag{11.32}$$
$$x \geq 0.$$

Then matrix A has m rows and $n + m$ columns, the last m of which form the $m \times m$ identity matrix I_m. Vector x has length $n + m$ and vector b has length m. Vector c has length $n + m$, with its last m components being zeros.

Next, we will apply the simplex method to this problem, however, this time the computations will be done using a matrix representation of the data at each step. The resulting method is referred to as the *revised simplex method*. To facilitate the discussion, we will illustrate the underlying ideas using the following LP:

$$\begin{aligned}
\text{maximize} \quad & 4x_1 + 3x_2 + 5x_3 \\
\text{subject to} \quad & x_1 + 2x_2 + 2x_3 \leq 4 \\
& 3x_1 + 4x_3 \leq 6 \\
& 2x_1 + x_2 + 4x_3 \leq 8 \\
& x_1, x_2, x_3 \geq 0.
\end{aligned} \tag{11.33}$$

For this problem, $n = m = 3$. Assume that this LP represents a production planning problem, where x_j is the number of units of product j to be produced, the objective coefficient c_j is the profit obtained from a unit of product j. The constraints represent limitations on 3 different resources used in the production process, with the coefficient a_{ij} for variable x_j in the constraint i standing for the amount of resource i used in the production of 1 unit of product j, and the right-hand side b_i in the constraint i representing the amount of resource i available; $i, j = 1, 2, 3$.

Following the general framework, we introduce the slack variables x_4, x_5, x_6 and put the LP in the standard form (11.32), where

$$c^T = [4, 3, 5, 0, 0, 0], \quad A = \begin{bmatrix} 1 & 2 & 2 & 1 & 0 & 0 \\ 3 & 0 & 4 & 0 & 1 & 0 \\ 2 & 1 & 4 & 0 & 0 & 1 \end{bmatrix}, \quad b = \begin{bmatrix} 4 \\ 6 \\ 8 \end{bmatrix}. \tag{11.34}$$

Each basic feasible solution $\tilde{x}$ partitions $x_1, x_2, \ldots, x_{n+m}$ into m basic and n nonbasic variables, which can be listed in two separate vectors x_B and x_N.

Taking into account the correspondence of each variable to a coefficient in c, we can naturally partition c into two vectors, c_B and c_N, containing the objective function coefficients for the basic and nonbasic variables, respectively. Similarly, the columns of A can be split to form two different matrices, B and N, containing basic and nonbasic variable columns, respectively. As a result, the objective function $c^T x$ and the constraints system $Ax = b$ can be equivalently written as

$$z = c^T x = c_B^T x_B + c_N^T x_N \tag{11.35}$$

and

$$Ax = Bx_B + Nx_N = b, \tag{11.36}$$

respectively, where

$x_B = $ the vector of basic variables listed in the increasing order of their indices;

$x_N = $ the vector of nonbasic variables listed in the increasing order of their indices;

$c_B = $ the vector consisting of the components of c that correspond to the basic variables, listed in the increasing order of their indices;

$c_N^T = $ the vector consisting of the components of c that correspond to the nonbasic variables, listed in the increasing order of their indices;

$B = $ the matrix whose columns are the columns of A that correspond to the basic variables, listed in the increasing order of their indices;

$N = $ the matrix whose columns are the columns of A that correspond to the nonbasic variables, listed in the increasing order of their indices.

For example, for LP (11.33) at step 0 of the simplex method we have:

$$x_B = \begin{bmatrix} x_4 \\ x_5 \\ x_6 \end{bmatrix}, \quad c_B = \begin{bmatrix} 0 \\ 0 \\ 0 \end{bmatrix}, \quad B = \begin{bmatrix} 1 & 0 & 0 \\ 0 & 1 & 0 \\ 0 & 0 & 1 \end{bmatrix}, \quad \tilde{x}_B = \begin{bmatrix} 4 \\ 6 \\ 8 \end{bmatrix},$$

$$x_N = \begin{bmatrix} x_1 \\ x_2 \\ x_3 \end{bmatrix}, \quad c_N = \begin{bmatrix} 4 \\ 3 \\ 5 \end{bmatrix}, \quad N = \begin{bmatrix} 1 & 2 & 2 \\ 3 & 0 & 4 \\ 2 & 1 & 4 \end{bmatrix}.$$

Note that B is given by the 3×3 identity matrix and is clearly nonsingular. Next we show that matrix B will remain nonsingular for any basis obtained during the execution of the simplex method.

Theorem 11.6 *Consider LP (11.31) and its standard form representation (11.32). Let B denote a matrix consisting of the columns of A*

> that correspond to the basic variables obtained at an arbitrary step of the
> simplex method applied to (11.32). Then B is nonsingular.

Proof. Let S denote the $m \times m$ matrix consisting of the last m columns of A. Note that each column of S corresponds to a slack variable, and S is just the $m \times m$ identity matrix I_m. Next, we analyze what happens to B and S as we proceed with the execution of the simplex method. The only operation applied to matrix A at each simplex step is a pivot, which is a set of elementary row operations aiming to eliminate the coefficients of the entering variable in all rows but one, for which the entering variable is basic. Let A' be the matrix that is obtained from A after applying all the elementary row operations used by the simplex method to arrive at the current basis, and let B' and S' be the matrices resulting from B and S as a result of applying the same set of elementary row operations. Then the rows of A' can always be arranged in the order such that B' becomes the identity matrix. Let A'', B'', and S'' be the matrices obtained from A', B', and S' as a result of applying such row ordering. Then we have

$$[B|S] = [B|I_m] \sim [B''|S''] = [I_m|S''],$$

hence $S'' = B^{-1}$ and B is nonsingular. □

Note that the proof above is constructive and allows one to easily extract B^{-1} from the simplex tableau, as illustrated in the following example.

Example 11.8 *Consider the LP solved in Section 11.2:*

$$
\begin{aligned}
maximize \quad & 15x_1 + 25x_2 \\
subject\ to \quad & x_1 + x_2 \leq 450 \\
& x_2 \leq 300 \\
& 4x_1 + 5x_2 \leq 2{,}000 \\
& x_1 \leq 350 \\
& x_1, x_2 \geq 0.
\end{aligned}
$$

Its step 0 tableau is given by:

z	x_1	x_2	s_1	s_2	s_3	s_4	rhs	$Basis$
1	-15	-25	0	0	0	0	0	z
0	1	1	1	0	0	0	450	s_1
0	0	1	0	1	0	0	300	s_2
0	4	5	0	0	1	0	2,000	s_3
0	1	0	0	0	0	1	350	s_4

(11.37)

After two steps of the simplex method we had the following tableau:

z	x_1	x_2	s_1	s_2	s_3	s_4	rhs	$Basis$
1	0	0	0	25/4	15/4	0	9,375	z
0	0	0	1	1/4	−1/4	0	25	s_1
0	0	1	0	1	0	0	300	x_2
0	1	0	0	−5/4	1/4	0	125	x_1
0	0	0	0	5/4	−1/4	1	225	s_4

(11.38)

Rearranging the rows of this tableau so that the matrix comprised of the columns corresponding to the basic variables x_1, x_2, s_1, and s_4 is the 4×4 identity matrix, we obtain:

z	x_1	x_2	s_1	s_2	s_3	s_4	rhs	$Basis$
1	0	0	0	25/4	15/4	0	9,375	z
0	1	0	0	−5/4	1/4	0	125	x_1
0	0	1	0	1	0	0	300	x_2
0	0	0	1	1/4	−1/4	0	25	s_1
0	0	0	0	5/4	−1/4	1	225	s_4

(11.39)

Matrix B consisting of columns of the basic variables x_1, x_2, s_1, and s_4 in step 0 tableau (11.37) is given by

$$B = \begin{bmatrix} 1 & 1 & 1 & 0 \\ 0 & 1 & 0 & 0 \\ 4 & 5 & 0 & 0 \\ 1 & 0 & 0 & 1 \end{bmatrix},$$

(11.40)

and its inverse can be read from the columns for s_1, s_2, s_3, and s_4 in (11.39):

$$B^{-1} = \begin{bmatrix} 0 & -5/4 & 1/4 & 0 \\ 0 & 1 & 0 & 0 \\ 1 & 1/4 & -1/4 & 0 \\ 0 & 5/4 & -1/4 & 1 \end{bmatrix}.$$

(11.41)

This is easy to verify by checking that the product of the matrices in (11.40) and (11.41) gives the 4×4 identity matrix.

Knowing that B^{-1} always exists, we can express x_B through x_N by multiplying both sides of (11.36) by B^{-1}:

$$x_B = B^{-1}b - B^{-1}Nx_N.$$

Then, using (11.35), z can be expressed through nonbasic variables only as follows:

$$z = c_B^T x_B + c_N^T x_N = c_B^T B^{-1} b + (c_N^T - c_B^T B^{-1} N) x_N.$$

In summary, the corresponding revised dictionary is given by

$$\begin{array}{rcl} z & = & c_B^T B^{-1} b + (c_N^T - c_B^T B^{-1} N) x_N \\ \hline x_B & = & B^{-1} b - B^{-1} N x_N. \end{array} \tag{RD}$$

Writing the tableau for the same basis, we obtain the revised tableau:

z	x_B	x_N	rhs
1	0	$-(c_N^T - c_B^T B^{-1} N)$	$c_B^T B^{-1} b$
0	I_m	$B^{-1} N$	$B^{-1} b$

$$\tag{RT}$$

Note that the formulas in (RD) and (RT) are valid for any LP in the standard form (11.32), as long as it has a basic solution, which ensures that B is nonsingular for every basic solution.

We will use the matrix representation of the dictionary (RD) or of the tableau (RT) to carry out the steps of the revised simplex method. Let us consider the tableau form (RT) to motivate the method. With the "usual" simplex method, at each step of the method we update the data in (RT) by applying certain elementary row operations to the tableau computed at the previous step. However, note that we do not need to know every single entry of the tableau in order to determine the entering and leaving variables, which is the only information needed to update the basis. The main reason for computing the whole tableau at each step was due to the fact that parts of the tableau that may not be used at the current step may become necessary at further steps. But, as we can see from (RT), the information required to perform a pivot may be obtained using the input data given by A, b, and c directly, rather than using the output from the previous step. We will demonstrate how the corresponding computations can be carried out using the LP given in (11.33), with c, A, and b as in (11.34):

$$c^T = [4, 3, 5, 0, 0, 0], \quad A = \begin{bmatrix} 1 & 2 & 2 & 1 & 0 & 0 \\ 3 & 0 & 4 & 0 & 1 & 0 \\ 2 & 1 & 4 & 0 & 0 & 1 \end{bmatrix}, \quad b = \begin{bmatrix} 4 \\ 6 \\ 8 \end{bmatrix}. \tag{11.42}$$

In the beginning, the basic variables are given by $BV = \{x_4, x_5, x_6\}$, so we have the following step 0 data:

$$x_B = \begin{bmatrix} x_4 \\ x_5 \\ x_6 \end{bmatrix}, \quad c_B = \begin{bmatrix} 0 \\ 0 \\ 0 \end{bmatrix}, \quad B = \begin{bmatrix} 1 & 0 & 0 \\ 0 & 1 & 0 \\ 0 & 0 & 1 \end{bmatrix}, \quad \tilde{x}_B = \begin{bmatrix} 4 \\ 6 \\ 8 \end{bmatrix},$$

$$x_N = \begin{bmatrix} x_1 \\ x_2 \\ x_3 \end{bmatrix}, \quad c_N = \begin{bmatrix} 4 \\ 3 \\ 5 \end{bmatrix}, \quad N = \begin{bmatrix} 1 & 2 & 2 \\ 3 & 0 & 4 \\ 2 & 1 & 4 \end{bmatrix}.$$

Here $\tilde{x}_B$ is the vector storing the values of the basic variables in the current basic feasible solution.

Step 1. Since the initial values of B and c_B are given by the 3×3 identity matrix and the 3-dimensional 0 vector, the formulas in (RT) simplify to the following:

z	x_B	x_N	rhs	
1	0	$-c_N^T$	0	(RT-1)
0	I_m	N	b	

Choosing the entering variable. At the first step, selecting the entering variable is easy; we just take a variable corresponding to the highest positive coefficient in c_N, which is x_3 in our case.

Choosing the leaving variable. We select the leaving variable based on the ratio test, which is performed by considering the components of the vector of right-hand sides given by $\tilde{x}_B$ that correspond to positive entries of the column of N representing the entering variable, x_3. Let us denote this column by N_{x_3}. Then we have:

$$
\begin{array}{cccc}
N_{x_3} & \tilde{x}_B & ratios & x_B \\
\begin{bmatrix} 2 \\ 4 \\ 4 \end{bmatrix} & \begin{bmatrix} 4 \\ 6 \\ 8 \end{bmatrix} & \begin{matrix} 2 \\ \mathbf{3/2} \\ 2 \end{matrix} & \begin{matrix} x_4 \\ x_5 \\ x_6 \end{matrix} \Rightarrow x_5 \text{ wins.}
\end{array}
$$

Updating the basic feasible solution. The information used for the ratio test is also sufficient for determining the values of the new vector $\tilde{x}_B$. For the entering variable, $\tilde{x}_3$ is given by the minimum ratio value, i.e., $\tilde{x}_3 = 3/2$. As for the remaining components of $\tilde{x}_B$, for each basic variable x_j they are computed based on the following observation. To perform the pivot, we would apply the elementary row operation which, in order to eliminate x_3 from the x_j-row, multiplies the x_5-row (i.e., the row where x_5 was basic) by the coefficient for x_3 in the x_j-row divided by 4 and then subtracts the result from the x_j-row. When this elementary row operation is applied to the right-hand side column, we obtain:

$$
\begin{aligned}
\tilde{x}_3 &= 3/2 \\
\tilde{x}_4 &= 4 - 2(6/4) = 1 \\
\tilde{x}_6 &= 8 - 4(6/4) = 2.
\end{aligned}
$$

In summary, we have the following step 1 output:

$$
x_B = \begin{bmatrix} x_3 \\ x_4 \\ x_6 \end{bmatrix}, \quad c_B = \begin{bmatrix} 5 \\ 0 \\ 0 \end{bmatrix}, \quad B = \begin{bmatrix} 2 & 1 & 0 \\ 4 & 0 & 0 \\ 4 & 0 & 1 \end{bmatrix}, \quad \tilde{x}_B = \begin{bmatrix} 3/2 \\ 1 \\ 2 \end{bmatrix},
$$

$$x_N = \begin{bmatrix} x_1 \\ x_2 \\ x_5 \end{bmatrix}, \quad c_N = \begin{bmatrix} 4 \\ 3 \\ 0 \end{bmatrix}, \quad N = \begin{bmatrix} 1 & 2 & 0 \\ 3 & 0 & 1 \\ 2 & 1 & 0 \end{bmatrix}.$$

Step 2. In order to carry out the necessary computations, we will use the formulas in (RT) written as follows:

z	x_B	x_N	rhs	
1	0	$-\tilde{c}_N = -(c_N^T - c_B^T B^{-1} N)$	$\tilde{z} = c_B^T B^{-1} b$	(RT-2)
0	I_m	$\tilde{N} = B^{-1} N$	$\tilde{x}_B = B^{-1} b$	

Choosing the entering variable. To find the entering variable, we need to compute

$$\tilde{c}_N^T = c_N^T - c_B^T B^{-1} N.$$

Instead of computing $c_B^T B^{-1}$ directly, we can denote by $u^T = c_B^T B^{-1}$ and then find u by solving the system $u^T B = c_B^T$, or, equivalently,

$$B^T u = c_B.$$

We have

$$\begin{bmatrix} 2 & 4 & 4 \\ 1 & 0 & 0 \\ 0 & 0 & 1 \end{bmatrix} \begin{bmatrix} u_1 \\ u_2 \\ u_3 \end{bmatrix} = \begin{bmatrix} 5 \\ 0 \\ 0 \end{bmatrix} \quad \Leftrightarrow \quad u = \begin{bmatrix} 0 \\ 5/4 \\ 0 \end{bmatrix},$$

so,

$$\tilde{c}_N^T = c_N^T - u^T N = [4, 3, 0] - [0, 5/4, 0] \begin{bmatrix} 1 & 2 & 0 \\ 3 & 0 & 1 \\ 2 & 1 & 0 \end{bmatrix} = [1/4, 3, -5/4],$$

and the entering variable is the second nonbasic variable, which is x_2.

Choosing the leaving variable. To perform the ratio test, we only need $\tilde{x}_B$ and the second column $\tilde{N}_{x_2}$ of the matrix $\tilde{N} = B^{-1} N$, which is equal to

$$\tilde{N}_{x_2} = B^{-1} \times N_{x_2}.$$

We can find $v = \tilde{N}_{x_2}$ by solving the system $Bv = N_{x_2}$ for v:

$$\begin{bmatrix} 2 & 1 & 0 \\ 4 & 0 & 0 \\ 4 & 0 & 1 \end{bmatrix} \begin{bmatrix} v_1 \\ v_2 \\ v_3 \end{bmatrix} = \begin{bmatrix} 2 \\ 0 \\ 1 \end{bmatrix} \quad \Leftrightarrow \quad v = \begin{bmatrix} 0 \\ 2 \\ 1 \end{bmatrix}.$$

We compare the ratios of components of $\tilde{x}_B$ and $\tilde{N}_{x_2} = v$:

$$
\begin{array}{cccc}
\tilde{N}_{x_2} & \tilde{x}_B & \text{ratios} & x_B \\
\begin{bmatrix} 0 \\ 2 \\ 1 \end{bmatrix} & \begin{bmatrix} 3/2 \\ 1 \\ 2 \end{bmatrix} & \begin{matrix} - \\ \mathbf{1/2} \\ 2 \end{matrix} & \begin{bmatrix} x_3 \\ x_4 \\ x_6 \end{bmatrix} \Rightarrow x_4 \text{ wins.}
\end{array}
$$

Updating the basic feasible solution. New $\tilde{x}_B$ is computed similarly to the previous step:

$$
\begin{aligned}
\tilde{x}_2 &= 1/2 \ (\text{min ratio value}) \\
\tilde{x}_3 &= 3/2 - 0(1/2) = 3/2 \\
\tilde{x}_6 &= 2 - 1(1/2) = 3/2.
\end{aligned}
$$

In summary, step 2 produces the following output, which is also the input for step 3.

$$
x_B = \begin{bmatrix} x_2 \\ x_3 \\ x_6 \end{bmatrix}, \quad c_B = \begin{bmatrix} 3 \\ 5 \\ 0 \end{bmatrix}, \quad B = \begin{bmatrix} 2 & 2 & 0 \\ 0 & 4 & 0 \\ 1 & 4 & 1 \end{bmatrix}, \quad \tilde{x}_B = \begin{bmatrix} 1/2 \\ 3/2 \\ 3/2 \end{bmatrix}
$$

$$
x_N = \begin{bmatrix} x_1 \\ x_4 \\ x_5 \end{bmatrix}, \quad c_N = \begin{bmatrix} 4 \\ 0 \\ 0 \end{bmatrix}, \quad N = \begin{bmatrix} 1 & 1 & 0 \\ 3 & 0 & 1 \\ 2 & 0 & 0 \end{bmatrix}.
$$

Step 3. We use the formulas as in (RT-2), which we mention again for convenience:

z	x_B	x_N	rhs	
1	0	$-\tilde{c}_N = -(c_N^T - c_B^T B^{-1} N)$	$\tilde{z} = c_B^T B^{-1} b$	(RT-3)
0	I_m	$\tilde{N} = B^{-1} N$	$\tilde{x}_B = B^{-1} b$	

Choosing the entering variable. To find the entering variable, we need to compute $\tilde{c}_N^T = c_N^T - c_B^T B^{-1} N$. Again, we denote by $u^T = c_B^T B^{-1}$ and then find u by solving the system $B^T u = c_B$. We have

$$
\begin{bmatrix} 2 & 0 & 1 \\ 2 & 4 & 4 \\ 0 & 0 & 1 \end{bmatrix} \begin{bmatrix} u_1 \\ u_2 \\ u_3 \end{bmatrix} = \begin{bmatrix} 3 \\ 5 \\ 0 \end{bmatrix} \Leftrightarrow u = \begin{bmatrix} 3/2 \\ 1/2 \\ 0 \end{bmatrix}
$$

and

$$
\tilde{c}_N^T = c_N^T - u^T N = [4, 0, 0] - [3/2, 1/2, 0] \begin{bmatrix} 1 & 1 & 0 \\ 3 & 0 & 1 \\ 2 & 0 & 0 \end{bmatrix} = [1, -3/2, -1/2]
$$

and the entering variable is the first nonbasic variable, which is x_1.

Choosing the leaving variable. To perform the ratio test, we only need $\tilde{x}_B$ and the first column $\tilde{N}_{x_1}$ of the matrix $\tilde{N} = B^{-1}N$, which is equal to

$$\tilde{N}_{x_1} = B^{-1} \times N_{x_1}.$$

We can find $v = \tilde{N}_{x_1}$ by solving the system $Bv = N_{x_1}$ for v:

$$\begin{bmatrix} 2 & 2 & 0 \\ 0 & 4 & 0 \\ 1 & 4 & 1 \end{bmatrix} \begin{bmatrix} v_1 \\ v_2 \\ v_3 \end{bmatrix} = \begin{bmatrix} 1 \\ 3 \\ 2 \end{bmatrix} \Leftrightarrow v = \begin{bmatrix} -1/4 \\ 3/4 \\ -3/4 \end{bmatrix}.$$

We compare the ratios of components of $\tilde{x}_B$ and $\tilde{N}_{x_1} = v$:

$$\begin{array}{cccc} \tilde{N}_{x_1} & \tilde{x}_B & \text{ratios} & x_B \\ \begin{bmatrix} -1/4 \\ 3/4 \\ -3/4 \end{bmatrix} & \begin{bmatrix} 1/2 \\ 3/2 \\ 3/2 \end{bmatrix} & \begin{array}{c} - \\ \mathbf{2} \\ - \end{array} & \begin{array}{c} x_2 \\ x_3 \\ x_6 \end{array} \Rightarrow x_3 \text{ wins.} \end{array}$$

Updating the basic feasible solution. Next, we update $\tilde{x}_B$:

$$\tilde{x}_1 = 2 \text{ (min ratio value)}$$
$$\tilde{x}_2 = 1/2 + (1/4)2 = 1$$
$$\tilde{x}_6 = 3/2 + (3/4)2 = 3.$$

The output of step 3 is given by:

$$x_B = \begin{bmatrix} x_1 \\ x_2 \\ x_6 \end{bmatrix}, \quad c_B = \begin{bmatrix} 4 \\ 3 \\ 0 \end{bmatrix}, \quad B = \begin{bmatrix} 1 & 2 & 0 \\ 3 & 0 & 0 \\ 2 & 1 & 1 \end{bmatrix}, \quad \tilde{x}_B = \begin{bmatrix} 2 \\ 1 \\ 3 \end{bmatrix},$$

$$x_N = \begin{bmatrix} x_3 \\ x_4 \\ x_5 \end{bmatrix}, \quad c_N = \begin{bmatrix} 5 \\ 0 \\ 0 \end{bmatrix}, \quad N = \begin{bmatrix} 2 & 1 & 0 \\ 4 & 0 & 1 \\ 4 & 0 & 0 \end{bmatrix}.$$

Step 4. We proceed similarly to the previous steps. To find the entering variable, we compute $\tilde{c}_N^T = c_N^T - c_B^T B^{-1} N$ by first solving the system $B^T u = c_B$ for u to find $c_B^T B^{-1}$,

$$\begin{bmatrix} 1 & 3 & 2 \\ 2 & 0 & 1 \\ 0 & 0 & 1 \end{bmatrix} \begin{bmatrix} u_1 \\ u_2 \\ u_3 \end{bmatrix} = \begin{bmatrix} 4 \\ 3 \\ 0 \end{bmatrix} \Leftrightarrow u = \begin{bmatrix} 3/2 \\ 5/6 \\ 0 \end{bmatrix},$$

and then use this information to find $\tilde{c}_N^T$:

$$\tilde{c}_N^T = c_N^T - u^T N = [5, 0, 0] - [3/2, 5/6, 0] \begin{bmatrix} 2 & 1 & 0 \\ 4 & 0 & 1 \\ 4 & 0 & 0 \end{bmatrix} = [-4/3, -3/2, -5/6].$$

Since none of the components of $\tilde{c}_N^T$ is positive, this is a basic optimal solution. We have

$$x_B^* = \tilde{x}_B = \begin{bmatrix} 2 \\ 1 \\ 3 \end{bmatrix},$$

so,

$$x^* = [2, 1, 0, 0, 0, 3]^T, \ z^* = c^T x^* = 11.$$

The final answer for the original variables is

$$x^* = [2, 1, 0]^T, \ z^* = 11.$$

11.7   Complexity of the Simplex Method

In this section we discuss the worst-case behavior of the simplex method in terms of the number of steps it has to execute before finding an optimal solution. Consider the problem in the form

$$\begin{aligned} \text{maximize} \quad & \sum_{j=1}^{n} c_j x_j \\ \text{subject to} \quad & \sum_{j=1}^{n} a_{ij} x_j \ \leq \ b_i, \quad i = 1, \ldots, m \\ & x_j \ \geq \ 0, \quad j = 1, \ldots, n. \end{aligned} \tag{11.43}$$

After introducing the slack variables $x_{n+1}, x_{n+2}, \ldots, x_{n+m}$ we write this problem in the standard form:

$$\begin{aligned} \text{maximize} \quad & c^T x \\ \text{subject to} \quad & Ax \ = \ b \\ & x \ \geq \ 0. \end{aligned} \tag{11.44}$$

The matrix A has m rows and $n + m$ columns, and we may potentially have

$$K = \binom{m+n}{m} = \frac{(m+n)!}{m!\,n!}$$

different basic feasible solutions. If $n = m$, taking into account that $\frac{n+i}{i} \geq 2$ for $i = 1, \ldots, n$, we have:

$$K = \frac{(2n)!}{(n!)^2} = \frac{(n+1)(n+2) \cdot \ldots \cdot 2n}{1 \cdot 2 \cdot \ldots \cdot n} = \prod_{i=1}^{n} \frac{n+i}{i} \geq 2^n.$$

Thus, the number of basic feasible solutions may be exponential with respect to the problem input size. If there existed an LP with K basic feasible solutions, and if the simplex method would have to visit each basic feasible solution before terminating, this would imply that the simplex method requires an exponential number of steps in the worst case. Unfortunately, such examples have been constructed for various strategies for selecting leaving and entering variables in the simplex method. The first such example was constructed by Klee and Minty in 1972, and is widely known as the *Klee-Minty problem*, which can be formulated as follows:

$$
\begin{array}{lrl}
\text{maximize} & \displaystyle\sum_{j=1}^{m} 10^{m-j} x_j & \\
\text{subject to} & x_i + 2 \displaystyle\sum_{j=1}^{i-1} 10^{i-j} x_j \leq 100^{i-1}, & i = 1, \ldots, m \\
& x_j \geq 0, & j = 1, \ldots, m.
\end{array}
\tag{11.45}
$$

If the entering variable is always selected to be the nonbasic variable with the highest coefficient in the objective (written in the dictionary format), then the simplex method will visit each of the 2^m basic feasible solutions of this LP.

It is still unknown whether a variation of the simplex method can be developed that would be guaranteed to terminate in a number of steps bounded by a polynomial function of the problem's input size. It should be noted that all currently known polynomial time algorithms for linear optimization problems, such as the *ellipsoid method* and the *interior point methods*, are based on the approaches that treat linear programs as continuous convex optimization problems, whereas the simplex method exploits the discrete nature of an LP's extreme points along with the fact that any optimal LP has an optimal solution at an extreme point.

Exercises

11.1. Write the following problems as LPs in the standard form:

(a)

$$
\begin{array}{lrcrcrcl}
\text{maximize} & 2x_1 & + & 3x_2 & - & 3x_3 & & \\
\text{subject to} & x_1 & + & x_2 & + & x_3 & \leq & 7 \\
& & & x_2 & - & x_3 & \leq & 5 \\
& -x_1 & + & x_2 & + & 4x_3 & \geq & 4 \\
& & & & & x_1 \in \mathbb{R}, x_2, x_3 & \geq & 0
\end{array}
$$

(b)

$$
\begin{array}{lrcrcl}
\text{minimize} & |x_1| & + & 2|x_2| & & \\
\text{subject to} & 2x_1 & + & 3x_2 & \leq & 4 \\
& 5|x_1| & - & 6x_2 & \leq & 7 \\
& 8x_1 & + & 9x_2 & \geq & 10 \\
& & & x_1, x_2 & \in & \mathbb{R}.
\end{array}
$$

11.2. Solve the following LPs using the simplex method:

(a)   maximize $3x_1 + 5x_2$
       subject to
$$
\begin{aligned}
x_1 + x_2 &\leq 5 \\
x_2 &\leq 3 \\
3x_1 + 5x_2 &\leq 18 \\
x_1 &\leq 4 \\
x_1, x_2 &\geq 0
\end{aligned}
$$

(b)   minimize $x_1 + 5x_2$
       subject to
$$
\begin{aligned}
x_1 + 2x_2 &\leq 3 \\
x_1 + x_2 &\leq 2 \\
5x_1 + x_2 &\leq 5 \\
x_1, x_2 &\geq 0.
\end{aligned}
$$

Illustrate the simplex steps graphically (that is, draw the feasible region, the iso-profit/iso-cost line corresponding to the optimal solution(s), and indicate the basic feasible solutions explored by the algorithm).

11.3. Solve the following LPs using the simplex method:

(a)   maximize $5x_1 - 2x_2 + 3x_3$
       subject to
$$
\begin{aligned}
3x_1 + 2x_2 - 2x_3 &\leq 6 \\
x_1 + x_2 + x_3 &\leq 3 \\
x_1, x_2, x_3 &\geq 0
\end{aligned}
$$

(b)   maximize $x_1 + 2x_2 + 3x_3$
       subject to
$$
\begin{aligned}
x_1 + x_2 + x_3 &\leq 3 \\
3x_1 + 2x_2 + 2x_3 &\leq 6 \\
2x_1 + x_2 + 3x_3 &\leq 6 \\
x_1, x_2, x_3 &\geq 0
\end{aligned}
$$

(c)   minimize $-x_1 + x_2 + x_3$
       subject to
$$
\begin{aligned}
- x_2 + x_3 &\leq 2 \\
x_1 - 2x_2 - 2x_3 &\leq 6 \\
-2x_1 + x_2 + x_3 &\leq 2 \\
x_1, x_2, x_3 &\geq 0
\end{aligned}
$$

(d)   maximize $2x_1 + 3x_2 + 4x_3 + 5x_4$
       subject to
$$
\begin{aligned}
x_1 + 2x_2 \qquad\quad + x_4 &\leq 5 \\
x_1 + x_2 + x_3 + 5x_4 &\leq 1 \\
x_1, x_2, x_3, x_4 &\geq 0.
\end{aligned}
$$

11.4. Find the basic solutions corresponding to points G, H, I, J, K, L, and M in Figure 11.1 at page 252 (Section 11.3).

11.5. Use the simplex method to solve the Klee-Minty LP (11.45) at page 277 for $m = 3$. Always select the nonbasic variable with the highest coefficient in the objective as the entering variable. Illustrate your steps geometrically.

11.6. Explain how the simplex algorithm can be used for finding the second-best basic feasible solution to an LP. Then find the second best bfs for problems (a) and (b) in Exercise 11.3.

11.7. Solve the LP given in Eq. (11.25) at page 255 using

(a) the two-phase simplex method;

(b) the big-M method.

11.8. Solve the following LPs using

(a) the two-phase simplex method;

(b) the big-M method.

$$
\begin{array}{lrrl}
\text{(i)} \quad \text{maximize} & 4x_1 & - & 2x_2 \\
\text{subject to} & x_1 & - & 2x_2 & = & 2 \\
& 2x_1 & + & x_2 & \leq & 7 \\
& x_1 & + & 2x_2 & \leq & 4 \\
& & & x_1, x_2 & \geq & 0
\end{array}
$$

$$
\begin{array}{lrrl}
\text{(ii)} \quad \text{minimize} & x_1 & + & x_2 \\
\text{subject to} & x_1 & - & 3x_2 & = & 3 \\
& -x_1 & + & 5x_2 & \geq & 2 \\
& & & x_1, x_2 & \geq & 0
\end{array}
$$

$$
\begin{array}{lrrl}
\text{(iii)} \quad \text{minimize} & 3x_1 & + & 2x_2 \\
\text{subject to} & x_1 & - & 3x_2 & \geq & 3 \\
& -2x_1 & + & 5x_2 & \geq & 3 \\
& & & x_1, x_2 & \geq & 0
\end{array}
$$

$$
\begin{array}{lrrl}
\text{(iv)} \quad \text{maximize} & 2x_1 & + & x_2 \\
\text{subject to} & x_1 & - & 2x_2 & = & 2 \\
& 2x_1 & + & x_2 & \leq & 7 \\
& x_1 & + & 2x_2 & \geq & 5 \\
& & & x_1, x_2 & \geq & 0.
\end{array}
$$

11.9. Use the revised simplex method to solve LPs (a) and (b) in Exercise 11.3.

Chapter 12

Duality and Sensitivity Analysis in Linear Programming

12.1 Defining the Dual LP

Consider an instance of the production planning problem discussed in Section 10.2.6. SuperCleats Inc. is a small business manufacturing 4 models of soccer shoes, Dynamo, Spartacus, Torpedo, and Kanga. Manufacturing one pair of each model is associated with the material, labor, and overhead costs given in the table below, which also lists the company's annual budget for each kind of resource.

	Dynamo	Spartacus	Torpedo	Kanga	Annual Budget
Price	100	80	60	50	
Material Cost	15	10	8	5	45,000
Labor Cost	35	22	18	10	100,000
Overhead Cost	5	5	4	4	20,000

In addition, the annual demand for Dynamo and Spartacus shoes is limited to 1,000 pairs each. The LP that SuperCleats Inc. solves in order to decide on the quantity of each shoe model to manufacture so that its overall profit is maximized is given by

$$
\begin{aligned}
\text{maximize} \quad & 100x_1 + 80x_2 + 60x_3 + 50x_4 \\
\text{subject to} \quad & 15x_1 + 10x_2 + 8x_3 + 5x_4 \leq 45,000 \\
& 35x_1 + 22x_2 + 18x_3 + 10x_4 \leq 100,000 \\
& 5x_1 + 5x_2 + 4x_3 + 4x_4 \leq 20,000 \qquad (12.1) \\
& x_1 \leq 1,000 \\
& x_2 \leq 1,000 \\
& x_1, x_2, x_3, x_4 \geq 0.
\end{aligned}
$$

All Shoes United (ASU) is a bigger shoe manufacturer looking to acquire SuperCleats. They want to prepare an offer that would minimize ASU's annual expense to cover SuperCleats' operations, while making sure that the offer is attractive for SuperCleats Inc. ASU's management assumes that for the offer to be attractive, the annual payment to SuperCleats must be no less than the profit SuperCleats Inc. expects to generate.

First, assume that ASU managers are partially aware of the information presented in the table above. They know the prices, the cost of materials required to manufacture one pair of shoes of each type, and SuperCleats' budget for materials. They want to use this information to get an estimate of how much the potential acquisition may cost. Based on the information in the first two rows of the table, we can write down the objective and the first constraint of LP (12.1). We have

$$z = \quad 100x_1 + 80x_2 + 60x_3 + 50x_4;$$
$$15x_1 + 10x_2 + 8x_3 + 5x_4 \leq 45{,}000.$$

If we multiply both sides of the constraint by 10,

$$150x_1 + 100x_2 + 80x_3 + 50x_4 \leq 450{,}000,$$

the coefficient of each variable in the constraint will be no less than the coefficient of the same variable in the objective, thus, we can easily obtain an upper bound on z by just taking the right-hand side of the constraint:

$$\begin{aligned} z = \quad & 100x_1 + \quad 80x_2 + 60x_3 + 50x_4 \\ \leq \quad & 150x_1 + 100x_2 + 80x_3 + 50x_4 \\ \leq \quad & 450{,}000. \end{aligned}$$

Thus, $z \leq 450{,}000$, and ASU obtains an upper bound on how much they should pay annually for the acquisition.

To get a better estimate of the minimum price that would be attractive for SuperCleats, we can use all the constraints in (12.1). Namely, we can multiply both sides of each constraint i by some nonnegative value y_i for $i = 1, \ldots, 5$, so that the inequalities are preserved,

$$\begin{array}{rrrrll} 15x_1 + & 10x_2 + & 8x_3 + & 5x_4 \leq 45{,}000 & \times y_1 \\ 35x_1 + & 22x_2 + & 18x_3 + & 10x_4 \leq 100{,}000 & \times y_2 \\ 5x_1 + & 5x_2 + & 4x_3 + & 4x_4 \leq 20{,}000 & \times y_3 \\ x_1 & & & \leq 1{,}000 & \times y_4 \\ & x_2 & & \leq 1{,}000 & \times y_5 \end{array}$$

and then sum the resulting inequalities to obtain:

$$\begin{aligned} (15x_1 + 10x_2 &+ 8x_3 + 5x_4)y_1 + (35x_1 + 22x_2 + 18x_3 + 10x_4)y_2 \\ &+ (5x_1 + 5x_2 + 4x_3 + 4x_4)y_3 + x_1 y_4 + x_2 y_5 \\ &= (15y_1 + 35y_2 + 5y_3 + y_4)x_1 + (10y_1 + 22y_2 + 5y_3 + y_5)x_2 \\ &\quad + (8y_1 + 18y_2 + 4y_3)x_3 + (5y_1 + 10y_2 + 4y_3)x_4 \\ &\leq 45{,}000y_1 + 100{,}000y_2 + 20{,}000y_3 + 1{,}000y_4 + 1{,}000y_5. \end{aligned}$$

If we select $y_1, \ldots, y_5$ so that the coefficients for $x_j, j = 1, \ldots, 4$ in the last expression are no less than the corresponding coefficients in

$$z = 100x_1 + 80x_2 + 60x_3 + 50x_4,$$

that is,

$$
\begin{aligned}
15y_1 + 35y_2 + 5y_3 + y_4 &\geq 100 \\
10y_1 + 22y_2 + 5y_3 \quad\quad + y_5 &\geq 80 \\
8y_1 + 18y_2 + 4y_3 &\geq 60 \\
5y_1 + 10y_2 + 4y_3 &\geq 50,
\end{aligned}
$$

then we have

$$
z = 100x_1 + 80x_2 + 60x_3 + 50x_4 \leq 1,000(45y_1 + 100y_2 + 20y_3 + y_4 + y_5).
$$

To get the lowest possible upper bound on z this way, we need to solve the following LP:

$$
\begin{aligned}
\text{minimize} \quad & 1,000(45y_1 + 100y_2 + 20y_3 + y_4 + y_5) \\
\text{subject to} \quad & 15y_1 + 35y_2 + 5y_3 + y_4 \geq 100 \\
& 10y_1 + 22y_2 + 5y_3 \quad\quad + y_5 \geq 80 \\
& 8y_1 + 18y_2 + 4y_3 \geq 60 \\
& 5y_1 + 10y_2 + 4y_3 \geq 50 \\
& y_1, y_2, y_3, y_4, y_5 \geq 0.
\end{aligned}
\tag{12.2}
$$

This LP is called the *dual* to the SuperCleats' maximization problem (12.1), which is called the *primal LP*. If we compare the primal LP (12.1) to the dual LP (12.2), we can see that the coefficients in the rows of one LP appear in the columns of the other LP, and vice versa.

More generally, if the primal LP is given by

$$
\begin{aligned}
\text{maximize} \quad & c_1x_1 + \cdots + c_nx_n \\
\text{subject to} \quad & a_{11}x_1 + \cdots + a_{1n}x_n \leq b_1 \\
& \quad\vdots \quad \ddots \quad \vdots \\
& a_{m1}x_1 + \cdots + a_{mn}x_n \leq b_m \\
& x_1, \ldots, x_n \geq 0
\end{aligned}
$$

or

$$
\begin{aligned}
\text{maximize} \quad & \sum_{j=1}^{n} c_jx_j \\
\text{subject to} \quad & \sum_{j=1}^{n} a_{ij}x_j \leq b_i, \; i = 1, \ldots, m \\
& x_1, \ldots, x_n \geq 0,
\end{aligned}
$$

then its dual LP is

$$
\begin{aligned}
\text{minimize} \quad & b_1y_1 + \cdots + b_my_m \\
\text{subject to} \quad & a_{11}y_1 + \cdots + a_{m1}y_m \geq c_1 \\
& \quad\vdots \quad \ddots \quad \vdots \quad\quad \vdots \\
& a_{1n}y_1 + \cdots + a_{mn}y_m \geq c_n \\
& y_1, \ldots, y_m \geq 0
\end{aligned}
$$

or

$$\text{minimize} \quad \sum_{i=1}^{m} b_i y_i$$
$$\text{subject to} \quad \sum_{i=1}^{m} a_{ij} y_i \geq c_j, \; j = 1, \ldots, n$$
$$y_1, \ldots, y_m \geq 0.$$

Using matrix representation, we have the following correspondence between the primal and dual LPs:

Primal LP	Dual LP
maximize $c^T x$	minimize $b^T y$
subject to $Ax \leq b$	subject to $A^T y \geq c$
$x \geq 0$	$y \geq 0$

The variables $x_j, j = 1, \ldots, n$ of the primal LP are called the *primal variables*, and the variables $y_i, i = 1, \ldots, m$ of the dual LP are called the *dual variables*.

12.1.1 Forming the dual of a general LP

We discussed how to form the dual of an LP with "$\leq$"-type inequality constraints only. Next we discuss how to form the dual for LPs involving different types of constraints.

If we have a "$\geq$"-type inequality constraint, we can multiply both sides by -1 and convert it into an equivalent "$\leq$"-type inequality. Assume now that we have a primal problem with an equality constraint:

$$\text{maximize} \quad -x_1 - 2x_2 - 3x_3$$
$$\text{subject to} \quad 7x_1 - 8x_2 + 9x_3 = 10$$
$$x_1, x_2, x_3 \geq 0.$$

In order to use the rules for forming the dual that we already know, we equivalently represent this equality by two inequality constraints,

$$7x_1 - 8x_2 + 9x_3 \leq 10$$
$$7x_1 - 8x_2 + 9x_3 \geq 10,$$

and then convert the second inequality into "$\leq$"-type:

$$7x_1 - 8x_2 + 9x_3 \leq 10$$
$$-7x_1 + 8x_2 - 9x_3 \leq -10,$$

Let y' and y'' be the dual variable for the first and the second constraints above, respectively. Then the corresponding dual LP is given by

$$
\begin{array}{rl}
\text{minimize} & 10y' - 10y'' \\
\text{subject to} & 7y' - 7y'' \geq -1 \\
& -8y' + 8y'' \geq -2 \\
& 9y' - 9y'' \geq -3 \\
& y', y'' \geq 0,
\end{array}
$$

or, equivalently,

$$
\begin{array}{rl}
\text{minimize} & 10(y' - y'') \\
\text{subject to} & 7(y' - y'') \geq -1 \\
& -8(y' - y'') \geq -2 \\
& 9(y' - y'') \geq -3 \\
& y', y'' \geq 0.
\end{array}
$$

Since y' and y'' always appear together, with the same absolute value but opposite sign coefficients, we can make the following change of variables:

$$
y = y' - y''.
$$

Then we obtain an equivalent one-variable problem:

$$
\begin{array}{rl}
\text{minimize} & 10y \\
\text{subject to} & 7y \geq -1 \\
& -8y \geq -2 \\
& 9y \geq -3 \\
& y \in \mathbb{R}.
\end{array}
$$

Thus, an equality constraint in the primal problem corresponds to a free variable in the dual problem.

Next we show that if we take the dual of the dual LP, we obtain the primal LP we started with.

Consider the primal LP

$$
\begin{array}{rll}
\text{maximize} & \sum_{j=1}^{n} c_j x_j & \\
\text{subject to} & \sum_{j=1}^{n} a_{ij} x_j \leq b_i, & i = 1, \ldots, m \\
& x_1, \ldots, x_n \geq 0 &
\end{array} \tag{12.3}
$$

and the corresponding dual LP

$$
\begin{array}{rll}
\text{minimize} & \sum_{i=1}^{m} b_i y_i & \\
\text{subject to} & \sum_{i=1}^{m} a_{ij} y_i \geq c_j, & j = 1, \ldots, n \\
& y_1, \ldots, y_m \geq 0.
\end{array} \tag{12.4}
$$

We can write (12.4) as an equivalent maximization problem,

$$
\begin{aligned}
\text{maximize} \quad & \sum_{i=1}^{m}(-b_i)y_i \\
\text{subject to} \quad & \sum_{i=1}^{m}(-a_{ij})y_i \leq -c_j, \ j=1,\ldots,n \\
& y_1,\ldots,y_m \geq 0
\end{aligned}
\tag{12.5}
$$

and then form its dual:

$$
\begin{aligned}
\text{minimize} \quad & \sum_{j=1}^{n}(-c_j)x_j \\
\text{subject to} \quad & \sum_{j=1}^{n}(-a_{ij})x_j \geq -b_i, \ i=1,\ldots,m \\
& x_1,\ldots,x_n \geq 0.
\end{aligned}
\tag{12.6}
$$

Representing (12.6) as an equivalent maximization problem,

$$
\begin{aligned}
\text{maximize} \quad & \sum_{j=1}^{n}c_j x_j \\
\text{subject to} \quad & \sum_{j=1}^{n}a_{ij}x_j \leq b_i, \ i=1,\ldots,m \\
& x_1,\ldots,x_n \geq 0,
\end{aligned}
\tag{12.7}
$$

we obtain an LP that is identical to the original primal LP (12.3).

Given an LP $\mathcal{P}$, let LP $\mathcal{D}$ be its dual. Then the dual of $\mathcal{D}$ is given by $\mathcal{P}$.

Taking into account that the dual of the dual gives the primal LP, we have the following summary of the correspondence between the constraints and variables of the primal and dual LPs, which can be used as rules for forming the dual for a given primal LP. We assume that the primal LP is a maximization problem and the dual LP is a minimization problem.

Primal LP		Dual LP
Equality constraint	$\leftrightarrow$	Free variable
Inequality constraint ($\leq$)	$\leftrightarrow$	Nonnegative variable
Free variable	$\leftrightarrow$	Equality constraint
Nonnegative variable	$\leftrightarrow$	Inequality constraint ($\geq$)

Example 12.1 *The dual of the LP*

$$
\begin{array}{rrcrcl}
\textit{maximize} & 2x_1 & + & x_2 & & \\
\textit{subject to} & x_1 & + & x_2 & = & 2 \\
& 2x_1 & - & x_2 & \geq & 3 \\
& x_1 & - & x_2 & \leq & 1 \\
& x_1 \geq 0, & & x_2 & \in & \mathbb{R}
\end{array}
$$

is given by

$$
\begin{array}{rrcrcrcl}
\textit{minimize} & 2y_1 & - & 3y_2 & + & y_3 & & \\
\textit{subject to} & y_1 & - & 2y_2 & + & y_3 & \geq & 2 \\
& y_1 & + & y_2 & - & y_3 & = & 1 \\
& & & y_1 \in \mathbb{R}, & & y_2, y_3 & \geq & 0.
\end{array}
$$

12.2 Weak Duality and the Duality Theorem

Let z and w denote the objective function of the primal and the dual LP, respectively. Then we have:

$$
\sum_{j=1}^{n} c_j x_j \leq \sum_{j=1}^{n} \left(\sum_{i=1}^{m} a_{ij} y_i \right) x_j = \sum_{i=1}^{m} \left(\sum_{j=1}^{n} a_{ij} x_j \right) y_i \leq \sum_{i=1}^{m} b_i y_i. \quad (12.8)
$$

Hence, for any feasible solution x to the primal LP and any feasible solution y to the dual LP we have the property of *weak duality*:

$$
z = \sum_{j=1}^{n} c_j x_j \leq \sum_{i=1}^{m} b_i y_i = w.
$$

Therefore, if we find a feasible solution x^* for the primal LP and a feasible solution y^* for the dual LP such that

$$
\sum_{j=1}^{n} c_j x_j^* = \sum_{i=1}^{m} b_i y_i^*,
$$

then x^* is an optimal solution of the primal problem, and y^* is an optimal solution of the dual problem. The following theorem establishes the result, known as *strong duality*.

> **Theorem 12.1 (The duality theorem)** *If the primal problem has an optimal solution x^*, then the dual problem also has an optimal solution y^* such that*
>
> $$\sum_{j=1}^{n} c_j x_j^* = \sum_{i=1}^{m} b_i y_i^*.$$

Proof. Assume that the primal problem is a general maximization problem with "$\leq$" and "$=$" constraints only. Let

$$\begin{array}{rl} \text{maximize} & c^T x \\ \text{subject to} & Ax = b \\ & x \geq 0 \end{array} \qquad (12.9)$$

be the associated big-M problem, which has $n + m' + m'' + m'''$ variables and $m = m' + m''$ constraints, where m' is the number of inequality constraints in the original LP, m'' is the number of equality constraints in the original LP, and m''' is the number of inequality constraints in the original LP that required an artificial variable in addition to the slack variable in order to formulate (12.9). Assume that the columns of A correspond to the variables in the following natural order: the first n columns correspond to the original n variables, the next m' columns to the slack variables, followed by m'' and m''' columns corresponding to the artificial variables for m'' equality and m''' inequality constraints, respectively. The dual to the big-M problem is given by

$$\begin{array}{rl} \text{minimize} & b^T y \\ \text{subject to} & A^T y \geq c \\ & y \in \mathbb{R}^n. \end{array} \qquad (12.10)$$

We will use a matrix representation of the optimal tableau obtained using the big-M method for the primal LP given by equation (RT) at page 271:

z	x_B	x_N	rhs
1	0	$-\bar{c}_N^T = -(c_N^T - c_B^T B^{-1} N)$	$c_B^T B^{-1} b$
0	I_m	$B^{-1} N$	$B^{-1} b$

$\qquad (12.11)$

We claim that the vector $y^* = c_B^T B^{-1}$ is an optimal solution to the dual LP (12.10). To prove this, we first show that y^* is feasible for (12.10), i.e., $A^T y^* \geq c$. This inequality can be equivalently represented as a system of two inequalities, $(y^*)^T B \geq c_B^T$ and $(y^*)^T N \geq c_N^T$. We have:

$$(y^*)^T B = c_B^T B^{-1} B = c_B^T,$$

so the first inequality is satisfied at equality. As for the second inequality, we have

$$(y^*)^T N = c_B^T B^{-1} N \geq c_N^T,$$

since $\tilde{c}_N^T = c_N^T - c_B^T B^{-1} N$ must be nonpositive due to optimality of the tableau (12.11).

Knowing that y^* is feasible for the dual LP, it suffices to show that the corresponding objective function value, $b^T y^*$, coincides with the optimal objective value, $c_B^T B^{-1} b$, of the primal big-M LP (12.9). Since $y^* = c_B^T B^{-1}$, we trivially have $b^T y^* = c_B^T B^{-1} b$. Thus, y^* must be optimal for the dual LP (12.10). $\qquad\qquad\square$

12.3 Extracting an Optimal Solution of the Dual LP from an Optimal Tableau of the Primal LP

The proof of the duality theorem above is constructive and provides an easy way of extracting an optimal solution of the dual LP from an optimal tableau of the primal LP. To see this, we first analyze the row 0 coefficients in (12.11) for the slack variables and the artificial variables corresponding to the equality constraints in (12.9).

First, assume that a slack or an artificial variable $x_{n+i}, i = 1, \ldots, m$ is basic in the optimal tableau (12.11). Then, obviously, the coefficient in row 0 of (12.11) is 0 for this variable. We show that in this case the ith component of $c_B^T B^{-1}$ equals 0 (that is, $y_i^* = 0$). Indeed, if we denote by $u^T = c_B^T B^{-1}$, then, multiplying both sides by B from the right, we have $B^T u = c_B$. Since x_{n+i} is the slack or artificial variable for the ith row of the LP, the corresponding column of B has 1 in the ith component and 0 in the remaining components. Also, the entry of c_B corresponding to x_{n+i} is 0. Hence, the row of the system $B^T u = c_B$ corresponding to x_{n+i} gives $u_i = 0$.

Second, assume that a slack variable $x_{n+i}, i = 1, \ldots, m'$ is nonbasic, then $c_{n+i} = 0$, and the column corresponding to x_{n+i} in N has the ith component equal to 1, and the rest components equal to 0. Hence, $\tilde{c}_{n+i} = 0 - y_i = -y_i$, where y_i is the ith component of $c_B^T B^{-1}$, $i = 1, \ldots, m'$. As for an artificial variable $x_{n+i}, i = m' + 1, \ldots, m$, we have $c_{n+i} = -M$, and if it is nonbasic then $\tilde{c}_{n+i} = -M - y_i$, where, again, y_i is the ith component of $c_B^T B^{-1}$, $i = m' + 1, \ldots, m$. But $c_B^T B^{-1}$ is exactly the dual optimal solution y^* we used to prove the duality theorem.

Thus, the coefficient for $x_{n+i}, i = 1, \ldots, m$ in row 0 of the optimal tableau (12.11) is always given by y_i^* if x_{n+i} is a slack variable and by $M + y_i^*$ if x_{n+i} is the artificial variable representing the ith constraint. This observation can be summarized as follows.

We consider a maximization LP with "$\leq$," "$\geq$," and "$=$" constraints. For convenience, we convert each "$\geq$" constraint into an equivalent "$\leq$" constraint by multiplying both sides by -1. Hence, we only need to deal with "$\leq$" and "$=$" constraints. Let s_i be the slack variable for the ith constraint if the ith

constraint is a "$\leq$" constraint, and let a_i be the artificial variable for the ith constraint if the ith constraint is a "$=$" constraint used in the big-M method. Then the following table represents the correspondence between the optimal value y_i^* of the dual variable corresponding to the ith constraint and the coefficients of s_i or a_i in row 0 of the optimal tableau.

i^{th} constraint type in the primal LP		y_i* (found in row 0 of the optimal primal LP tableau)
$\leq$	$\rightarrow$	coefficient of s_i
$=$	$\rightarrow$	(coefficient of a_i)$-M$

(Note that this correspondence is for the tableau format; sign changes would have to be made in the dictionary format).

12.4 Correspondence between the Primal and Dual LP Types

From the duality theorem we know that if the primal LP is optimal, then so is its dual. Next we investigate the correspondence between the primal and dual LP types in all possible cases. More specifically, the question we are addressing here is, what can we say about the type of the dual (primal) LP if we know the primal (dual) LP type? Recall than an LP can either be optimal or infeasible, or unbounded.

To establish the correspondence sought, we use the fact that the dual of the dual LP is the primal LP. Consider a primal LP $\mathcal{P}$ with the objective function denoted by $z(x)$ and its dual $\mathcal{D}$ with the objective function denoted by $w(y)$. Then according to the duality theorem, $\mathcal{P}$ is an optimal LP if and only if $\mathcal{D}$ is an optimal LP. If $\mathcal{P}$ is unbounded, then $\mathcal{D}$ must be infeasible, since if $\mathcal{D}$ had a feasible solution $\bar{y}$, then $z(x) \leq w(\bar{y})$ for any feasible solution x of $\mathcal{P}$, contradicting the unboundedness of $\mathcal{P}$. Finally, if $\mathcal{P}$ is infeasible, then since the dual of the dual is primal and the dual of an unbounded LP is infeasible, it is possible that $\mathcal{D}$ is unbounded. It is also possible that $\mathcal{D}$ is infeasible, as shown in the following example.

Example 12.2 (An infeasible LP with infeasible dual) *Consider the LP*

$$
\begin{aligned}
maximize \quad & 5x_1 - 2x_2 \\
subject\ to \quad & 2x_1 - x_2 \leq 2 \\
& -2x_1 + x_2 \leq -3 \\
& x_1, x_2 \geq 0.
\end{aligned}
$$

This LP is, clearly, infeasible. The dual of this LP is

$$
\begin{aligned}
minimize \quad & 2y_1 \; - \; 3y_2 \\
subject\ to \quad & 2y_1 \; - \; 2y_2 \; \geq \; 5 \\
& -y_1 \; + \; y_2 \; \geq \; -2 \\
& x_1, x_2 \; \geq \; 0,
\end{aligned}
$$

which is also infeasible.

The correspondence between the primal and dual LP types is summarized below.

		Dual LP		
		Optimal	Infeasible	Unbounded
Primal LP	Optimal	possible	impossible	impossible
	Infeasible	impossible	possible	possible
	Unbounded	impossible	possible	impossible

12.5   Complementary Slackness

In this section, we explore the relationship between the constraints of the primal LP and the variables of its dual when both problems are optimal. Let the primal LP be given by

$$
\begin{aligned}
maximize \quad & \sum_{j=1}^{n} c_j x_j \\
subject\ to \quad & \sum_{j=1}^{n} a_{ij} x_j \; \leq \; b_i, \; i = 1, \dots, m \\
& x_1, \dots, x_n \; \geq \; 0.
\end{aligned}
\tag{P}
$$

Then the corresponding dual LP is given by

$$
\begin{aligned}
minimize \quad & \sum_{i=1}^{m} b_i y_i \\
subject\ to \quad & \sum_{i=1}^{m} a_{ij} y_i \; \geq \; c_j, \; j = 1, \dots, n \\
& y_1, \dots, y_m \; \geq \; 0.
\end{aligned}
\tag{D}
$$

Let x^* and y^* be optimal solutions to the primal problem (P) and the dual problem (D), respectively. Then the strong duality holds,

$$\sum_{j=1}^{n} c_j x_j^* = \sum_{i=1}^{m} b_i y_i^*,$$

and from the derivation of the weak duality property we have

$$\sum_{j=1}^{n} c_j x_j^* \overset{(a)}{\leq} \sum_{j=1}^{n} \left(\sum_{i=1}^{m} a_{ij} y_i^* \right) x_j^* = \sum_{i=1}^{m} \left(\sum_{j=1}^{n} a_{ij} x_j^* \right) y_i^* \overset{(a)}{\leq} \sum_{i=1}^{m} b_i y_i^*. \qquad (12.12)$$

Note that the equality in the inequality (a) above is possible if and only if for every $j = 1, \ldots, n$, either $x_j^* = 0$ or $c_j = \sum_{i=1}^{m} a_{ij} y_i^*$. Similarly, the equality in the inequality (b) above is possible if and only if for every $i = 1, \ldots, m$, either $y_i^* = 0$ or $b_i = \sum_{j=1}^{n} a_{ij} x_j^*$. Thus we obtain the following result.

Theorem 12.2 (Complementary Slackness) *Let $x^* = [x_1^*, \ldots, x_n^*]^T$ be a feasible solution of the primal LP (P) and $y^* = [y_1^*, \ldots, y_m^*]^T$ be a feasible solution of the dual LP (D). Then x^* is an optimal solution of (P) and y^* is an optimal solution of (D) simultaneously if and only if both of the following statements hold:*

$$x_j^* = 0 \ or \ \sum_{i=1}^{m} a_{ij} y_i^* = c_j \ for \ all \ j = 1, \ldots, n \qquad (12.13)$$

and

$$y_i^* = 0 \ or \ \sum_{j=1}^{n} a_{ij} x_j^* = b_i \ for \ all \ i = 1, \ldots, m. \qquad (12.14)$$

The theorem above implies that a feasible solution $x^* = [x_1^*, \ldots, x_n^*]^T$ of (P) is optimal if and only if there exist numbers $y_1^*, \ldots, y_m^*$ such that

- if $x_j^* > 0$ then $\sum_{i=1}^{m} a_{ij} y_i^* = c_j$; $j = 1, \ldots, n$,

- if $\sum_{j=1}^{n} a_{ij} x_j^* < b_j$ then $y_i^* = 0$; $i = 1, \ldots, m$,

- $\sum_{i=1}^{m} a_{ij} y_i^* \geq c_j$, $j = 1, \ldots, n$, and

- $y_i^* \geq 0$, $i = 1, \ldots, m$.

We can use these conditions to test optimality of a given solution to an LP.

Example 12.3 *Check if $\tilde{x} = [9, 0, 11, 5, 0, 4, 3]^T$ is an optimal solution of the following LP:*

$$
\begin{array}{llllllll}
\text{minimize} & x_1 & +x_2 & +x_3 & +x_4 & +x_5 & +x_6 & +x_7 \\
\text{subject to} & x_1 & & & & & +x_6 & +x_7 & \geq & 16 \\
& x_1 & +x_2 & & & & & +x_7 & \geq & 12 \\
& x_1 & +x_2 & +x_3 & & & & & \geq & 18 \\
& & x_2 & +x_3 & +x_4 & & & & \geq & 13 \\
& & & x_3 & +x_4 & +x_5 & & & \geq & 15 \\
& & & & x_4 & +x_5 & +x_6 & & \geq & 9 \\
& & & & & x_5 & +x_6 & +x_7 & \geq & 7 \\
& x_1 & +x_2 & +x_3 & -x_4 & -x_5 & -x_6 & -x_7 & \geq & 0 \\
& & & & & & x_1, x_2, x_3, x_4, x_5, x_6, x_7 & \geq & 0.
\end{array}
$$

The dual of this LP is given by

$$
\begin{array}{llllllll}
\text{maximize} & 16y_1 + 12y_2 + 18y_3 + 13y_4 + 15y_5 + 9y_6 + 7y_7 \\
\text{subject to} & y_1 & +y_2 & +y_3 & & & & & +y_8 & \leq & 1 \\
& & y_2 & +y_3 & +y_4 & & & & +y_8 & \leq & 1 \\
& & & y_3 & +y_4 & +y_5 & & & +y_8 & \leq & 1 \\
& & & & y_4 & +y_5 & +y_6 & & -y_8 & \leq & 1 \\
& & & & & y_5 & +y_6 & +y_7 & -y_8 & \leq & 1 \\
& y_1 & & & & & +y_6 & +y_7 & -y_8 & \leq & 1 \\
& y_1 & +y_2 & & & & & +y_7 & -y_8 & \leq & 1 \\
& & & & & y_1, y_2, y_3, y_4, y_5, y_6, y_7, y_8 & \geq & 0.
\end{array}
$$

First, we determine which of the constraints are not binding at $\tilde{x}$:

$$
\begin{array}{llllllllll}
\tilde{x}_1 & & & & & +\tilde{x}_6 & +\tilde{x}_7 & = & 16 \\
\tilde{x}_1 & +\tilde{x}_2 & & & & & +\tilde{x}_7 & = & 12 \\
\tilde{x}_1 & +\tilde{x}_2 & +\tilde{x}_3 & & & & & = & 20 & > & 18 \\
& \tilde{x}_2 & +\tilde{x}_3 & +\tilde{x}_4 & & & & = & 16 & > & 13 \\
& & \tilde{x}_3 & +\tilde{x}_4 & +\tilde{x}_5 & & & = & 16 & > & 15 \\
& & & \tilde{x}_4 & +\tilde{x}_5 & +\tilde{x}_6 & & = & 9 \\
& & & & \tilde{x}_5 & +\tilde{x}_6 & +\tilde{x}_7 & = & 7 \\
\tilde{x}_1 & +\tilde{x}_2 & +\tilde{x}_3 & -\tilde{x}_4 & -\tilde{x}_5 & -\tilde{x}_6 & -\tilde{x}_7 & = & 8 & > & 0
\end{array}
$$

We see that the 3rd, 4th, 5th, and 8th constraints are not binding, thus, due to the complementary slackness, if $\tilde{x}$ is optimal for the primal problem, we must have

$$\tilde{y}_3 = \tilde{y}_4 = \tilde{y}_5 = \tilde{y}_8 = 0$$

in an optimal solution $\tilde{y}$ of the dual problem.

Also observe that $\tilde{x}_1, \tilde{x}_3, \tilde{x}_4, \tilde{x}_6, \tilde{x}_7$ are all positive, thus the 1st, 3rd, 4th, 6th, and 7th constraints of the dual LP must be binding at an optimal point $\tilde{y}$. However, this is impossible since taking $\tilde{y}_3 = \tilde{y}_4 = \tilde{y}_5 = \tilde{y}_8 = 0$ in the third constraint of the dual problem as binding gives $0 = 1$. Therefore, $\tilde{x}$ is not optimal for the considered LP. Noting that if we reduce $\tilde{x}_3$ by 1 then the resulting vector $x^* = [9, 0, 10, 5, 0, 4, 3]^T$ is still feasible for the primal LP and its objective function value (31) is better than that of $\tilde{x}$ (32), we see that $\tilde{x}$ is indeed not optimal.

Next, we will use the complementary slackness to show that x^* is, in fact, an optimal solution for the given primal LP. For x^*, the 5th constraint becomes binding, whereas the 3rd, 4th, and 8th constraints remain nonbinding. Thus,

$$y_3^* = y_4^* = y_8^* = 0$$

in an optimal solution y^* of the dual problem. As with $\tilde{x}$, since the same components of x^* are positive, we must have the 1st, 3rd, 4th, 6th, and 7th constraints of the dual LP binding at an optimal point y^*. Taking into account that $y_3^* = y_4^* = y_8^* = 0$, this gives the following system:

$$
\begin{aligned}
y_1^* + y_2^* & & & = 1 \\
y_5^* & & & = 1 \\
y_5^* + y_6^* & & & = 1 \\
y_1^* + y_6^* + y_7^* & & & = 1 \\
y_1^* + y_2^* + y_7^* & & & = 1.
\end{aligned}
$$

Solving this system, we find $y_1^* = y_5^* = 1$, $y_2^* = y_6^* = y_7^* = 0$, so

$$y^* = [1, 0, 0, 0, 1, 0, 0, 0]^T.$$

It remains to check that y^* satisfies the 2nd and 5th constraints of the dual problem, and that the dual objective function value at y^* is 31, which is the same as the primal objective function value at x^*. We conclude that x^* is optimal for the primal problem and y^* is optimal for the dual problem.

Finally, note that the problem we considered in this example is the scheduling LP formulated in Section 10.2.3 (page 215).

12.6   Economic Interpretation of the Dual LP

Consider the following problem. Romeo Winery produces two types of wines, Bordeaux and Romerlot, by blending the Merlot and Cabernet Sauvignon grapes. Making one barrel of Bordeaux blend requires 250 pounds of Merlot and 250 pounds of Cabernet Sauvignon, whereas making one barrel of Romerlot requires 450 pounds of Merlot and 50 pounds of Cabernet Sauvignon. The profit received from selling Bordeaux is $800 per barrel, and from selling Romerlot, $600 per barrel. Romeo Winery has 9,000 pounds of Merlot and 5,000 pounds of Cabernet Sauvignon available.

We formulate an LP model aiming to maximize the winery's revenue. Let x_1 and x_2 be the amount (in barrels) of Bordeaux and Romerlot made, respectively. Then we have:

$$\begin{aligned}
\text{maximize} \quad & 800x_1 + 600x_2 \\
\text{subject to} \quad & 250x_1 + 450x_2 \leq 9,000 \quad \text{(Merlot constraint)} \\
& 250x_1 + 50x_2 \leq 5,000 \quad \text{(Cabernet constraint)} \\
& x_1, x_2 \geq 0.
\end{aligned}$$

$$(12.15)$$

The dual of this LP is given by

$$\begin{aligned}
\text{minimize} \quad & 9,000y_1 + 5,000y_2 \\
\text{subject to} \quad & 250y_1 + 250y_2 \geq 800 \\
& 450y_1 + 50y_2 \geq 600 \\
& y_1, y_2 \geq 0.
\end{aligned}$$

$$(12.16)$$

The optimal solution of the primal LP (12.15) is $x_1^* = 18, x_2^* = 10$. The optimal solution of the dual LP (12.16) is $y_1^* = 1.1, y_2^* = 2.1$. The optimal objective value of both the primal and dual LP is 20,400.

We start the analysis of the economic meaning of the dual LP by determining the units of measure for the dual variables y_1 and y_2. Consider the first constraint of the dual LP:

$$250y_1 + 250y_2 \geq 800.$$

The coefficient for both y_1 and y_2 is 250 pounds/barrel, and the right-hand side is 800 dollars/barrel. Thus, for the constraint to make a physical sense, the units for both dual variables must be dollars/pound, meaning that y_1 and y_2 express the cost of grapes.

Note that multiplying the first and the second constraint of the primal LP (12.15) by $y_1^* = 1.1$ and $y_2^* = 2.1$, respectively, and then summing up the left-hand sides of the resulting inequalities gives the primal objective function:

$$z = 800x_1 + 600x_2 = (250x_1 + 450x_2)1.1 + (250x_1 + 50x_2)2.1.$$

Now, assume that the winery is considering purchasing p_1 pounds of Merlot grapes in addition to 9,000 pounds already available. The new primal LP is then

$$\begin{aligned}
\text{maximize} \quad & 800x_1 + 600x_2 \\
\text{subject to} \quad & 250x_1 + 450x_2 \leq 9,000 + p_1 \\
& 250x_1 + 50x_2 \leq 5,000 \\
& x_1, x_2 \geq 0.
\end{aligned}$$

$$(12.17)$$

If we multiply the constraints by y_1^* and y_2^*, respectively,

$$\begin{aligned}
\text{maximize} \quad & 800x_1 + 600x_2 & & \\
\text{subject to} \quad & 250x_1 + 450x_2 \leq 9,000 + p_1 & & \times y_1^* = 1.1 \\
& 250x_1 + 50x_2 \leq 5,000 & & \times y_2^* = 2.1 \\
& x_1, x_2 \geq 0,
\end{aligned}$$

then for any feasible solution we have:

$$
\begin{aligned}
z = 800x_1 + 600x_2 &= (250x_1 + 450x_2)1.1 + (250x_1 + 50x_2)2.1 \\
&\leq (9,000 + p_1)1.1 + 5,000 \cdot 2.1 \\
&= 20,400 + 1.1p_1.
\end{aligned}
$$

Thus, we got the following upper bound

$$
z \leq 20,400 + 1.1p_1
$$

on the objective, in which the optimal value of the first dual variable $y_1^* = 1.1$ is the coefficient for the variable representing the extra Merlot grapes. The extra profit added to the currently optimal profit of \$20,400 will never exceed $1.1p_1$, thus \$1.1/pound is the maximum *extra* amount the winery should be willing to pay for additional Merlot grapes.

Similarly, if the winery looks to purchase p_2 more pounds of Cabernet Sauvignon grapes, we obtain the upper bound

$$
\begin{aligned}
z = 800x_1 + 600x_2 &= (250x_1 + 450x_2)1.1 + (250x_1 + 50x_2)2.1 \\
&\leq 9,000 \cdot 1.1 + (5,000 + p_2)2.1 \\
&= 20,400 + 2.1p_2,
\end{aligned}
$$

i.e.,

$$
z \leq 20,400 + 2.1p_2,
$$

implying that the company should not pay more than \$2.1/pound in addition to what they already pay for Cabernet Sauvignon grapes. This quantity is sometimes referred to as the *shadow price*.

> **Definition 12.1** *The shadow price for the ith resource constraint of an LP is defined as the amount by which the optimal objective function value is improved if the right-hand side of this constraint is increased by 1.*

12.7 Sensitivity Analysis

Sensitivity analysis studies how changes in the problem's input data impact its solution. We will first illustrate the idea geometrically using the Heavenly Pouch LP formulated in Section 10.1 and solved graphically in Section 10.4

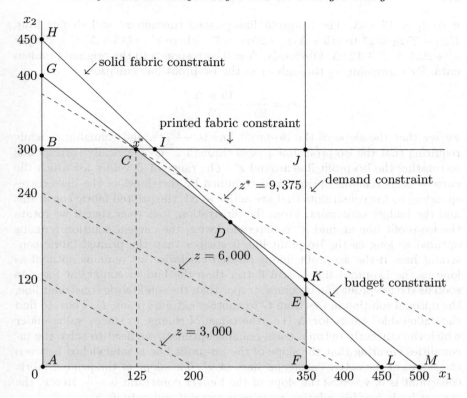

FIGURE 12.1: Graphical solution of the Heavenly Pouch LP.

(see also Sections 11.2 and 11.3). The Heavenly Pouch LP is given by:

$$
\begin{array}{lrcrcll}
\text{maximize} & 15x_1 & + & 25x_2 \\
\text{subject to} & x_1 & + & x_2 & \leq & 450 & \text{(solid color fabric constraint)} \\
& & & x_2 & \leq & 300 & \text{(printed fabric constraint)} \\
& 4x_1 & + & 5x_2 & \leq & 2{,}000 & \text{(budget constraint)} \\
& x_1 & & & \leq & 350 & \text{(demand constraint)} \\
& & & x_1, x_2 & \geq & 0 & \text{(nonnegativity constraints)},
\end{array}
$$

and its graphical solution is shown in Figure 12.1. The optimal solution is $x* = [125, 300]^T$, $z^* = 9,375$. What happens to this solution if some of the coefficients of the LP change? What changes to the problem parameters are allowed if we require that the currently optimal basis must still remain optimal after the changes take effect? These are some of the questions addressed using sensitivity analysis.

First, we consider a situation when there is a change in one of the objective function coefficients, say c_1, which corresponds to changing the profit obtained from each non-reversible carrier. Currently, $c_1 = 15$, and we consider changing it to $\bar{c}_1 = 15 + \Delta$. The iso-profit line passing through x^* will change from $15x_1 + 25x_2 = z^*$ to $(15 + \Delta)x_1 + 25x_2 = \bar{z}^*$, where $\bar{z}^* = (15 + \Delta)x_1^* + 25x_2^* = z^* + \Delta x_1^* = z^* + 125\Delta$. Obviously, $\Delta = 0$ corresponds to the original problem data. By expressing x_2 through x_1 in the iso-profit line equation,

$$x_2 = \frac{\bar{z}^*}{25} - \frac{15 + \Delta}{25}x_1,$$

we see that the slope of the iso-profit line is $-\frac{15+\Delta}{25}$, and changing Δ while requiring that the iso-profit line passes through x^* geometrically corresponds to rotating the iso-profit line around x^*. The range of Δ values for which the current basis remains optimal is determined by the slopes of the lines corresponding to the constraints that are active at x^*, the printed fabric constraint and the budget constraint. From the illustration, it is clear that if we rotate the iso-profit line around x^* counter-clockwise, the current solution remains optimal as long as the iso-profit line is steeper than the printed fabric constraint line. If the iso-profit line is rotated clockwise, x^* remains optimal as long as the iso-profit line is still flatter than the budget constraint line. As soon as the iso-profit line becomes steeper than the solid fabric constraint line, the optimal solution moves from C to another extreme point, D. Thus, to find the "allowable" range for Δ (i.e., the range of change in the c_1 value under which the currently optimal basis remains optimal), we need to solve the inequalities ensuring that the slope of the iso-profit line is sandwiched between the slopes of the active constraint lines at x^*. The slope of the printed fabric constraint is 0, whereas the slope of the budget constraint is $-\frac{4}{5}$. Hence, the current basic feasible solution remains optimal if and only if

$$-\frac{4}{5} \leq -\frac{15 + \Delta}{25} \leq 0 \quad \Leftrightarrow \quad -15 \leq \Delta \leq 5.$$

Thus, with $-15 \leq \Delta \leq 5$, x^* remains optimal. Recall that the case of $\Delta = 5$ was solved graphically in Figure 10.4 (page 227).

Next we analyze an example where there is a change in a constraint's right-hand side. To be specific, assume that the right-hand side of the budget constraint is changed from $2,000$ to $2,000 + \Delta$. Change in the right-hand side does not affect the line's slope, and geometrically, such a change corresponds to a move of the line on parallel to its original position. We are interested in what is happening to the optimal point C, which is defined by the intersection of the printed fabric constraint and the budget constraint lines.

If Δ increases from 0 to a positive value, point C moves to the right, toward I $(x_1 = 150, x_2 = 300)$, at which point the total budget is $4 \times 150 + 5 \times 300 = 2,100$, i.e., $\Delta = 2,100 - 2,000 = 100$. If we keep increasing Δ beyond that point, C becomes infeasible, and I is the optimal point. Thus, the highest allowable increase for the right-hand side of the budget constraint is $\Delta = 100$.

If, on the other hand, Δ is negative, then as Δ grows in absolute value, the budget constraint line moves down, point C moves to the left, and D moves toward E along the solid fabric constraint line. When $\Delta = -100$, D turns

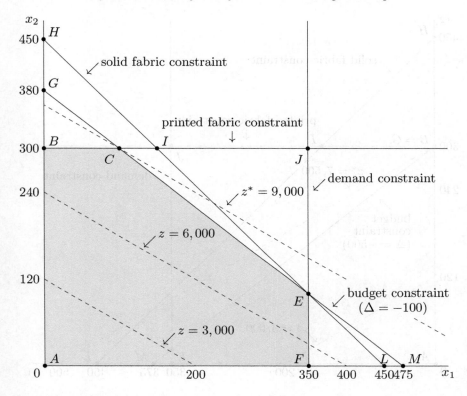

FIGURE 12.2: Graphical solution of the Heavenly Pouch LP with changed right-hand side for the budget constraint ($\Delta = -100$).

into E (see Figure 12.2). At this point, C is still the optimal solution, with $x_1^* = 100, x_2^* = 300, z^* = 9,000$. When $\Delta = -500$, both C and G converge to B (see Figure 12.3), which becomes the optimal point with $x_1^* = 0, x_2^* = 300, z^* = 7,500$. If Δ keeps increasing in the absolute value after that, the optimum will relocate to point G, which is the point on the intersection of the budget constraint line and x_2 axis. Thus, $\Delta = -500$ gives the highest decrease in the budget constraint's right-hand side that does not alter the optimality of the basis.

In summary, based on the graphical sensitivity analysis we conclude that $-500 \leq \Delta \leq 100$ is the allowable range of change for the budget constraint's right-hand side.

Next, we will discuss how the sensitivity analysis can be performed in algebraic form. Unlike the graphical approach, these techniques can be applied to LPs with an arbitrary number of variables. Therefore, we consider a general

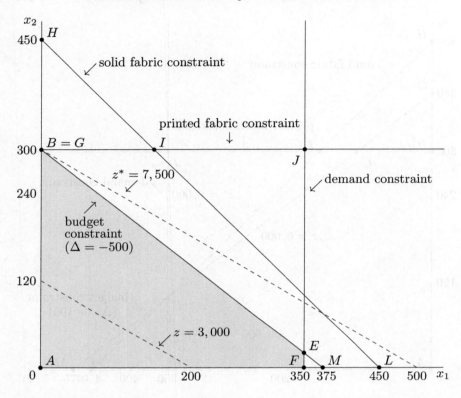

FIGURE 12.3: Graphical solution of the Heavenly Pouch LP with changed right-hand side for the budget constraint ($\Delta = -500$).

LP in the standard form

$$\text{maximize} \quad c^T x$$
$$\text{subject to} \quad Ax = b \qquad (12.18)$$
$$x \geq 0.$$

The cases we discuss include:

- changing a coefficient in the objective function,
- changing the column of a nonbasic variable,
- changing the right-hand side,
- introducing a new variable, and
- introducing a new constraint.

The analysis will be based on observations made regarding the simplex tableaux in the matrix form, as introduced in Section 11.6:

z	x_B	x_N	rhs	
1	0	$-(c_N^T - c_B^T B^{-1} N)$	$c_B^T B^{-1} b$	(12.19)
0	I_m	$B^{-1} N$	$B^{-1} b$	

More specifically, we will use the formulas in (12.19) to answer the following important questions:

(1) Will the current basis still remain feasible after a given change in the problem input data?

(2) Will the current basis still remain optimal after a given change in the problem input data?

(3) How can we find a new optimal solution if the feasibility or optimality of the optimal basis for the original LP is altered by the change in the input data?

The first question can be addressed by checking whether the vector of the right-hand sides of the constraints given by $B^{-1}b$ in (12.19) has only nonnegative entries. To answer the second question, we need to verify the signs of the entries of nonbasic variable coefficients in the objective given by $c_N^T - c_B^T B^{-1} N$. To address the third question, note that the types of changes we analyze can be grouped into two categories, the first one being the changes that impact optimality (determined based on $c_N^T - c_B^T B^{-1} N$) of the current solution but do not impact its feasibility (i.e., the sign of $B^{-1}b$ entries), and the second one representing the changes that may alter feasibility, but do not impact the vector of nonbasic variable coefficients in row 0 of the optimal tableau. The first group includes changing a coefficient in the objective function, introducing a new variable, changing the column of a nonbasic variable, and the second group includes changing the right-hand side and introducing a new constraint. Based on this classification, we will have two different recipes for dealing with the third question.

We will illustrate the underlying ideas using the following LP, which was solved using the revised simplex method in Section 11.6:

$$
\begin{array}{lrcrcrcl}
\text{maximize} & 4x_1 & + & 3x_2 & + & 5x_3 & & \\
\text{subject to} & x_1 & + & 2x_2 & + & 2x_3 & \leq & 4 \\
& 3x_1 & & & + & 4x_3 & \leq & 6 \\
& 2x_1 & + & x_2 & + & 4x_3 & \leq & 8 \\
& & & & x_1, x_2, x_3 & \geq & 0.
\end{array}
\tag{12.20}
$$

Recall that this LP represents a production planning problem, where x_j is the number of units of product j to be produced; the objective coefficient c_j is the profit obtained from selling a unit of product j. The constraints represent limitations on three different resources used in the production process, with the coefficient a_{ij} for variable x_j in the constraint i standing for the amount of resource i used in the production of 1 unit of product j, and the right-hand side b_i in the constraint i representing the amount of resource i available; $i, j = 1, 2, 3$. As before, let x_4, x_5, and x_6 be the slack variables for (12.20).

The optimal basis for this problem consists of variables x_1, x_2, and x_6, and the corresponding basic optimal solution is given by

$$x^* = [2, 1, 0, 0, 0, 3]^T,\ z^* = 11.$$

We will refer to this solution and the corresponding basis as current. The corresponding tableau is given by

z	x_B	x_N	rhs	
1	0	$-\tilde{c}_N = -(c_N^T - c_B^T B^{-1} N)$	$c_B^T B^{-1} b$	(12.21)
0	I_m	$B^{-1} N$	$\tilde{x}_B = B^{-1} b$	

Assume that (12.21) represents the optimal tableau for LP (12.20), with

$$\begin{aligned} x_B^T &= [x_1, x_2, x_6], \\ \tilde{x}_B^T &= [2, 1, 3], \\ \tilde{c}_N^T &= [-4/3, -3/2, -5/6], \end{aligned}$$

(12.22)

as was shown in Section 11.6 (page 275).

We consider the types of changes in the problem data outlines above and analyze how sensitive the current optimal basis is to these changes.

12.7.1 Changing the objective function coefficient of a basic variable

Assume that the objective function coefficient of a basic variable x_1 is changed from $c_1 = 4$ to $c_1 = 4 + \Delta$. Then the modified problem is

$$\begin{aligned} \text{maximize}\quad & (4+\Delta)x_1 + 3x_2 + 5x_3 \\ \text{subject to}\quad & x_1 + 2x_2 + 2x_3 \leq 4 \\ & 3x_1 \qquad\quad + 4x_3 \leq 6 \\ & 2x_1 + x_2 + 4x_3 \leq 8 \\ & x_1, x_2, x_3 \geq 0, \end{aligned}$$

(12.23)

and (12.19) is modified as follows:

z	x_B	x_N	rhs	
1	0	$-\bar{c}_N = -(c_N^T - c_B^T B^{-1}N)$	$c_B^T B^{-1}b$	(12.24)
0	I_m	$B^{-1}N$	$B^{-1}b$	

Here, c_B is the vector resulting from c_B after changing c_1 to c_1, i.e.,

$$c_B^T = c_B^T + [\Delta, 0, 0] = [c_1, c_2, c_6] + [\Delta, 0, 0] = [4 + \Delta, 3, 0].$$

Thus, the objective function coefficients for nonbasic variables in the tableau of the modified problem with the same basis as current are given by

$$\bar{c}_N^T = c_N^T - c_B^T B^{-1}N.$$

To compute $c_B^T B^{-1}N$, we first compute $u^T = c_B^T B^{-1}$ by solving the system $B^T u = c_B$ and then computing $u^T N$.

$$B^T u = \begin{bmatrix} 1 & 3 & 2 \\ 2 & 0 & 1 \\ 0 & 0 & 1 \end{bmatrix} \begin{bmatrix} u_1 \\ u_2 \\ u_3 \end{bmatrix} = \begin{bmatrix} 4 + \Delta \\ 3 \\ 0 \end{bmatrix} \Leftrightarrow \begin{bmatrix} u_1 \\ u_2 \\ u_3 \end{bmatrix} = \begin{bmatrix} 3/2 \\ 5/6 + \Delta/3 \\ 0 \end{bmatrix}.$$

We have:

$$u^T N = [3/2, 5/6 + \Delta/3, 0] \begin{bmatrix} 2 & 1 & 0 \\ 4 & 0 & 1 \\ 4 & 0 & 0 \end{bmatrix} = [(19 + 4\Delta)/3, \ 3/2, \ 5/6 + \Delta/3].$$

So,

$$\begin{aligned} \bar{c}_N^T = c_N^T - c_B^T B^{-1}N &= [5, 0, 0] - [(19 + 4\Delta)/3, \ 3/2, \ 5/6 + \Delta/3] \\ &= [-4(1 + \Delta)/3, \ -3/2, \ -5/6 - \Delta/3]. \end{aligned}$$

The current solution will remain optimal if and only if

$$[-4(1 + \Delta)/3, \ -3/2, \ -5/6 - \Delta/3] \leq 0,$$

which is the case if and only if $\Delta \geq -1$.

12.7.2 Changing the objective function coefficient of a nonbasic variable

Assume that the objective function coefficient of a nonbasic variable x_3 is changed from $c_3 = 5$ to $c_3 = 5 + \Delta$. Then the modified problem is

$$\begin{aligned} \text{maximize} \quad & 4x_1 + 3x_2 + (5 + \Delta)x_3 \\ \text{subject to} \quad & x_1 + 2x_2 + 2x_3 \leq 4 \\ & 3x_1 + 4x_3 \leq 6 \\ & 2x_1 + x_2 + 4x_3 \leq 8 \\ & x_1, x_2, x_3 \geq 0, \end{aligned} \tag{12.25}$$

and the tableau (12.19) becomes:

z	x_B	x_N	rhs	
1	0	$-\bar{c}_N = -(\boldsymbol{c}_N^T - c_B^T B^{-1} N)$	$c_B^T B^{-1} b$	(12.26)
0	I_m	$B^{-1}N$	$B^{-1}b$	

Hence, the objective function coefficients for nonbasic variables in the tableau of the modified problem with the same basis as current are given by

$$\bar{c}_N^T = \boldsymbol{c}_N^T - c_B^T B^{-1} N,$$

where

$$\boldsymbol{c}_N^T = c_N^T + [\Delta, 0, 0] = [c_3, c_4, c_5] + [\Delta, 0, 0] = [5 + \Delta, 0, 0].$$

Recalling from (12.22) that

$$\tilde{c}_N^T = c_N^T - c_B^T B^{-1} N = [-4/3, -3/2, -5/6],$$

we have:

$$\begin{aligned}
\bar{c}_N^T &= c_N^T - c_B^T B^{-1} N \\
&= c_N^T + [\Delta, 0, 0] - c_B^T B^{-1} N \\
&= c_N^T - c_B^T B^{-1} N + [\Delta, 0, 0] \\
&= [-4/3, -3/2, -5/6] + [\Delta, 0, 0] \\
&= [-4/3 + \Delta, -3/2, -5/6].
\end{aligned}$$

The current solution will remain optimal if and only if

$$[-4/3 + \Delta, -3/2, -5/6] \leq 0,$$

which is true if and only if $\Delta \leq 4/3$.

Thus, if $\Delta \leq 4/3$, the current solution is still optimal, and the new optimal objective function value can be easily computed. If $\Delta > 4/3$, then the current basis is still feasible, and we can use the corresponding tableau to compute a new optimal solution with the simplex method.

Definition 12.2 *The absolute value of the coefficient of a nonbasic variable in row 0 of an optimal simplex tableau is called the reduced cost of this variable.*

In other words, the reduced cost of a nonbasic variable x_j is the minimum amount by which c_j needs to be increased in order for x_j to become a basic variable. For example, to make producing product 3 reasonable, we need to either increase its price or reduce its manufacturing cost by at least 4/3.

12.7.3 Changing the column of a nonbasic variable

Assume that in addition to changing c_3 to $\boldsymbol{c}_3$, we also change the column N_{x_3} of N corresponding to x_3 to $\boldsymbol{N}_{x_3}$. The resulting tableau corresponding to the current basis is then given by:

z	x_B	x_N	rhs	
1	0	$-\bar{c}_N = -(c_N^T - c_B^T B^{-1}N)$	$c_B^T B^{-1}b$	(12.27)
0	I_m	$B^{-1}N$	$B^{-1}b$	

where N is the matrix resulting from N as a result of changing N_{x_3} to $\boldsymbol{N}_{x_3}$. The current basis may not be optimal anymore as a result of the change, however, the modification does not impact the feasibility of the current basis. We could apply the simplex method starting with the feasible tableau above to obtain a solution to the modified problem.

12.7.4 Changing the right-hand side

Assume that the right-hand side vector $b = [4, 6, 8]^T$ is changed to

$$\boldsymbol{b} = b + [\Delta, 0, 0]^T = [4 + \Delta, 6, 8]^T.$$

Then the modified problem is

$$
\begin{array}{llllllll}
\text{maximize} & 4x_1 & + & 3x_2 & + & 5x_3 & & \\
\text{subject to} & x_1 & + & 2x_2 & + & 2x_3 & \leq & 4 + \Delta \\
& 3x_1 & & & + & 4x_3 & \leq & 6 \\
& 2x_1 & + & x_2 & + & 4x_3 & \leq & 8 \\
& & & & x_1, x_2, x_3 & \geq & 0.
\end{array}
\tag{12.28}
$$

Recall that (12.21) represents the optimal tableau for the original LP (12.20). Then the tableau with the same basis for problem (12.28) is given by:

z	x_B	x_N	rhs	
1	0	$-(c_N^T - c_B^T B^{-1}N)$	$c_B^T B^{-1}\boldsymbol{b}$	(12.29)
0	I_m	$\bar{x} = B^{-1}N$	$B^{-1}\boldsymbol{b}$	

The corresponding values for basic variables are given by $\bar{x} = B^{-1}\boldsymbol{b}$. Thus, the basic solution corresponding to the tableau (12.29) will be feasible if and only if $B^{-1}\boldsymbol{b} \geq 0$. Note that if this is the case, then this tableau is optimal (since the coefficients for nonbasic variables are all nonpositive due to the optimality of tableau (12.21) for problem (12.20)).

To compute $v = B^{-1}b$, we solve the system $Bv = b$ for v:

$$\begin{bmatrix} 1 & 2 & 0 \\ 3 & 0 & 0 \\ 2 & 1 & 1 \end{bmatrix} \begin{bmatrix} v_1 \\ v_2 \\ v_3 \end{bmatrix} = \begin{bmatrix} 4 + \Delta \\ 6 \\ 8 \end{bmatrix} \Leftrightarrow \begin{bmatrix} v_1 \\ v_2 \\ v_3 \end{bmatrix} = \begin{bmatrix} 2 \\ 1 + \Delta/2 \\ 3 - \Delta/2 \end{bmatrix}, \quad (12.30)$$

and $v \geq 0$ if and only if $-2 \leq \Delta \leq 6$. Thus, the basis consisting of x_1, x_2, and x_6 will remain optimal if and only if $-2 \leq \Delta \leq 6$. Next we consider two concrete examples of the right-hand side change, one with Δ chosen within the allowable interval, and the other one outside the interval.

First, consider $\Delta = -1$ (which is within the interval $[-2, 6]$). Then from (12.30) we have

$$x_B^* = \begin{bmatrix} x_1 \\ x_2 \\ x_6 \end{bmatrix} = B^{-1}b = \begin{bmatrix} 2 \\ 1/2 \\ 7/2 \end{bmatrix},$$

so the optimal solution to (12.28) with $\Delta = -1$ is $[x_1, x_2, x_3]^T = [2, 1/2, 0]$ with $z^* = 4 \times 2 + 3 \times 1/2 + 5 \times 0 = 19/2$.

Now consider $\Delta = 8$ (which is outside the interval $[-2, 6]$). Then from (12.30) we have

$$x_B^* = \begin{bmatrix} x_1 \\ x_2 \\ x_6 \end{bmatrix} = B^{-1}b = \begin{bmatrix} 2 \\ 5 \\ -1 \end{bmatrix},$$

so the tableau corresponding to the current basis is infeasible. Therefore we will use the corresponding dual tableau. First, we compute the primal tableau (12.29) for the basis x_1, x_2, x_6. We obtain

z	x_1	x_2	x_3	x_4	x_5	x_6	rhs	$Basis$
1	0	0	$\frac{4}{3}$	$\frac{3}{2}$	$\frac{5}{6}$	0	23	z
0	1	0	$\frac{4}{3}$	0	$\frac{1}{3}$	0	2	x_1
0	0	1	$\frac{1}{3}$	$\frac{1}{2}$	$-\frac{1}{6}$	0	5	x_2
0	0	0	1	$-\frac{1}{2}$	$-\frac{1}{2}$	1	-1	x_6

(12.31)

The corresponding dual tableau is

$-w$	y_1	y_2	y_3	y_4	y_5	y_6	rhs	$Basis$
1	0	0	-1	2	5	0	-23	$-w$
0	1	0	$\frac{1}{2}$	0	$-\frac{1}{2}$	0	$\frac{3}{2}$	y_1
0	0	1	$\boxed{\frac{1}{2}}$	$-\frac{1}{3}$	$\frac{1}{6}$	0	$\frac{5}{6}$	y_2
0	0	0	-1	$-\frac{4}{3}$	$-\frac{1}{3}$	1	$\frac{4}{3}$	y_6

(12.32)

We will apply the simplex method to find the optimal dual tableau. The entering variable is y_3, the leaving variable is y_2, and the tableau after the pivot is given by

$-w$	y_1	y_2	y_3	y_4	y_5	y_6	rhs	$Basis$
1	0	2	0	$\frac{4}{3}$	$\frac{16}{3}$	0	$-\frac{64}{3}$	$-w$
0	1	-1	0	$\frac{1}{3}$	$-\frac{2}{3}$	0	$\frac{2}{3}$	y_1
0	0	2	1	$-\frac{2}{3}$	$\frac{1}{3}$	0	$\frac{5}{3}$	y_3
0	0	2	0	-2	0	1	3	y_6

(12.33)

This tableau is optimal with the solution $y_1^* = 2/3; y_2^* = 0; y_3^* = 5/3$. The optimal solution for the primal problem (12.28) with $\Delta = 8$ can be read from row 0 of the optimal dual tableau:

$$x_1^* = 4/3, \quad x_2^* = 16/3, \quad x_3^* = 0, \quad z^* = 64/3.$$

12.7.5 Introducing a new variable

Next we introduce a new variable, x_4 in problem (12.20), to obtain the modified problem:

$$
\begin{array}{llllllllll}
\text{maximize} & 4x_1 & + & 3x_2 & + & 5x_3 & + & 5x_4 \\
\text{subject to} & x_1 & + & 2x_2 & + & 2x_3 & + & 3x_4 & \leq & 4 \\
& 3x_1 & & & + & 4x_3 & + & 4x_4 & \leq & 6 \\
& 2x_1 & + & x_2 & + & 4x_3 & + & 5x_4 & \leq & 8 \\
& & & & & & & x_1, x_2, x_3, x_4 & \geq & 0.
\end{array}
$$

(12.34)

Introducing a new variable in our production planning problem can be thought of as considering a new activity, such as producing a new product. Then x_4 represents the quantity of this new product 4 to be produced, its coefficient in the objective function, $c_4 = 5$, is the profit obtained from 1 unit of product 4, and the corresponding coefficients in the constraints, given by the vector $N_{x_4} = [3, 4, 5]^T$, give the amount of the first, the second, and the third resource, respectively, required to produce one unit of product 4.

The new slack variables are x_5, x_6, x_7 (in the original problem, the slack variables were x_4, x_5, x_6), so the basis of (12.34) corresponding to the optimal basis x_1, x_2, x_6 of the original LP (12.20) now consists of variables x_1, x_2, x_7. The newly introduced variable x_4 is treated as a nonbasic variable and its inclusion results in the following changes to c_N and N:

$$c_N = [c_N, c_4]^T = [c_3, c_5, c_6, c_4]^T = [5, 0, 0, 5]^T;$$

$$N = [N, a_4] = \begin{bmatrix} 2 & 1 & 0 & 3 \\ 4 & 0 & 1 & 4 \\ 4 & 0 & 0 & 5 \end{bmatrix}.$$

The corresponding tableau with the current basis x_1, x_2, x_7 is now given by:

z	x_B	x_N	rhs	
1	0	$-(c_N^T - c_B^T B^{-1} N)$	$c_B^T B^{-1} b$	(12.35)
0	I_m	$B^{-1} N$	$B^{-1} b$	

Thus, the coefficient $\bar{c}_4$ of x_4 in row 0 of the tableau is

$$\bar{c}_4 = c_4 - c_B^T B^{-1} a_4 = 5 - [3/2, 5/6, 0] \begin{bmatrix} 3 \\ 4 \\ 5 \end{bmatrix} = -17/6 < 0.$$

This means that the basis x_1, x_2, x_7 remains optimal, and the optimal solution is the same as for the original problem. Thus, $x_4 = 0$, meaning that we should not produce the new product.

In general, even if $\bar{c}_4$ was positive, the tableau with the same basis would still be feasible, even though it would not be optimal anymore. We could apply the simplex method starting with this feasible tableau to obtain a solution to the modified problem.

12.7.6 Introducing a new constraint

We add the constraint $x_1 + 2x_2 + 4x_3 \leq 6$ to (12.20) to obtain the following LP:

$$\begin{array}{rlrlrlr}
\text{maximize} & 4x_1 & + & 3x_2 & + & 5x_3 & \\
\text{subject to} & x_1 & + & 2x_2 & + & 2x_3 & \leq & 4 \\
& 3x_1 & & & + & 4x_3 & \leq & 6 \\
& 2x_1 & + & x_2 & + & 4x_3 & \leq & 8 \\
& 1x_1 & + & 2x_2 & + & 4x_3 & \leq & 6 \\
& & & & x_1, x_2, x_3 & \geq & 0.
\end{array}$$
(12.36)

This results in additional entries in B, b, c_B, and N. Since $x_B = B^{-1} b$ in the solution corresponding to the tableau (12.19), the changes in B and b may lead to an infeasible tableau. However, we will show that the corresponding dual tableau will still be feasible. We have the basic variables x_1, x_2, x_6, x_7 (x_7 is the slack variable for the new constraint) and let B, b, c_B, and N denote the corresponding problem data after the new constraint was introduced. Then

$$B = \begin{bmatrix} 1 & 2 & 0 & 0 \\ 3 & 0 & 0 & 0 \\ 2 & 1 & 1 & 0 \\ 1 & 2 & 0 & 1 \end{bmatrix}, \quad b = \begin{bmatrix} 4 \\ 6 \\ 8 \\ 6 \end{bmatrix}, \quad c_B = \begin{bmatrix} 4 \\ 3 \\ 0 \\ 0 \end{bmatrix}, \quad N = \begin{bmatrix} 2 & 1 & 0 \\ 4 & 0 & 1 \\ 4 & 0 & 0 \\ 4 & 0 & 0 \end{bmatrix},$$

and the current tableau is given by

	z	x_B	x_N	rhs	
	1	0	$-(c_N^T - c_B^T B^{-1} N)$	$c_B^T B^{-1} b$	(12.37)
	0	I_m	$B^{-1} N$	$B^{-1} b$	

From (12.37) it appears that introducing the new constraint causes changes to both the vector of nonbasic variable coefficients and the vector of right-hand sides. However, next we show that, in fact, $c_N^T - c_B^T B^{-1} N = c_N^T - c_B^T B^{-1} N$, so only the feasibility can be impacted by adding a new constraint. To compute $c_B^T B^{-1} N$, we first compute $u^T = c_B^T B^{-1}$ by solving the system $B^T u = c_B$ and then computing $u^T N$.

$$B^T u = \begin{bmatrix} 1 & 3 & 2 & 1 \\ 2 & 0 & 1 & 2 \\ 0 & 0 & 1 & 0 \\ 0 & 0 & 0 & 1 \end{bmatrix} \begin{bmatrix} u_1 \\ u_2 \\ u_3 \\ u_4 \end{bmatrix} = \begin{bmatrix} 4 \\ 3 \\ 0 \\ 0 \end{bmatrix} \Leftrightarrow \begin{bmatrix} u_1 \\ u_2 \\ u_3 \\ u_4 \end{bmatrix} = \begin{bmatrix} 3/2 \\ 5/6 \\ 0 \\ 0 \end{bmatrix},$$

and

$$u^T N = [3/2, 5/6, 0, 0] \begin{bmatrix} 2 & 1 & 0 \\ 4 & 0 & 1 \\ 4 & 0 & 0 \\ 4 & 0 & 0 \end{bmatrix} = [19/3, 3/2, 5/6].$$

Recall that $c_N^T = [5, 0, 0]$, so,

$$c_N^T - c_B^T B^{-1} N = [5, 0, 0] - [19/3, \ 3/2, \ 5/6] = [-4/3, \ -3/2, \ -5/6],$$

which is the same as the vector of coefficients $\tilde{c}_N = c_N^T - c_B^T B^{-1} N$ for nonbasic variables in the optimal tableau of the original problem. Note that this will always be the case if we add a new constraint, i.e., the vector of the coefficients for nonbasic variables in the optimal tableau of the original problem and the vector of the coefficients for nonbasic variables in the corresponding tableau of the modified problem will be the same:

$$c_N^T - c_B^T B^{-1} N = c_N^T - c_B^T B^{-1} N.$$

Hence, (12.37) is equivalent to

	z	x_B	x_N	rhs	
	1	0	$-(c_N^T - c_B^T B^{-1} N)$	$c_B^T B^{-1} b$	(12.38)
	0	I_m	$B^{-1} N$	$\bar{x} = B^{-1} b$	

and since row 0 is not affected by the change, the corresponding dual tableau is feasible and we can use it to solve the dual of the modified problem to optimality.

12.7.7 Summary

While performing the sensitivity analysis above, we saw that with the following modifications:

- changing a coefficient in the objective,

- introducing a new variable, and

- changing the column of a nonbasic variable,

the optimal basis still remains feasible. Then, if the basis is not optimal, we can apply the simplex method starting with this basis and find a new optimal solution.

With the following modifications:

- changing the right-hand side,

- introducing a new constraint

the *dual* optimal basis still remains feasible. Then, if the dual tableau is not optimal, we can apply the simplex method starting with this basis to solve the dual problem. We can then extract the optimal solution of the primal problem from the optimal dual tableau.

Exercises

12.1. Write the dual for each of the following LPs:

(a)
$$
\begin{array}{lrcrcrcl}
\text{maximize} & x_1 & + & 2x_2 & - & 3x_3 & & \\
\text{subject to} & 4x_1 & + & 5x_2 & + & x_3 & \leq & 6 \\
& 7x_1 & + & x_2 & & & \leq & 8 \\
& & & 9x_2 & + & x_3 & = & 10 \\
& \multicolumn{6}{l}{x_1 \geq 0,\ x_2,\ x_3\ \in\ \mathbb{R}}
\end{array}
$$

(b)
$$
\begin{array}{lrcrcrcl}
\text{minimize} & 8x_1 & + & 4x_2 & - & 2x_3 & & \\
\text{subject to} & x_1 & + & 3x_2 & & & \leq & 5 \\
& 6x_1 & - & 7x_2 & + & 8x_3 & = & 7 \\
& \multicolumn{6}{l}{x_1 \in \mathbb{R},\ x_2, x_3\ \geq\ 0}
\end{array}
$$

(c)
$$
\begin{array}{lrcrcrcl}
\text{maximize} & 9x_1 & - & 7x_2 & + & 5x_3 & & \\
\text{subject to} & 3x_1 & + & x_2 & - & x_3 & \leq & 2 \\
& 4x_1 & - & 6x_2 & + & 3x_3 & \geq & 8 \\
& x_1 & + & 2x_2 & + & x_3 & = & 11 \\
& \multicolumn{6}{l}{x_1 \in \mathbb{R},\ x_2, x_3\ \geq\ 0.}
\end{array}
$$

12.2. Write down the dual of the LP formulated for the diet problem in Example 10.2 (page 213). Provide an economic interpretation of the dual LP.

12.3. Prove or disprove each of the following statements concerning a primal LP,

$$\text{maximize } z = c^T x \quad \text{subject to } Ax \leq b, \ x \geq 0$$

and the corresponding dual LP,

$$\text{minimize } w = b^T y \quad \text{subject to } A^T y \geq c, \ y \geq 0.$$

(a) If the primal LP is infeasible, the corresponding dual LP must be unbounded.

(b) If the primal LP has a nonempty, bounded feasible region, then the corresponding dual LP must have a global minimum.

(c) If $b < 0$, then the primal LP is infeasible.

(d) If the primal LP has a feasible solution, then the dual LP must also have a feasible solution.

(e) Even if the primal LP has an unbounded feasible region, the corresponding dual LP can still have an optimal solution.

(f) If the primal LP has an optimal solution, then altering the entries of right-hand side vector b cannot result in an unbounded LP.

12.4. Consider the following LP:

$$\begin{array}{rrrrrl}
\text{maximize} & 2x_1 & + & 3x_2 & - & x_3 \\
\text{subject to} & 3x_1 & + & 3x_2 & - & 2x_3 & \leq & 6 \\
& -x_1 & - & 2x_2 & + & 3x_3 & \leq & 6 \\
& & & & x_1, x_2, x_3 & \geq & 0.
\end{array}$$

Determine the missing entries in the z-row of its optimal tableau:

z	x_1	x_2	x_3	s_1	s_2	rhs
1	8/5	0	?	7/5	3/5	?

12.5. Consider the following LP:

$$\begin{array}{rrrrrl}
\text{maximize} & 5x_1 & + & 3x_2 & - & x_3 \\
\text{subject to} & 3x_1 & + & 3x_2 & - & 2x_3 & \leq & 4 \\
& -x_1 & - & 2x_2 & + & 3x_3 & \leq & 6 \\
& & & & x_1, x_2, x_3 & \geq & 0.
\end{array}$$

Find the missing entries in the z-row of its optimal tableau:

z	x_1	x_2	x_3	s_1	s_2	rhs
1	?	1	0	2	?	14

12.6. Consider the following primal LP:

$$
\begin{aligned}
\text{maximize} \quad & 5x_1 - 2x_2 + 3x_3 \\
\text{subject to} \quad & 3x_1 + 2x_2 - 3x_3 \leq 6 \\
& x_1 + x_2 + x_3 \leq 6 \\
& x_1, x_2, x_3 \geq 0.
\end{aligned}
$$

(a) Formulate its dual and solve it graphically.

(b) Use your solution to the dual LP and complementary slackness to solve the primal LP.

12.7. Use the complementary slackness theorem to determine whether x^* is an optimal solution of the LP

$$
\begin{aligned}
\text{maximize} \quad & x_1 + 3x_2 + 3x_3 \\
\text{subject to} \quad & x_1 + 3x_2 - 2x_3 \leq 3 \\
& 5x_1 - 2x_2 + 4x_3 \leq 2 \\
& 4x_1 - 2x_2 + 5x_3 \leq 4 \\
& x_1, x_2, x_3 \geq 0,
\end{aligned}
$$

where

(a) $x^* = [1, 2, 1/2]^T$.

(b) $x^* = [0, 2, 3/2]^T$.

12.8. Consider the following LP:

$$
\begin{aligned}
\text{maximize} \quad & x_1 + 2x_2 \\
\text{subject to} \quad & x_1 + 5x_2 \leq 5 \\
& 3x_1 + 2x_2 \leq 6 \\
& 5x_1 + x_2 \leq 5 \\
& x_1, x_2 \geq 0.
\end{aligned}
$$

(a) Solve the LP graphically.

(b) Graphically determine the range of values of the objective function coefficient for x_1 for which the optimal basis does not change.

(c) Graphically determine the range of values of the right-hand side of the first constraint for which the optimal basis does not change.

12.9. For the following LP,

$$
\begin{aligned}
\text{maximize} \quad & 2x_1 + 2x_2 - 3x_3 \\
\text{subject to} \quad & -3x_1 + 2x_2 + x_3 \leq 3 \\
& 2x_1 + x_2 + x_3 \leq 5 \\
& x_1, x_2, x_3 \geq 0,
\end{aligned}
$$

consider its optimal tableau:

z	x_1	x_2	x_3	s_1	s_2	rhs
1	0	0	33/7	2/7	10/7	8
0	0	1	5/7	2/7	3/7	3
0	1	0	1/7	−1/7	2/7	1

(a) Write the dual to this LP and use the above tableau to find the dual optimal solution.

(b) Find the range of values of the objective function coefficient of x_2 for which the current basis remains optimal.

(c) Find the range of values of the objective function coefficients of x_3 for which the current basis remains optimal.

(d) Find the range of values of the right-hand side of the first constraint for which the current basis remains optimal.

(e) What will be the optimal solution to this LP if the right-hand side of the first constraint is changed from 3 to 8?

12.10. Use the LP and its optimal tableau given in Exercise 12.9 to solve the following LPs:

(a)
$$\begin{array}{rl}
\text{maximize} & 2x_1 + 2x_2 - 2x_3 \\
\text{subject to} & -3x_1 + 2x_2 + 2x_3 \le 3 \\
& 2x_1 + x_2 + 2x_3 \le 5 \\
& x_1, x_2, x_3 \ge 0
\end{array}$$

(b)
$$\begin{array}{rl}
\text{maximize} & 5x_1 + 2x_2 - 3x_3 \\
\text{subject to} & -3x_1 + 2x_2 + x_3 \le 3 \\
& 2x_1 + x_2 + x_3 \le 5 \\
& x_1, x_2, x_3 \ge 0
\end{array}$$

(c)
$$\begin{array}{rl}
\text{maximize} & 2x_1 + 2x_2 - 3x_3 + 3x_4 \\
\text{subject to} & -3x_1 + 2x_2 + x_3 + 3x_4 \le 3 \\
& 2x_1 + x_2 + x_3 + 2x_4 \le 5 \\
& x_1, x_2, x_3, x_4 \ge 0.
\end{array}$$

12.11. Use the LP and its optimal tableau given in Exercise 12.9 to solve the following LPs:

(a)
$$\begin{array}{rl}
\text{maximize} & 2x_1 + 2x_2 - 3x_3 \\
\text{subject to} & -3x_1 + 2x_2 + x_3 \le 3 \\
& 2x_1 + x_2 + x_3 \le 1 \\
& x_1, x_2, x_3 \ge 0,
\end{array}$$

(b)
$$\begin{array}{rl}
\text{maximize} & 2x_1 + 2x_2 - 3x_3 \\
\text{subject to} & -3x_1 + 2x_2 + x_3 \le 3 \\
& 2x_1 + x_2 + x_3 \le 5 \\
& -x_1 + 2x_2 + x_3 \le 4 \\
& x_1, x_2, x_3 \ge 0.
\end{array}$$

12.12. For the following LP,

$$
\begin{array}{llrcrcrcl}
\text{maximize} & & x_1 & + & 3x_2 & + & 3x_3 \\
\text{subject to} & & 2x_1 & + & 3x_2 & - & 2x_3 & \leq & 1 \\
& & 4x_1 & - & 2x_2 & + & 4x_3 & \leq & 2 \\
& & & & & & x_1, x_2, x_3 & \geq & 0,
\end{array}
$$

consider its optimal tableau:

z	x_1	x_2	x_3	s_1	s_2	rhs
1	11	0	0	9/4	15/8	6
0	2	1	0	1/2	1/4	1
0	2	0	1	1/4	3/8	1

(a) Write the dual to this LP and use the above tableau to find the dual optimal solution.

(b) Find the range of values of the objective function coefficient of x_1 for which the current basis remains optimal.

(c) Find the range of values of the objective function coefficients of x_2 for which the current basis remains optimal.

(d) Find the range of values of the right-hand side of the second constraint for which the current basis remains optimal.

(e) Extract B^{-1} from the optimal tableau.

12.13. A manufacturing company makes three products, 1, 2, and 3, using two kinds of resources, I and II. The resource requirements and selling price for the three products are shown in the following table:

	Product		
Resource	1	2	3
I (units)	3	2	4
II (units)	4	2	3
Price (\$)	10	7	12

Currently, 100 units of Resource I are available. Up to 120 units of Resource II can be purchased at \$1 per unit. Then the following LP can be formulated to maximize the company's profits:

$$
\begin{array}{llrcrcrcrcl}
\text{maximize} & & 10x_1 & + & 7x_2 & + & 12x_3 & - & r \\
\text{subject to} & & 3x_1 & + & 2x_2 & + & 4x_3 & - & r & \leq & 0 \\
& & 4x_1 & + & 2x_2 & + & 3x_3 & & & \leq & 100 \\
& & & & & & & & r & \leq & 120 \\
& & & & & & x_1, x_2, x_3, r & \geq & 0,
\end{array}
$$

where $x_i =$ units of product i produced, $i = 1, 2, 3$, and $r =$ number of units of resource II purchased.

The optimal solution to this LP is $x_1^* = 0, x_2^* = 20, x_3^* = 20, r^* = 120$ with the optimal objective function value $z^* = 260$. Sensitivity analysis yielded results summarized in the tables below.

Decision Variable	Optimal Value	Reduced Cost	Objective Coefficient	Allowable Increase	Allowable Decrease
x_1	0	$-5/2$	10	$5/2$	∞
x_2	20	0	7	$1/3$	$5/7$
x_3	20	0	12	2	$1/2$
r	120	0	-1	∞	$1/2$

Constraint	Shadow Price	Constraint RHS	Allowable Increase	Allowable Decrease
1	$3/2$	0	$40/3$	20
2	2	100	20	10
3	$1/2$	120	$40/3$	20

Use these results to answer the following questions.

(a) Write down the dual of this LP. What is its optimal solution?

(b) What is the most the company should be willing to pay for an additional unit of resource I?

(c) What is the most the company should be willing to pay for another unit of resource II?

(d) What selling price of product 1 would make it reasonable for the company to manufacture it?

(e) What would the company's profit be if 110 units of resource I was available?

(f) What would the company's profit be if 130 units of resource II could be purchased?

(g) Find the new optimal solution if product 3 sold for $13.

Chapter 13

Unconstrained Optimization

In this chapter we deal with *unconstrained optimization* problems in the form

$$\text{minimize } f(x),$$

where $f : \mathbb{R}^n \to \mathbb{R}$ is a continuously differentiable function. We first develop *optimality conditions*, which characterize analytical properties of local optima and are fundamental for developing solution methods. Since computing not only global but even local optimal solutions is a very challenging task in general, we set a more realistic goal of finding a point that satisfies the *first-order necessary conditions*, which are the conditions that the first-order derivative (gradient) of the objective function must satisfy at a point of local optimum. We discuss several classical numerical methods converging to such a point.

13.1 Optimality Conditions

Optimality conditions characterize analytical properties of local optima and are fundamental for developing numerical methods. *Necessary* optimality conditions aim to characterize a point x^*, assuming that x^* is a local minimizer of the considered problem. Such conditions allow one to determine a set of candidate points that may need to be further analyzed in order to find out if they are local minimizers. In some cases, such analysis can be done using *sufficient* optimality conditions, providing properties that, if satisfied, guarantee that a candidate point x^* is a local minimizer.

13.1.1 First-order necessary conditions

The first-order necessary conditions (FONC) for unconstrained optimization can be viewed as a special case of the following optimality conditions for a more general set-constrained optimization problem.

Theorem 13.1 (FONC for a Set-Constrained Problem)
If $f : X \to \mathbb{R}$, where $X \subseteq \mathbb{R}^n$, is a continuously differentiable function on an open set containing X, and x^* is a point of local minimum for the set-constrained problem

$$
\begin{aligned}
\text{minimize} \quad & f(x) \\
\text{subject to} \quad & x \in X,
\end{aligned}
\tag{13.1}
$$

then for any feasible direction d at x^*, the rate of increase of f at x^* in the direction d is nonnegative: $\nabla f(x^*)^T d \geq 0$.

Proof. Let $x^* \in X$ be a local minimizer of (13.1). Consider a feasible direction d at x^* such that $\|d\| = 1$. Using Taylor's theorem for f, x^*, and $x = x^* + \alpha d$, where $\alpha > 0$, we have:

$$
\begin{aligned}
f(x) &= f(x^*) + \nabla f(x^*)^T \alpha d + o(\|\alpha d\|) \\
&= f(x^*) + \alpha \nabla f(x^*)^T d + o(\alpha).
\end{aligned}
\tag{13.2}
$$

Since x^* is a local minimizer of (13.1), from the above equation we have:

$$
\frac{f(x) - f(x^*)}{\alpha} = \nabla f(x^*)^T d + \frac{o(\alpha)}{\alpha}.
\tag{13.3}
$$

Since $f(x) - f(x^*) \geq 0$, $\alpha > 0$ and $\frac{o(\alpha)}{\alpha} \to 0$ as $\alpha \to 0$, we obtain

$$
\nabla f(x^*)^T d \geq 0.
$$

Indeed, if we assume $f(x^*)^T d < 0$, then selecting ϵ such that $f(x^*) \leq f(x)$ for any $x \in B(x^*, \epsilon)$ and the error term $\frac{o(\alpha)}{\alpha}$ in (13.3) is less than $|f(x^*)^T d|$ (such ϵ exists since x^* is a local minimizer and by definition of $o(\alpha)$) would result in $f(x) - f(x^*) < 0$. We obtain a contradiction with the assumption that $f(x^*) \leq f(x)$ for any $x \in B(x^*, \epsilon)$. $\quad\square$

As a corollary of Theorem 13.1, we get the following result.

Corollary 13.1 *If x^* is an interior point of X and a local minimizer of (13.1), then $\nabla f(x^*) = 0$.*

Proof. If x^* is an interior point of X, then any $d \in \mathbb{R}^n$ is a feasible direction at x^*. Thus, using Theorem 13.1 for an arbitrary direction $d \neq 0$ and its opposite direction $-d$ we have: $\nabla f(x^*)^T d \geq 0, \nabla f(x^*)^T (-d) \geq 0$, therefore, $\nabla f(x^*)^T d = 0$. In particular, if we use $d = [d_1, \ldots d_n]^T$ with $d_j = 1$ for an arbitrary $j \in \{1, \ldots, n\}$ and $d_i = 0$ for all $i \neq j$, this implies that the jth component of $\nabla f(x^*)$ equals 0. Since this is the case for each j, we have $\nabla f(x^*) = 0$. $\quad\square$

In the case of an unconstrained problem, $X = \mathbb{R}^n$ and any point $x \in \mathbb{R}^n$ is an interior point of X. Thus, we obtain the following FONC for unconstrained optimization.

Theorem 13.2 (FONC for an Unconstrained Problem)
If x^* is a point of local minimum of the unconstrained problem

$$\text{minimize} \quad f(x), \tag{13.4}$$

where $f : \mathbb{R}^n \to \mathbb{R}$ is continuously differentiable, then $\nabla f(x^*) = 0$.

Definition 13.1 (Stationary Point) *A point x^* satisfying the FONC for a given problem is called a stationary point for this problem.*

The following example shows that the FONC is not sufficient for a local minimizer.

Example 13.1 *Applying the FONC to $f(x) = x^3$, we have*

$$f'(x) = 3x^2 = 0 \quad \Leftrightarrow \quad x = 0.$$

But, obviously, $x = 0$ is not a local minimizer of $f(x)$. In fact, for any given point x and any small $\epsilon > 0$, there always exist $x_*, x^* \in B(x^*, \epsilon)$ such that $f(x_*) < f(x) < f(x^*)$, so $f(x)$ does not have any local or global minimum or maximum.

Next, we prove that a point satisfying the FONC is, in fact, a global minimizer if a problem is convex. The proof is based on the first-order characterization of a convex function. Consider a convex problem $\min_{x \in X} f(x)$. For any $x, y \in X$, a differentiable convex function satisfies

$$f(y) \geq f(x) + \nabla f(x)^T (y - x),$$

hence, if for $x = x^*$: $\nabla f(x^*) = 0$, we obtain

$$f(y) \geq f(x^*)$$

for any $y \in X$. Thus, x^* is a point of global minimum for this problem. This implies that the FONC in the case of an unconstrained convex problem becomes a sufficient condition for a global minimizer.

Theorem 13.3 (Optimality Condition for a Convex Problem)
For a convex unconstrained problem minimize $f(x)$, where $f : \mathbb{R}^n \to$

$\mathbb{R}$ *is differentiable and convex, x^* is a global minimizer if and only if*
$\nabla f(x^*) = 0$.

Consider a quadratic problem

$$\max_{x \in \mathbb{R}^n} q(x),$$

where $q(x) = \frac{1}{2}x^T Q x + c^T x$ for a given $n \times n$-matrix Q and n-vector c. If Q is a positive semidefinite matrix, then $q(x)$ is convex and any point satisfying the FONC is a global minimizer. We have

$$\nabla q(x) = 0 \quad \Leftrightarrow \quad Qx = -c.$$

If Q is positive definite, then this system has a unique solution $x^* = -Q^{-1}c$, which is the only global minimizer of the considered problem.

Example 13.2 *Find the minimum of the function*

$$q(x) = \frac{1}{2}x^T Q x + c^T x,$$

where

$$Q = \begin{bmatrix} 3 & -1 & -1 \\ -1 & 2 & 0 \\ -1 & 0 & 4 \end{bmatrix} \quad and \quad c = \begin{bmatrix} 1 \\ -2 \\ 3 \end{bmatrix}.$$

Since

$$q_{11} = 3 > 0,$$

$$\begin{vmatrix} q_{11} & q_{12} \\ q_{21} & q_{22} \end{vmatrix} = \begin{vmatrix} 3 & -1 \\ -1 & 2 \end{vmatrix} = 5 > 0$$

and

$$\det(Q) = 18 > 0,$$

the matrix Q is positive definite. Thus $q(x)$ is convex, and it has a unique global minimum which can be found by solving the system

$$Qx = -c.$$

The solution to this system is

$$x^* = [-1/3, 5/6, -5/6]^T, \quad and \quad q(x^*) = -9/4.$$

13.1.2   Second-order optimality conditions

Next, we derive the second-order necessary conditions (SONC) and second-order sufficient conditions (SOSC) for a local minimizer of an unconstrained problem. We assume that the objective function $f \in C^{(2)}(\mathbb{R}^n)$, i.e., f is twice continuously differentiable.

Theorem 13.4 (SONC for an Unconstrained Problem)
If $x^* \in \mathbb{R}^n$ is a local minimizer for the problem $\min\limits_{x \in \mathbb{R}^n} f(x)$, where $f(x) \in \mathcal{C}^{(2)}(\mathbb{R}^n)$, then $\nabla^2 f(x^*)$ is positive semidefinite.

Proof. Let x^* be a point of local minimum of $f(x)$. Given an arbitrary $d \in \mathbb{R}^n$ such that $\|d\| = 1$ and a scalar $\alpha > 0$, using Taylor's theorem we have

$$f(x^* + \alpha d) = f(x^*) + \alpha \nabla f(x^*)^T d + \frac{1}{2}\alpha^2 d^T \nabla^2 f(x^*)d + o(\alpha^2). \qquad (13.5)$$

Since x^* is a local minimizer, $\nabla f(x^*) = 0$, and (13.5) can be written as

$$f(x^* + \alpha d) - f(x^*) = \frac{1}{2}\alpha^2 d^T \nabla^2 f(x^*)d + o(\alpha^2),$$

or, dividing both sides by α^2, as

$$\frac{f(x^* + \alpha d) - f(x^*)}{\alpha^2} = \frac{1}{2}d^T \nabla^2 f(x^*)d + \frac{o(\alpha^2)}{\alpha^2}. \qquad (13.6)$$

Due to the local minimality of x^*, there exists $\epsilon > 0$ such that

$$f(x^* + \alpha d) - f(x^*) \geq 0 \quad \text{for any} \quad \alpha \in (0, \epsilon). \qquad (13.7)$$

If we assume that $d^T \nabla^2 f(x^*)d < 0$, then, since $\lim\limits_{\alpha \to 0} \frac{o(\alpha^2)}{\alpha^2} = 0$, there exists a sufficiently small $\alpha \in (0, \epsilon)$ such that

$$\frac{1}{2}d^T \nabla^2 f(x^*)d + \frac{o(\alpha^2)}{\alpha^2} < 0. \qquad (13.8)$$

However, (13.7) and (13.8) combined contradict (13.6). Thus, the assumption that $d^T \nabla^2 f(x^*)d < 0$ is incorrect and $d^T \nabla^2 f(x^*)d \geq 0$ for any d, implying that $\nabla^2 f(x^*)$ is positive semidefinite. $\qquad \square$

Next we derive second-order sufficient conditions (SOSC) for a local minimizer of an unconstrained problem.

Theorem 13.5 (SOSC for an Unconstrained Problem)
If x^* satisfies the FONC and SONC for an unconstrained problem $\min\limits_{x \in \mathbb{R}^n} f(x)$ and $\nabla^2 f(x^*)$ is positive definite, then x^* is a point of strict local minimum for this problem.

Proof. We assume that x^* satisfies the FONC and SONC. Then, for any $d \in \mathbb{R}^n$ with $\|d\| = 1$ and any $\alpha > 0$ we have

$$\frac{f(x^* + \alpha d) - f(x^*)}{\alpha^2} = \frac{1}{2}d^T \nabla^2 f(x^*)d + \frac{o(\alpha^2)}{\alpha^2}.$$

If we additionally assume that $\nabla^2 f(x^*)$ is positive definite, then by Rayleigh's inequality,

$$\frac{1}{2}d^T \nabla^2 f(x^*)d \geq \lambda_{\min}\|d\|^2 = \lambda_{\min} > 0.$$

Here $\lambda_{\min}$ denotes the smallest eigenvalue of $\nabla^2 f(x^*)$. Thus, there exists $\epsilon > 0$ such that for any $\alpha \in (0, \epsilon)$ we have $f(x^* + \alpha d) - f(x^*) > 0$. Since d is an arbitrary direction in $\mathbb{R}^n$, x^* is a point of strict local minimum by definition. Thus, the FONC, SONC, together with the positive definiteness of $\nabla^2 f(x^*)$, constitute the sufficient conditions for a strict local minimizer.   □

Example 13.3 *Consider the function $f(x) = x_1^3 - x_2^3 + 3x_1 x_2$. We apply the optimality conditions above to find its local optima. FONC system for this problem is given by*

$$\nabla f(x) = \left[\begin{array}{c} 3x_1^2 + 3x_2 \\ -3x_2^2 + 3x_1 \end{array} \right] = \left[\begin{array}{c} 0 \\ 0 \end{array} \right].$$

From the second equation of this system we have $x_1 = x_2^2$. Substituting for x_1 in the first equation gives $x_2^4 + x_2 = 0$, which yields $x_2 = 0$ or $x_2 = -1$. The corresponding stationary points are $\hat{x} = [0, 0]^T$ and $\tilde{x} = [1, -1]^T$, respectively.
    The Hessian of $f(x)$ is

$$\nabla^2 f(x) = \left[\begin{array}{cc} 6x_1 & 3 \\ 3 & -6x_2 \end{array} \right] \Rightarrow \nabla^2 f(\hat{x}) = \left[\begin{array}{cc} 0 & 3 \\ 3 & 0 \end{array} \right]; \quad \nabla^2 f(\tilde{x}) = \left[\begin{array}{cc} 6 & 3 \\ 3 & 6 \end{array} \right].$$

Since the determinant of $\nabla^2 f(\hat{x})$ equals -9, the Hessian is indefinite at $\hat{x}$, and from the SONC, $\hat{x}$ cannot be a local optimum. On the other hand, $\nabla^2 f(\tilde{x})$ is positive definite, hence $\tilde{x}$ is a strict local minimum by the SOSC. Note that if we fix $x_2 = 0$, the function f is then given by x_1^3, which is unbounded from below or above. Thus, f has no global minima and no local or global maxima. Its only local minimum is $\tilde{x} = [1, -1]^T$.

13.1.3 Using optimality conditions for solving optimization problems

    As we have seen in examples above, in some cases optimality conditions can be used to find points of local or global optima. In particular, a search for local optima of a given function f can be carried out by first finding all points satisfying FONC (stationary points), and then applying second order optimality conditions to analyze the nature of the stationary points. If the SONC condition is not satisfied at a stationary point x^*, then x^* is not a local optimizer. A stationary point that is not a local optimum is called a *saddle point*. If, on the other hand, a stationary point x^* satisfies the SOSC, then it is a local optimum of f. In case a stationary point x^* satisfies the SONC but does not satisfy the SOSC, the second-order optimality conditions are inconclusive and cannot be used to decide whether x^* is a local optimum or a saddle point. In addition, if the problem of optimizing f is known to have a global optimal solution, its global optima can potentially be computed by considering all stationary points and comparing the corresponding objective function values.

On a negative side, verifying the existence of a global optimum and finding the stationary points are difficult problems in general, which limits the use of optimality conditions for the purpose of finding local and global optimal solutions. Given the intractability of these problems, when designing numerical methods for solving the problem of optimizing $f(x)$, it is reasonable to settle for the goal of computing a point x^* such that $\nabla f(x^*) \approx 0$, i.e., a point where the FONC is approximately satisfied. This is the premise of the methods for unconstrained optimization that we discuss below in this chapter. However, before proceeding to outlining such methods, we first consider some simple derivative-free techniques for solving problems involving a single variable.

13.2 Optimization Problems with a Single Variable

Consider a function $f : \mathbb{R} \to \mathbb{R}$ which has a unique minimizer c over a closed interval $[a, b]$. Also, assume that f is strictly decreasing on $[a, c]$ and strictly increasing on $[c, b]$. Such a function is referred to as a *unimodal* function. The golden section search and Fibonacci search are methods designed to find the minimizer of a unimodal function over a closed interval.

13.2.1 Golden section search

Assume that we are given a unimodal function $f : [a_0, b_0] \to \mathbb{R}$ with the minimizer x^*. We aim to reduce the search space by locating a smaller interval containing the minimizer. To do so, we evaluate f at two points $a_1, b_1 \in (a_0, b_0)$. We choose these points in the way that

$$a_1 - a_0 = b_0 - b_1 = \rho(b_0 - a_0),$$

where $\rho < 1/2$. If $f(a_1) < f(b_1)$, then the minimizer is in the interval $[a_0, b_1]$. Otherwise, if $f(a_1) > f(b_1)$, then the minimizer is in $[a_1, b_0]$ (see Figure 13.1). Thus, the range of uncertainty will be reduced by the factor of $(1 - \rho)$, and we can continue the search using the same method over a smaller interval.

In the golden section search method, we want to reduce the number of function evaluations by using previously computed intermediate points. Consider the example shown in Figure 13.1. In this example, $f(a_1) < f(b_1)$, so the range of uncertainty reduces to the interval $[a_0, b_1]$. To continue the process, we need to choose two points in $[a_0, b_1]$ and evaluate f in these points. However, we know that $a_1 \in [a_0, b_1]$. So, a_1 can be chosen as one of the two

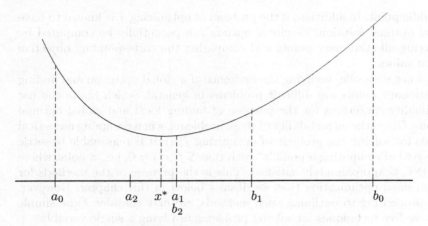

FIGURE 13.1: Golden section search.

points: $b_2 = a_1$, and it suffices to find only one new point a_2, which would be as far from a_0, as b_2 is from b_1. The advantage of such a choice of intermediate points is that in the next iteration we would have to evaluate f only in one new point, a_2. Now we need to compute the value of ρ which would result in having a_1 chosen as one of the intermediate points. Figure 13.2 illustrates this situation: If we assume that the length of $[a_0, b_0]$ is l, and the length of $[a_0, b_1]$ is d, then we have

$$d = (1 - \rho)l,$$

so

$$l = \frac{d}{1 - \rho}.$$

On the other hand, if we consider the interval $[a_0, a_1] = [a_0, b_2]$, then its length can be expressed in two different ways:

$$\rho l = (1 - \rho)d.$$

From the last two expressions, we obtain

$$\frac{\rho}{1 - \rho} = 1 - \rho.$$

This formula can be interpreted as follows. If we divide the interval of length 1 into two segments, one of length $\rho < 1/2$ and the other of length $1 - \rho > 1/2$, then the ratio of the shorter segment to the longer equals to the ratio of the longer to the sum of the two. In Ancient Greece, this division was referred to as the *golden ratio*, therefore the name of the method. We can now compute ρ by solving the quadratic equation

$$\rho^2 - 3\rho + 1 = 0.$$

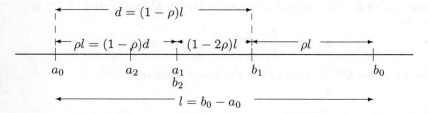

FIGURE 13.2: Finding ρ in the golden section search.

Taking into account that we are looking for $\rho < 1/2$, we obtain the solution

$$\rho = \frac{3 - \sqrt{5}}{2} \approx 0.382.$$

Note that the uncertainty interval is reduced by the factor of $1 - \rho \approx 0.618$ at each step. So, in N steps the reduction factor would be

$$(1 - \rho)^n \approx 0.618^N.$$

13.2.1.1 Fibonacci search

Fibonacci search is a generalization of the golden section search. Instead of using the same value of ρ at each step, we can vary it and use a different value ρ_k for each step k. Analogously to the golden section search, we want to select $\rho_k \in [0, 1/2]$ in such a way that only one new function evaluation is required at each step. Using reasonings similar to those for the choice of ρ in the golden section search (see also Figure 13.2), we obtain the following relations for the values of ρ_k:

$$\rho_{k+1} = 1 - \frac{\rho_k}{1 - \rho_k}, \quad k = 1, \ldots, n - 1.$$

In order to minimize the interval of uncertainty after N steps, we consider the following minimization problem:

$$\text{minimize} \quad (1 - \rho_1)(1 - \rho_2) \cdots (1 - \rho_N) \tag{13.9}$$

$$\text{subject to} \quad \rho_{k+1} = 1 - \frac{\rho_k}{1 - \rho_k}, \quad k = 1, \ldots, N - 1 \tag{13.10}$$

$$0 \le \rho_k \le \frac{1}{2}, \quad k = 1, \ldots, N. \tag{13.11}$$

To describe the solution to this problem, we need the following definition. The *Fibonacci sequence* $\{F_k, k \ge 0\}$ is defined by $F_0 = F_1 = 1$ and the recursive relation $F_{k+1} = F_k + F_{k-1}$.

> **Theorem 13.6** *The optimal solution to the problem (13.9)–(13.11) is given by*
>
> $$\rho_k = 1 - \frac{F_{N-k+1}}{F_{N-k+2}}, \quad k = 1, \dots, N,$$
>
> *where F_k is the kth element of the Fibonacci sequence.*

Proof. Note that using (13.10) we can recursively express all variables ρ_k, $k = 1, \dots, N$ in the objective function of (13.9) through one of the variables, say ρ_N. If we denote the resulting univariate function by $f_N(\rho_N)$, then

$$f_N(\rho_N) = \frac{1 - \rho_N}{F_N - F_{N-2}\rho_N}, \quad N \geq 2. \tag{13.12}$$

We will prove (13.12) using induction by N. For $N = 2$, from (13.10) we have $\rho_1 = \frac{1-\rho_2}{2-\rho_2}$, so $f_2(\rho_2) = \left(1 - \frac{1-\rho_2}{2-\rho_2}\right)(1 - \rho_2) = \frac{1-\rho_2}{2-\rho_2} = \frac{1-\rho_2}{F_2 - F_0\rho_2}$, and (13.12) is correct for $N = 2$. Assuming that (13.12) is correct for some $N = K - 1$, i.e., $f_{K-1}(\rho_{K-1}) = \frac{1-\rho_{K-1}}{F_{K-1}-F_{K-3}\rho_{K-1}}$, we need to show that it is also correct for $N = K$. From (13.10) we have $\rho_{K-1} = \frac{1-\rho_K}{2-\rho_K}$, so

$$
\begin{aligned}
f_K(\rho_K) &= f_{K-1}\left(\frac{1-\rho_K}{2-\rho_K}\right)(1 - \rho_K) \\
&= \frac{1 - \frac{1-\rho_K}{2-\rho_K}}{F_{K-1} - F_{K-3}\frac{1-\rho_K}{2-\rho_K}}(1 - \rho_K) \\
&= \frac{1 - \rho_K}{2F_{K-1} - F_{K-3} - (F_{K-1} - F_{K-3})\rho_K} \\
&= \frac{1 - \rho_K}{F_K - F_{K-2}\rho_K},
\end{aligned}
$$

so (13.12) is correct for any $N \geq 2$.

Next, we will show that $f_N(\rho_N)$ is a strictly decreasing function on $[0, \frac{1}{2}]$. We can do so by showing that the derivative $f'_N(\rho_N) < 0, \forall \rho_N \in [0, \frac{1}{2}]$. Indeed,

$$f'_N(\rho_N) = \frac{-F_N + F_{N-2}}{(F_N - F_{N-2}\rho_N)^2} = \frac{-F_{N-1}}{(F_N - F_{N-2}\rho_N)^2} < 0, \forall \rho_N \leq \frac{1}{2}.$$

Therefore,

$$\min_{\rho_N \in [0,1/2]} f_N(\rho_N) = f_N(1/2) = \frac{1 - 1/2}{F_N - F_{N-2}/2} = \frac{1}{F_{N+1}},$$

which means that the reduction factor after N steps of the Fibonacci search is

$$\frac{1}{F_{N+1}}.$$

Returning to the original problem (13.9)–(13.11), we have

$$\rho_N = 1/2 = 1 - \frac{F_1}{F_2};$$

$$\rho_{N-1} = \frac{1 - \rho_N}{2 - \rho_N} = \frac{F_1}{F_3} = 1 - \frac{F_2}{F_3};$$

$$\vdots$$

$$\rho_{k+1} = 1 - \frac{F_{N-k}}{F_{N-k+1}};$$

$$\rho_k = \frac{1 - \rho_{k+1}}{2 - \rho_{k+1}} = \frac{F_{N-k}}{F_{N-k+2}} = 1 - \frac{F_{N-k+1}}{F_{N-k+2}};$$

$$\vdots$$

$$\rho_1 = 1 - \frac{F_N}{F_{N+1}}.$$

Note that the above $\rho_k, k = 1, \ldots, N$ satisfy the conditions (13.10)–(13.11) and thus represent the (unique) optimal solution to (13.9)–(13.11). $\square$

13.3 Algorithmic Strategies for Unconstrained Optimization

Consider the problem

$$\text{minimize} f(x).$$

Classical algorithms for this problem usually aim to construct a sequence of points $\{x^{(k)} : k \geq 0\}$, such that $x^{(k)} \to x^*, k \to \infty$, where x^* is a stationary point of $f(x)$ (that is, $\nabla f(x^*) = 0$). Each next point in this sequence is obtained from the previous point by moving some distance along a direction $d^{(k)}$:

$$x^{(k+1)} = x^{(k)} + \alpha_k d^{(k)}, \quad k \geq 0,$$

where $\alpha_k \geq 0$ is a scalar representing the step size. Two popular strategies used are *line search* and *trust region*. In a line search strategy, we proceed as follows. Given the current solution $x^{(k)}$, we first select a direction $d^{(k)}$ and then search along this direction for a new solution with a lower objective value. To determine the step size α_k, we usually need to solve (approximately) the following one-dimensional problem:

$$\min_{\alpha \geq 0} f(x^{(k)} + \alpha d^{(k)}).$$

The direction $d^{(k)}$ in a line search iteration is typically selected based on the gradient $\nabla f(x^{(k)})$, leading to *gradient methods*, such as the method of steepest descent discussed in Section 13.4.

Alternatively, in a trust region strategy, the information collected about f is used to construct a model function $\hat{f}_k$, which approximates f in some neighborhood of $x^{(k)}$. Then, instead of unconstrained minimization of f, we deal with minimization of $\hat{f}_k$ over a *trust region* T (the region where $\hat{f}_k$ approximates f reasonably well):

$$\min_{d \in T} \hat{f}_k(x^{(k)} + d).$$

Let $d^{(k)}$ be a solution of this problem. If the decrease in the value of f going from $x^{(k)}$ to $x^{(k+1)} = x^{(k)} + d^{(k)}$ is not sufficient, we conclude that the approximation is not good enough, perhaps because the region T is too large. Therefore, we shrink the trust region and resolve the problem. We stop if we are not able to get a meaningful decrease in the objective after a certain number of attempts.

In the reminder of this chapter we discuss line search methods for unconstrained optimization.

13.4   Method of Steepest Descent

Consider the problem

$$\min_{x \in \mathbb{R}^n} f(x),$$

where $f(x)$ is a continuously differentiable function. Given $x^{(0)} \in \mathbb{R}^n$ and a direction $d \in \mathbb{R}^n$, the directional derivative of $f(x)$ at $x^{(0)}$ is

$$\nabla f(x^{(0)})^T d.$$

Recall that if $\|d\| = 1$, the directional derivative is interpreted as the rate of increase of $f(x)$ at $x^{(0)}$ in the direction d. The Cauchy-Schwartz inequality, stating that for any two vectors $u, v \in \mathbb{R}^n : u^T v \le \|u\|\|v\|$ with equality if and only if $u = \alpha v$ for some scalar $\alpha \ge 0$, allows one to find the direction with the largest possible rate of increase. Applying this inequality for d and $\nabla f(x^{(0)})$, we have

$$\nabla f(x^{(0)})^T d \le \|\nabla f(x^{(0)})\|\|d\|,$$

where equality is possible if and only if $d = \alpha \nabla f(x^{(0)})$ with $\alpha \ge 0$. So, the direction $d = \alpha \nabla f(x^{(0)})$ is the direction of the maximum rate of increase for f at $x^{(0)}$.

Similarly, for d and $-\nabla f(x^{(0)})$, we have

$$\nabla f(x^{(0)})^T d \ge -\|\nabla f(x^{(0)})\|\|d\|,$$

where equality is possible if and only if $d = -\alpha \nabla f(x^{(0)})$ with $\alpha \ge 0$. So, the direction $d = -\alpha \nabla f(x^{(0)})$ is the direction of the maximum rate of decrease at $x^{(0)}$. Thus, intuitively, the direction opposite to the gradient is the "best" direction to take in a minimization method. The general outline of a gradient method is

$$x^{(k+1)} = x^{(k)} - \alpha_k \nabla_k, \ \ k \geq 0,$$

where $\alpha_k \geq 0$ and $\nabla_k = \nabla f(x^{(k)})$. Different choices of $\alpha_k, k \geq 0$ result in different variations of gradient methods, but in general α_k is chosen so that the *descent property* is satisfied:

$$f(x^{(k+1)}) < f(x^{(k)}), \ \ k \geq 0.$$

Next we show that if $\nabla_k \neq 0$, then α_k can always be chosen such that the sequence $\{f(x^{(k)}) : k \geq 0\}$ possesses the descent property. Using Taylor's theorem, we have

$$f(x^{(k+1)}) = f(x^{(k)}) + \nabla_k^T(x^{(k+1)} - x^{(k)}) + o(\|x^{(k+1)} - x^{(k)}\|).$$

Since $x^{(k+1)} = x^{(k)} - \alpha_k \nabla_k$, we have

$$f(x^{(k+1)}) - f(x^{(k)}) = -\alpha_k \|\nabla_k\|^2 + o(\alpha_k \|\nabla_k\|).$$

This implies that there exists $\bar{\alpha} > 0$ such that for any positive $\alpha_k \leq \bar{\alpha}$:

$$f(x^{(k+1)}) - f(x^{(k)}) < 0.$$

In the steepest descent method, the step size α_k corresponds to the largest decrease in the objective while moving along the direction $\nabla f(x^{(k)})$ from point $x^{(k)}$:

$$\alpha_k : \ f(x^{(k)} + \alpha_k \nabla_k) = \min_{\alpha \geq 0} f(x^{(k)} - \alpha \nabla_k),$$

i.e., $\alpha_k = \arg \min_{\alpha \geq 0} f(x^{(k)} - \alpha \nabla_k)$. From the above, it is obvious that such choice of α_k will guarantee the descent property, since

$$f(x^{(k+1)}) \leq f(x^{(k)} - \bar{\alpha} \nabla_k) < f(x^{(k)}).$$

Theorem 13.7 *If $x^{(k)} \to x^*$, where $\{x^{(k)} : k \geq 0\}$ is the sequence generated by the steepest descent method, then $\nabla f(x^*) = 0$.*

Proof. Consider $\phi_k(\alpha) = f(x^{(k)} - \alpha \nabla_k)$. Since in the steepest descent method α_k minimizes $\phi_k(\alpha)$ for $\alpha > 0$, by the FONC

$$\phi_k'(\alpha_k) = 0.$$

On the other hand, using the chain rule:

$$\phi_k'(\alpha_k) = \left. \frac{df(x^{(k)} - \alpha \nabla_k)}{d\alpha} \right|_{\alpha = \alpha_k} = \nabla_{k+1}^T \nabla_k.$$

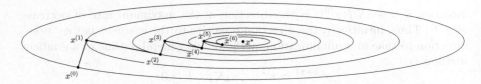

FIGURE 13.3: Illustration of steepest descent iterations.

So, $\nabla_{k+1}^T \nabla_k = 0$, $k \geq 0$ and $\|\nabla f(x^*)\|^2 = \lim_{k \to \infty} \nabla_{k+1}^T \nabla_k = 0$. $\qquad\square$

In the proof, we have shown that if we apply the method of steepest descent, then

$$\nabla_k^T \nabla_{k+1} = 0, \ k \geq 0.$$

Thus, the gradients of f in two consecutive points generated by the steepest descent are orthogonal to each other. Since the negative of the gradient represents the direction we move along at each iteration of the steepest descent, this means that the directions in the two consecutive steps are orthogonal as well. Indeed,

$$
\begin{aligned}
(x^{(k+1)} - x^{(k)})^T (x^{(k+2)} - x^{(k+1)}) &= (-\alpha_k \nabla_k)^T (-\alpha_{k+1} \nabla_{k+1}) \\
&= \alpha_k \alpha_{k+1} \nabla_k^T \nabla_{k+1} \\
&= 0.
\end{aligned}
$$

Figure 13.3 illustrates steepest descent iterations for optimizing a function of two variables geometrically. The figure shows six steps of the method starting from $x^{(0)}$. The direction used at each step is orthogonal to the level set of the objective function passing through the current point, as well as to the direction used at the previous step. The method will eventually converge to the optimal solution x^* in this example.

13.4.1 Convex quadratic case

Consider a convex quadratic function

$$f(x) = \frac{1}{2} x^T Q x + c^T x,$$

where Q is a positive definite matrix. Then, for some $x^{(k)}$,

$$\nabla_k = \nabla f(x^{(k)}) = Q x^{(k)} + c,$$

and the $(k+1)^{\text{st}}$ iteration of the steepest descent method is

$$x^{(k+1)} = x^{(k)} - \alpha_k \nabla_k,$$

with

$$\alpha_k = \arg\min_{\alpha \geq 0}\{\phi_k(\alpha)\},$$

where
$$\phi_k(\alpha) = f(x^{(k)} - \alpha\nabla_k); \quad \nabla_k = Qx^{(k)} + c.$$

We have

$$\begin{aligned}
\phi_k(\alpha) &= \frac{1}{2}(x^{(k)} - \alpha\nabla_k)^T Q(x^{(k)} - \alpha\nabla_k) + c^T(x^{(k)} - \alpha\nabla_k) \\
&= \alpha^2\left(\frac{1}{2}\nabla_k^T Q\nabla_k\right) - \alpha\left(\nabla_k^T \nabla_k\right) + f(x^{(k)}).
\end{aligned}$$

Since Q is positive definite, the coefficient for α^2 is positive, so $\phi_k(\alpha)$ is a convex quadratic function whose global minimizer is given by

$$\alpha_k = \frac{\nabla_k^T \nabla_k}{\nabla_k^T Q\nabla_k}.$$

Therefore, an iteration of the steepest descent method for the convex quadratic function $f(x) = \frac{1}{2}x^T Qx + c^T x$ is

$$x^{(k+1)} = x^{(k)} - \frac{\nabla_k^T \nabla_k}{\nabla_k^T Q\nabla_k}\nabla_k, \ k \geq 0.$$

Example 13.4 *For* $f(x) = \sum_{i=1}^{n} x_i^2 = x^T x = \|x\|^2$, *we have* $Q = 2I_n$, *where* I_n *is the* $n \times n$ *identity matrix,* $c = 0$, *and for any* $x^{(0)} \in \mathbb{R}^n$,

$$x^{(1)} = x^{(0)} - \frac{4(x^{(0)})^T x^{(0)}}{8(x^{(0)})^T x^{(0)}}2x^{(0)} = x^{(0)} - x^{(0)} = 0.$$

Thus, we get the global minimizer in one step in this case.

13.4.2 Global convergence of the steepest descent method

Recall that a numerical method is said to be globally convergent if it converges starting from any point. It can be shown that the steepest descent method globally converges to a stationary point. We discuss the global convergence analysis for the convex quadratic case only; however, the result holds in general.

Consider a convex quadratic function

$$f(x) = \frac{1}{2}x^T Qx + c^T x.$$

Then the global minimizer of $f(x)$ is $x^* = -Q^{-1}c$. For convenience, instead of $f(x)$ we will use the quadratic function in the form

$$q(x) = \frac{1}{2}(x - x^*)^T Q(x - x^*) \tag{13.13}$$

in our analysis. It is easy to check that $q(x) = f(x) + \frac{1}{2}x^{*T}Qx^*$, so the two functions differ only by a constant. We denote by ∇_k the gradient of $q(x)$ at point $x^{(k)}$:

$$\nabla_k = \nabla q(x^{(k)}) = Q(x^{(k)} - x^*). \qquad (13.14)$$

The steepest descent iteration for this function is

$$x^{(k+1)} = x^{(k)} - \alpha_k \nabla_k, \qquad (13.15)$$

where

$$\alpha_k = \frac{\nabla_k^T \nabla_k}{\nabla_k^T Q \nabla_k}. \qquad (13.16)$$

Next, we prove that

$$q(x^{(k+1)}) \le q(x^{(k)}) \left(1 - \frac{\lambda_{min}(Q)}{\lambda_{max}(Q)}\right),$$

where $\lambda_{min}(Q)$ and $\lambda_{max}(Q)$ are the smallest and the largest eigenvalues of Q, respectively. To show this, we first observe that

$$
\begin{aligned}
(x^{(k)} - x^*)^T Q (x^{(k)} - x^*) &= (x^{(k)} - x^*)^T Q^T Q^{-1} Q (x^{(k)} - x^*) \\
&= (Q(x^{(k)} - x^*))^T Q^{-1} (Q(x^{(k)} - x^*)) \\
&= \nabla_k^T Q^{-1} \nabla_k,
\end{aligned}
$$

and then express $q(x^{(k+1)})$ in terms of $q(x^{(k)})$ as follows:

$$
\begin{aligned}
q(x^{(k+1)}) &= \tfrac{1}{2}(x^{(k)} - \alpha_k \nabla_k - x^*)Q(x^{(k)} - \alpha_k \nabla_k - x^*) \\
&= q(x^{(k)}) - \alpha_k \nabla_k^T Q(x^{(k)} - x^*) + \tfrac{1}{2}\alpha_k^2 \nabla_k^T Q \nabla_k \\
&= q(x^{(k)}) \left(1 - \frac{\nabla_k^T \nabla_k}{\nabla_k^T Q \nabla_k} \frac{\nabla_k^T \nabla_k}{\tfrac{1}{2}(x^{(k)}-x^*)^T Q(x^{(k)}-x^*)} \right. \\
&\qquad\qquad \left. + (\frac{\nabla_k^T \nabla_k}{\nabla_k^T Q \nabla_k})^2 \frac{\tfrac{1}{2}\nabla_k^T Q \nabla_k}{\tfrac{1}{2}(x^{(k)}-x^*)^T Q(x^{(k)}-x^*)}\right) \\
&= q(x^{(k)}) \left(1 - \frac{2\|\nabla_k\|^4}{(\nabla_k^T Q \nabla_k)(\nabla_k^T Q^{-1}\nabla_k)} + \frac{\|\nabla_k\|^4}{(\nabla_k^T Q \nabla_k)(\nabla_k^T Q^{-1}\nabla_k)}\right) \\
&= q(x^{(k)}) \left(1 - \frac{\|\nabla_k\|^4}{(\nabla_k^T Q \nabla_k)(\nabla_k^T Q^{-1}\nabla_k)}\right).
\end{aligned}
$$

From Rayleigh's inequality,

$$\nabla_k^T Q \nabla_k \le \lambda_{max}(Q)\|\nabla_k\|^2$$

and

$$(\nabla_k^T Q^{-1}\nabla_k) \le \lambda_{max}(Q^{-1})\|\nabla_k\|^2 = (\lambda_{min}(Q))^{-1}\|\nabla_k\|^2.$$

Therefore,

$$q(x^{(k+1)}) = q(x^{(k)})\left(1 - \frac{\|\nabla_k\|^4}{(\nabla_k^T Q \nabla_k)(\nabla_k^T Q^{-1}\nabla_k)}\right) \le q(x^{(k)})\left(1 - \frac{\lambda_{min}(Q)}{\lambda_{max}(Q)}\right).$$

In summary, we have

$$q(x^{(k+1)}) \le q(x^{(0)}) \left(1 - \frac{\lambda_{min}(Q)}{\lambda_{max}(Q)} \right)^{k+1}, \tag{13.17}$$

and since $0 < \lambda_{min}(Q) \le \lambda_{max}(Q)$,

$$q(x^{(k)}) \to 0, \quad k \to \infty.$$

Note that $q(x) = 0 \iff x = x^*$, so $x^{(k)} \to x^*$, $k \to \infty$. Thus, the steepest descent method is globally convergent for a convex quadratic function. Note that the rate of convergence is linear.

From the above inequality (13.17), we also see that if $\lambda_{min}(Q) = \lambda_{max}(Q)$, then we will have convergence in one step (as for $f(x) = x_1^2 + x_2^2$ in Example 13.4). On the other hand, if $\lambda_{max}(Q)$ is much larger than $\lambda_{min}(Q)$, then $1 - \frac{\lambda_{min}(Q)}{\lambda_{max}(Q)} \approx 1$ and the convergence may be extremely slow in this case.

Recall that the ratio $k(Q) = \frac{\lambda_{max}(Q)}{\lambda_{min}(Q)} = \|Q\|\|Q^{-1}\|$ is called the *condition number* of matrix Q. When a matrix is poorly conditioned (i.e., it has a large condition number), we have "long, narrow" level sets, and the steepest descent may move back and forth ("zigzag") in search of the minimizer.

13.5   Newton's Method

As before, we consider the unconstrained problem

$$\min_{x \in \mathbb{R}^n} f(x).$$

Newton's method is based on minimizing a quadratic approximation of $f(x)$ obtained using Taylor's theorem instead of $f(x)$. We have

$$f(x) \approx f(x^{(k)}) + \nabla_k^T (x - x^{(k)}) + \frac{1}{2}(x - x^{(k)})^T \nabla_k^2 (x - x^{(k)}),$$

where $\nabla_k = \nabla f(x^{(k)})$ and $\nabla_k^2 = \nabla^2 f(x^{(k)})$. If ∇_k^2 is positive definite, then the global minimizer of the quadratic approximation

$$q(x) = f(x^{(k)}) + \nabla_k^T (x - x^{(k)}) + \frac{1}{2}(x - x^{(k)})^T \nabla_k^2 (x - x^{(k)})$$

is given by $x^{(k)} - (\nabla_k^2)^{-1} \nabla_k$. Setting

$$x^{(k+1)} = x^{(k)} - (\nabla_k^2)^{-1} \nabla_k,$$

we obtain an iteration of Newton's method:

$$x^{(k+1)} = x^{(k)} - (\nabla_k^2)^{-1} \nabla_k, \ k \ge 0.$$

Example 13.5 *Using Newton's iteration for a convex quadratic function*

$$f(x) = \frac{1}{2}x^T Q x + c^T x$$

with positive definite Q, we obtain

$$x^{(k+1)} = x^{(k)} - Q^{-1}(Qx^{(k)} + c) = -Q^{-1}c.$$

Thus, we get the global minimizer in one step.

13.5.1   Rate of convergence

Next we show that Newton's method has a quadratic rate of convergence under certain assumptions. We assume that $f \in C^{(3)}(\mathbb{R}^n)$, ∇_k^2 and $\nabla^2 f(x^*)$ are positive definite, where x^* is a stationary point of $f(x)$ and $x^{(k)}$ is a point close to x^*. Using Taylor's theorem for $\nabla f(x)$ and $x^{(k)}$, we have:

$$\nabla f(x) - \nabla_k - \nabla_k^2(x - x^{(k)}) = O(\|x - x^{(k)}\|^2).$$

Thus, by definition of $O(\cdot)$, there is a constant c_1 such that

$$\|\nabla f(x) - \nabla_k - \nabla_k^2(x - x^{(k)})\| \leq c_1 \|x - x^{(k)}\|^2 \qquad (13.18)$$

if x is sufficiently close to $x^{(k)}$. Consider a closed *epsilon*-ball $\bar{B}(x^*, \epsilon) = \{x : \|x - x^*\| \leq \epsilon\}$ centered at x^*, in which $\nabla^2 f(x)$ is nonsingular (such a neighborhood exists since $\nabla^2 f(x^*)$ is positive definite). Then, by the Weierstrass theorem, since $\|(\nabla^2 f(x))^{-1}\|$ is a continuous function, it has a maximizer over $\bar{B}(x^*, \epsilon)$, that is, there exists a constant c_2 such that for any $x \in \bar{B}(x^*, \epsilon)$

$$\|(\nabla^2 f(x))^{-1}\| \leq c_2. \qquad (13.19)$$

Consider $x^{(k+1)}$ obtained from $x^{(k)}$ using an iteration of Newton's method:

$$x^{(k+1)} = x^{(k)} - (\nabla_k^2)^{-1}\nabla_k.$$

Then,

$$
\begin{aligned}
\|x^{(k+1)} - x^*\| &= \|x^{(k)} - (\nabla_k^2)^{-1}\nabla_k - x^*\| \\
&= \|(\nabla_k^2)^{-1}(\nabla f(x^*) - \nabla_k - \nabla_k^2(x^* - x^{(k)}))\| \\
&\leq \|(\nabla_k^2)^{-1}\| \cdot \|(\nabla_k - \nabla f(x^*) - \nabla_k^2(x^{(k)} - x^*))\| \\
&\leq c_2 c_1 \|x^{(k)} - x^*\|^2.
\end{aligned}
$$

So, if we start close enough from the stationary point x^*, then Newton's method converges to x^* with a quadratic rate of convergence under the assumptions specified above.

13.5.2 Guaranteeing the descent

Consider an iteration of Newton's method:

$$x^{(k+1)} = x^{(k)} - (\nabla_k^2)^{-1}\nabla_k.$$

Here we assume that $\nabla_k \neq 0$. In general, Newton's method may not possess the descent property, that is, we may have $f(x^{(k+1)}) > f(x^{(k)})$. However, if ∇_k^2 is positive definite, the *Newton's direction* $d^{(k)} = -(\nabla_k^2)^{-1}\nabla_k$ is a descent direction in the sense that there exists $\bar{\alpha}$ such that for all $\alpha \in (0, \bar{\alpha}) : f(x^{(k)} + \alpha d^{(k)}) < f(x^{(k)})$. To show this, we introduce a function of one variable

$$\phi_k(\alpha) = f(x^{(k)} + \alpha d^{(k)})$$

and consider its derivative for $\alpha = 0$:

$$\phi_k'(0) = \nabla_k^T d^{(k)} = -\nabla_k^T (\nabla_k^2)^{-1}\nabla_k < 0.$$

The inequality above holds since $(\nabla_k^2)^{-1}$ is positive definite and $\nabla_k \neq 0$. Thus, there exists $\bar{\alpha}$ such that $\phi_k'(\alpha) < 0$ for $\alpha \leq \bar{\alpha}$, i.e., $\phi_k(\alpha)$ is decreasing on $[0, \bar{\alpha}]$ and for all $\alpha \in (0, \bar{\alpha}) : f(x^{(k)} + \alpha d^{(k)}) < f(x^{(k)})$.

Therefore, we can modify Newton's method to enforce the descent property by introducing a step size as follows:

- find $\alpha_k = \arg\min_{\alpha \geq 0}(f(x^{(k)} - \alpha(\nabla_k^2)^{-1}\nabla_k))$;

- set $x^{(k+1)} = x^{(k)} - \alpha_k(\nabla_k^2)^{-1}\nabla_k$.

Then the descent property follows from the observation that

$$f(x^{(k+1)}) \leq f(x^{(k)} - \alpha(\nabla_k^2)^{-1}\nabla_k) < f(x^{(k)})$$

for any $\alpha \in (0, \bar{\alpha}]$.

13.5.3 Levenberg-Marquardt method

We have assumed above that the Hessian is positive definite in Newton's method. Here we show that if the Hessian is not positive definite, it can still be modified so that the iteration resulting from this modification has the descent property.

Consider the matrix

$$M_k = \nabla_k^2 + \mu_k I_n,$$

where I_n is the $n \times n$ identity matrix and μ_k is a scalar. If we denote by $\lambda_i, i = 1, \ldots, n$ the eigenvalues of ∇_k^2, then the eigenvalues of M_k are given by

$$\lambda_i + \mu_k, \ i = 1, \ldots, n.$$

Indeed, if v_i is the eigenvector of ∇_k^2 corresponding to the eigenvalue λ_i, then

$$M_k v_i = (\nabla_k^2 + \mu_k I_n)v_i = \lambda_i v_i + \mu_k v_i = (\lambda_i + \mu_k)v_i.$$

Hence, if we choose $\mu_k > |\lambda_{min}(\nabla_k^2)|$, where $\lambda_{min}(\nabla_k^2)$ is the minimum eigenvalue of ∇_k^2, then all eigenvalues of M_k are positive, so M_k is a positive definite matrix.

To make sure that the descent property holds, we can use the direction $-M_k^{-1}\nabla_k$ instead of the direction $-(\nabla_k^2)^{-1}\nabla_k$ used in Newton's method. Including the step size, we obtain the following iteration:

$$x^{(k+1)} = x^{(k)} - \alpha_k M_k^{-1}\nabla_k,$$

where $\alpha_k = \arg\min_{\alpha \geq 0} f(x^{(k)} - \alpha M_k^{-1}\nabla_k)$. This method is referred to as the *Levenberg-Marquardt method*. With this modification of the Hessian matrix, the direction used becomes a descent direction. To see this, we write down the derivative of the function $\phi_k(\alpha) = f(x^{(k)} - \alpha M_k^{-1}\nabla_k)$. We have

$$\phi_k'(0) = -\nabla_k^T (M_k)^{-1}\nabla_k < 0,$$

since M_k^{-1} is positive definite.

The Levenberg-Marquardt method is in some sense intermediate between the steepest descent and Newton's methods. If $\mu_k = 0$, then it coincides with Newton's method. On the other hand, if μ_k is very large, then $M_k \approx C I_n$ for some very large $C > 0$, so $M_k^{-1} \approx \epsilon I_n$ for some small $\epsilon = 1/C > 0$, and the iteration is

$$x^{(k+1)} \approx x^{(k)} - \alpha_k \epsilon \nabla_k.$$

Thus, we obtain an approximation of the steepest descent iteration in this case.

13.6   Conjugate Direction Method

Next we discuss methods that search for a stationary point by exploiting a set of conjugate directions, which are defined next. Given a positive definite $n \times n$ matrix Q, the nonzero directions $d^{(0)}, d^{(1)}, \ldots, d^{(k)}$ are called Q-conjugate if

$$(d^{(i)})^T Q d^{(j)} = 0 \quad \text{for } i \neq j.$$

We first show that Q-conjugate directions $d^{(0)}, d^{(1)}, \ldots, d^{(k)}$ form a set of linearly independent vectors. Consider a linear combination of the given Q-conjugate directions that results in a zero vector:

$$c_0 d^{(0)} + c_1 d^{(1)} + \ldots + c_i d^{(i)} + \ldots + c_k d^{(k)} = 0.$$

If we premultiply both sides of this equation by $(d^{(i)})^T Q$, we obtain

$$c_0(d^{(i)})^T Q d^{(0)} + c_1(d^{(i)})^T Q d^{(1)} + \ldots + c_i(d^{(i)})^T Q d^{(i)} + \ldots + c_k(d^{(i)})^T Q d^{(k)} = 0.$$

But the directions $d^{(0)}, d^{(1)}, \ldots, d^{(k)}$ are Q-conjugate, so $(d^{(i)})^T Q d^{(j)} = 0$ for all $j \neq i$, and hence we have

$$c_i(d^{(i)})^T Q d^{(i)} = 0.$$

This means that $c_i = 0$, since Q is positive definite and $d^{(i)} \neq 0$. Note that the index i was chosen arbitrarily, hence $c_i = 0$ for all $i = 1, \ldots, k$.

The linear independence of the conjugate directions implies that one can choose at most n Q-conjugate directions in $\mathbb{R}^n$.

13.6.1 Conjugate direction method for convex quadratic problems

We consider a convex quadratic problem

$$\min_{x \in \mathbb{R}^n} f(x), \qquad (13.20)$$

where

$$f(x) = \frac{1}{2} x^T Q x + c^T x,$$

and Q is a positive definite matrix. Consider an iteration

$$x^{(k+1)} = x^{(k)} + \alpha_k d^{(k)},$$

where

$$\alpha_k = \arg \min_{\alpha \in \mathbb{R}} f(x^{(k)} + \alpha d^{(k)}).$$

In the conjugate direction method the directions $d^{(k)}$ are chosen so that

$$\{d^{(0)}, d^{(1)}, \ldots, d^{(n-1)}\}$$

is a set of Q-conjugate directions.

Denote by

$$\phi_k(\alpha) = f(x^{(k)} + \alpha d^{(k)}).$$

Then,

$$\begin{aligned}
\phi_k(\alpha) &= \frac{1}{2}(x^{(k)} + \alpha d^{(k)})^T Q(x^{(k)} + \alpha d^{(k)}) + c^T(x^{(k)} + \alpha d^{(k)}) \\
&= \alpha^2 \left(\frac{1}{2}(d^{(k)})^T Q d^{(k)} \right) + \alpha((x^{(k)})^T Q + c^T)d^{(k)} + f(x^{(k)}) \\
&= \left(\frac{1}{2}(d^{(k)})^T Q d^{(k)} \right) \alpha^2 + (\nabla_k^T d^{(k)})\alpha + f(x^{(k)}).
\end{aligned}$$

Solving $\phi_k'(\alpha) = 0$, we find that

$$\alpha_k = -\frac{\nabla_k^T d^{(k)}}{(d^{(k)})^T Q d^{(k)}},$$

so an iteration of the conjugate direction method in this case is

$$x^{(k+1)} = x^{(k)} - \frac{\nabla_k^T d^{(k)}}{(d^{(k)})^T Q d^{(k)}} d^{(k)}.$$

Next we establish some basic properties of the conjugate direction method in the convex quadratic case.

Lemma 13.1 *Let $x^{(k+1)}$ be the point obtained by applying $k + 1$ iterations of a conjugate direction method to the problem of minimizing $f(x)$ starting at $x^{(0)}$ and using a set of Q-conjugate directions $d^{(0)}, \ldots, d^{(k)}$. Then the gradient $\nabla_{k+1} = \nabla f(x^{(k+1)})$ satisfies*

$$\nabla_{k+1}^T d^{(i)} = 0, \ i = 0, \ldots, k.$$

Proof. We will use induction for the proof. We first show that for any k: $\nabla_{k+1}^T d^{(k)} = 0$. We know that α_k satisfies the FONC for $\phi_k(\alpha)$, hence

$$\phi_k'(\alpha_k) = 0.$$

On the other hand, using the chain rule

$$\phi_k'(\alpha_k) = \nabla f(x^{(k)} + \alpha_k d^{(k)})^T d^{(k)} = \nabla_{k+1}^T d^{(k)},$$

so $\nabla_{k+1}^T d^{(k)} = 0$.

Now, assume that the statement is correct for all $k = 1, \ldots, K$ for some K, i.e.,

$$\nabla_K^T d^{(i)} = 0, \ i = 0, \ldots, K - 1.$$

We need to show that it is correct for $k = K + 1$. For $i = 0, \ldots, K - 1$, consider

$$
\begin{aligned}
\nabla_{K+1}^T d^{(i)} &= (Qx^{(K+1)} + c)^T d^{(i)} \\
&= \left(Q(x^{(K)} + \alpha_K d^{(K)}) + c \right)^T d^{(i)} \\
&= \left(Qx^{(K)} + c + \alpha_K Q d^{(K)} \right)^T d^{(i)} \\
&= \left(\nabla_K + \alpha_K Q d^{(K)} \right)^T d^{(i)} \\
&= \nabla_K^T d^{(i)} + \alpha_K (d^{(K)})^T Q d^{(i)} \\
&= 0.
\end{aligned}
$$

We have already shown that the statement is correct for $i = K$. Thus, by induction, the statement is correct for any k and any $i = 0, \ldots, k$. $\qquad\square$

Lemma 13.2 *Denote by* $\alpha^{(k)} = [\alpha_0, \alpha_1, \ldots, \alpha_k]^T$ *the vector of step sizes obtained using k steps of the conjugate direction method, and by*

$$\Phi_k(a^{(k)}) = f(x^{(0)} + a_0 d^{(0)} + a_1 d^{(1)} + \ldots + a_k d^{(k)}) = f(x^{(0)} + D_k a^{(k)}),$$

where

$$D_k = [d^{(0)} \ d^{(1)} \ \cdots \ d^{(k)}] \ and \ a^{(k)} = [a_0, a_1, \ldots, a_k]^T.$$

Then we have

$$\alpha^{(k)} = \arg \min_{a^{(k)} \in \mathbb{R}^{k+1}} \Phi(a^{(k)}).$$

Proof. We have

$$\Phi_k(a^{(k)}) = (x^{(0)} + D_k a^{(k)})^T Q(x^{(0)} + D_k a^{(k)}) + c^T(x^{(0)} + D_k a^{(k)})$$
$$= \frac{1}{2}(a^{(k)})^T \left(D_k{}^T Q D_k\right) a^{(k)} + \left((x^{(0)})^T Q D_k + c^T D_k\right) a^{(k)} + f(x^{(0)}).$$

Since Q is a positive definite matrix and D_k is a matrix of full rank (due to linear independence of Q-conjugate directions), $D_k{}^T Q D_k$ is positive definite, so $\Phi_k(a^{(k)})$ is a convex quadratic function. The gradient of $\Phi_k(a^{(k)})$ is

$$\nabla \Phi_k(a^{(k)})^T = \nabla f(x^{(0)} + D_k a^{(k)})^T D_k.$$

Then for $a^{(k)} = \alpha^{(k)}$ we have

$$\nabla \Phi_k(\alpha^{(k)})^T = \nabla f(x^{(0)} + D_k \alpha^{(k)})^T D_k$$
$$= \nabla_{k+1}^T D_k$$
$$= \nabla_{k+1}^T [d^{(0)} \ d^{(1)} \ \cdots \ d^{(k)}]$$
$$= [\nabla_{k+1}^T d^{(0)} \ \nabla_{k+1}^T d^{(1)} \ \cdots \ \nabla_{k+1}^T d^{(k)}]$$
$$= 0^T.$$

Since $\Phi_k(a^{(k)})$ is a convex quadratic function, $a^{(k)} = \alpha^{(k)}$ is the global minimizer of $\Phi_k(a^{(k)})$. $\qquad \square$

The last property has an important implication concerning the convergence of the method. We state it in the following theorem.

Theorem 13.8 *The conjugate direction algorithm for a convex quadratic function converges to the global minimizer in no more than n steps for any starting point $x^{(0)}$.*

Proof. Note that

$$\{x = x^{(0)} + D_{n-1} a : a \in \mathbb{R}^n\} = \left\{x = x^{(0)} + \sum_{i=0}^{n-1} a_i d^{(i)} : a_i \in \mathbb{R} \ \forall i\right\} = \mathbb{R}^n.$$

Thus,

$$f(x^{(n)}) = f(x^{(0)} + D_{n-1}\alpha^{(n-1)}) = \min_{a^{(n-1)} \in \mathbb{R}^n} f(x^{(0)} + D_{n-1}a^{(n-1)}) = \min_{x \in \mathbb{R}^n} f(x).$$

$\square$

13.6.2   Conjugate gradient algorithm

To implement a conjugate direction algorithm, one needs to specify n Q-conjugate directions. This can be done using a procedure similar to the *Gram-Schmidt* orthogonalization (see Exercise 13.14). The conjugate gradient algorithm provides an alternative approach that allows us to find Q-conjugate directions one by one as the steps of the algorithm progress. The directions used in this algorithm are related to the gradient. More specifically, the directions are chosen so that

$$\begin{aligned} d^{(0)} &= -\nabla_0; \\ d^{(k)} &= -\nabla_k + \beta_k d^{(k-1)}, \ k = 1, \dots, n-1. \end{aligned}$$

Here β_k needs to be chosen so that $(d^{(k)})^T Q d^{(k-1)} = 0$. We have

$$(-\nabla_k + \beta_k d^{(k-1)})^T Q d^{(k-1)} = 0 \quad \Rightarrow \quad \beta_k = \frac{\nabla_k^T Q d^{(k-1)}}{(d^{(k-1)})^T Q d^{(k-1)}}.$$

It can be shown by induction that with such choice of β_k, $k = 1, \dots, n-1$, the directions $d^{(0)}, d^{(1)}, \dots, d^{(n-1)}$ are Q-conjugate. The conjugate gradient algorithm for a convex quadratic problem proceeds as described in Algorithm 13.1.

Algorithm 13.1 The conjugate gradient algorithm for minimizing a convex quadratic function $q(x) = \frac{1}{2}x^T Q x + c^T x$ on $\mathbb{R}^n$.

1: **Input:** $Q, c, x^{(0)}$
2: **Output:** x^*, the minimizer of $q(x)$
3: **for** $k = 0, \dots, n-1$ **do**
4: $\quad \alpha_k = -\frac{\nabla_k^T d^{(k)}}{(d^{(k)})^T Q d^{(k)}}$, where $\nabla_k = -(Qx^{(k)} + c)$
5: $\quad$ **if** $k = 0$ **then**
6: $\quad\quad d_0 = -\nabla_0$
7: $\quad$ **else**
8: $\quad\quad \beta_k = \frac{\nabla_k^T Q d^{(k-1)}}{(d^{(k-1)})^T Q d^{(k-1)}}$
9: $\quad\quad d^{(k)} = -\nabla_k + \beta_k d^{(k-1)}$
10: $\quad$ **end if**
11: $\quad x^{(k+1)} = x^{(k)} + \alpha_k d^{(k)}$
12: **end for**
13: **return** $x^{(n)}$

Figure 13.4 illustrates how the steepest descent method, conjugate gradient method, and Newton's method compare for a convex quadratic function.

(a) steepest descent

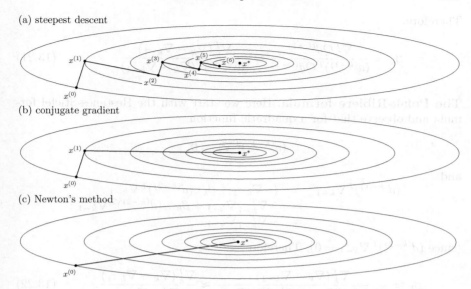

(b) conjugate gradient

(c) Newton's method

FIGURE 13.4: Comparison of the steepest descent method, conjugate gradient method, and Newton's method on a quadratic function.

13.6.2.1 Non-quadratic problems

If the minimized function is not quadratic, one can use its quadratic approximation given by Taylor's theorem:

$$f(x) \approx f(x^{(k)}) + \nabla f(x^{(k)})^T(x - x^{(k)}) + \frac{1}{2}(x - x^{(k)})^T \nabla^2 f(x^{(k)})(x - x^{(k)}).$$

We could replace $f(x)$ with this quadratic approximation and apply the conjugate gradient algorithm to find the minimizer of the quadratic function, which we would use as $x^{(k+1)}$. However, computing and evaluating the Hessian at each iteration is a computationally expensive procedure that we would like to avoid. In the conjugate gradient algorithm above, two operations involve the matrix Q, which would correspond to the Hessian. These operations are computing α_k and β_k. While α_k could be approximated using line search, we need to find a way to approximate β_k. Next we discuss several formulas designed for this purpose.

The Hestenes-Stiefel formula. We use the fact that for a quadratic function

$$\nabla_k - \nabla_{k-1} = Q(x^{(k)} - x^{(k-1)}) = \alpha_{k-1} Q d^{(k-1)},$$

so

$$Q d^{(k-1)} = \frac{\nabla_k - \nabla_{k-1}}{\alpha_{k-1}}.$$

Therefore,

$$\beta_k = \frac{\nabla_k^T Q d^{(k-1)}}{(d^{(k-1)})^T Q d^{(k-1)}} \approx \frac{\nabla_k^T (\nabla_k - \nabla_{k-1})}{(d^{(k-1)})^T (\nabla_k - \nabla_{k-1})}. \quad (13.21)$$

The Polak-Ribiere formula. Here we start with the Hestenes-Stiefel formula and observe that for a quadratic function

$$(d^{(k-1)})^T \nabla_k = 0$$

and

$$
\begin{aligned}
(d^{(k-1)})^T \nabla_{k-1} &= (-\nabla_{k-1} + \beta_{k-1} d^{(k-2)})^T \nabla_{k-1} \\
&= -\nabla_{k-1}^T \nabla_{k-1} + \beta_{k-1} (d^{(k-2)})^T \nabla_{k-1} \\
&= -\nabla_{k-1}^T \nabla_{k-1}
\end{aligned}
$$

(since $(d^{(k-2)})^T \nabla_{k-1} = 0$). Therefore,

$$\beta_k \approx \frac{\nabla_k^T (\nabla_k - \nabla_{k-1})}{(d^{(k-1)})^T \nabla_k - (d^{(k-1)})^T \nabla_{k-1}} \approx \frac{\nabla_k^T (\nabla_k - \nabla_{k-1})}{\nabla_{k-1}^T \nabla_{k-1}}. \quad (13.22)$$

The Fletcher-Reeves formula. Note that for a quadratic function

$$\nabla_k^T \nabla_{k-1} = \nabla_k^T (-d^{(k-1)} + \beta_{k-1} d^{(k-2)}) = 0.$$

So, starting with the Polak-Ribiere formula, we obtain

$$\beta_k \approx \frac{\nabla_k^T (\nabla_k - \nabla_{k-1})}{\nabla_{k-1}^T \nabla_{k-1}} = \frac{\nabla_k^T \nabla_k - \nabla_k^T \nabla_{k-1}}{\nabla_{k-1}^T \nabla_{k-1}} \approx \frac{\nabla_k^T \nabla_k}{\nabla_{k-1}^T \nabla_{k-1}}. \quad (13.23)$$

In the above derivations, each formula was obtained from the previous one using equalities that are exact for quadratic functions, but only approximate in general. Therefore, in the process of simplifying the formula for β_k, the quality of approximation of the original problem gradually decreases.

13.7   Quasi-Newton Methods

Recall an iteration of Newton's method with step size:

$$x^{(k+1)} = x^{(k)} - \alpha_k (\nabla_k^2)^{-1} \nabla_k, \ k \geq 0.$$

Quasi-Newton methods are obtained from Newton's method by approximating the inverse of the Hessian with a matrix that does not involve second-order derivatives. Thus, if we replace $(\nabla_k^2)^{-1}$ with its approximation H_k, we obtain a step of a quasi-Newton method:

$$x^{(k+1)} = x^{(k)} - \alpha_k H_k \nabla_k, \ k \geq 0.$$

Different choices of H_k result in different variations of the quasi-Newton methods. Matrix H_k will be chosen so that it satisfies some properties that the Hessian matrix satisfies in the case of a convex quadratic function $f(x) = \frac{1}{2} x^T Q x + c^T x$. In this case, since $\nabla_k = Q x^{(k)} + c$, we have

$$Q(x^{(k+1)} - x^{(k)}) = \nabla_{k+1} - \nabla_k.$$

If we denote by $p^{(k)} = x^{(k+1)} - x^{(k)}$ and by $g^{(k)} = \nabla_{k+1} - \nabla_k$, then

$$Qp^{(k)} = g^{(k)}, \ k \geq 0 \quad \Leftrightarrow \quad p^{(k)} = Q^{-1} g^{(k)}, \ k \geq 0.$$

Therefore, we choose H_k such that

$$p^{(i)} = H_k g^{(i)}, \ i = 0, \ldots, k-1,$$

then after $n-1$ steps we obtain

$$H_n g^{(i)} = p^{(i)}, \ i = 0, \ldots, n-1.$$

This can be written in matrix form as

$$H_n [g^{(0)} \ g^{(1)} \ \cdots \ g^{(n-1)}] = [p^{(0)} \ p^{(1)} \ \cdots \ p^{(n-1)}],$$

or, if we denote by $G_n = [g^{(0)} \ g^{(1)} \ \cdots \ g^{(n-1)}]$ and by $P_n = [p^{(0)} \ p^{(1)} \ \cdots \ p^{(n-1)}]$, then

$$H_n G_n = P_n.$$

If G_n is a nonsingular matrix, we have

$$H_n = P_n G_n^{-1}.$$

Similarly, we can show that

$$Q^{-1} G_n = P_n$$

and

$$Q^{-1} = P_n G_n^{-1}.$$

So, for a quadratic function, $H_n = Q^{-1}$. This means that after $n+1$ steps of a quasi-Newton method we get the same answer as we get after one step of Newton's method, which is the global minimizer of the convex quadratic function. Next, we show that the global minimizer of the convex quadratic function is, in fact, obtained in no more than n steps of a quasi-Newton method. More specifically, assuming that H_k is symmetric for any k, we show that the quasi-Newton directions

$$\{d^{(k)} = -H_k \nabla_k : \; k = 0, \ldots, n-1\}$$

are Q-conjugate. We use induction.

Assume that the statement we are proving is correct for some $K < n - 1$, i.e.,

$$\{d^{(k)} = -H_k \nabla_k : \; k = 0, \ldots, K\}$$

are Q-conjugate. Then for $k = K + 1$ and for any $i \le K$:

$$
\begin{aligned}
(d^{(K+1)})^T Q d^{(i)} &= (d^{(K+1)})^T Q p^{(i)}/\alpha_i & & \left(p^{(i)} = x^{(i+1)} - x^{(i)} = \alpha_i d^{(i)}\right) \\
&= (d^{(K+1)})^T g^{(i)}/\alpha_i & & (Q p^{(i)} = g^{(i)}) \\
&= -\nabla_{K+1}^T H_{K+1} g^{(i)}/\alpha_i & & (d^{(K+1)} = H_{K+1} \nabla_{K+1}) \\
&= -\nabla_{K+1}^T p^{(i)}/\alpha_i & & (H_{K+1} g^{(i)} = p^{(i)}) \\
&= -\nabla_{K+1}^T d^{(i)} & & (p^{(i)} = \alpha_i d^{(i)}) \\
&= 0
\end{aligned}
$$

(since $d^{(i)}$, $i = 0, \ldots, K$ are Q-conjugate by the induction assumption). Thus, by induction, the directions $d^{(0)}, d^{(1)}, \ldots, d^{(n-1)}$ in a quasi-Newton method are Q-conjugate.

Next, we discuss how to select H_k, $k \ge 0$ in quasi-Newton methods.

13.7.1   Rank-one correction formula

In the rank-one correction formula, we start with a symmetric positive definite matrix H_0 (say, $H_0 = I_n$). At step $k + 1$ we add a rank-one matrix to H_k to obtain H_{k+1}:

$$H_{k+1} = H_k + a_k z^{(k)} (z^{(k)})^T \tag{13.24}$$

for some vector $z^{(k)}$ and a scalar a_k (note that $z^{(k)} (z^{(k)})^T$ is an $n \times n$ matrix of rank 1). As before, we require that

$$H_{k+1} g^{(i)} = p^{(i)}, \; i = 0, \ldots, k.$$

For $i = k$, we have

$$H_{k+1} g^{(k)} = p^{(k)},$$

so,

$$(H_k + \alpha_k z^{(k)} (z^{(k)})^T) g^{(k)} = p^{(k)}, \tag{13.25}$$

which is equivalent to

$$H_k g^{(k)} + \alpha_k z^{(k)} (z^{(k)})^T g^{(k)} = p^{(k)}.$$

From here we find that

$$z^{(k)} = \frac{p^{(k)} - H_k g^{(k)}}{\alpha_k (z^{(k)})^T g^{(k)}},$$

thus,

$$\alpha_k z^{(k)} (z^{(k)})^T = \alpha_k \frac{(p^{(k)} - H_k g^{(k)})}{\alpha_k (z^{(k)})^T g^{(k)}} \frac{(p^{(k)} - H_k g^{(k)})^T}{\alpha_k (z^{(k)})^T g^{(k)}}$$

$$= \frac{(p^{(k)} - H_k g^{(k)})(p^{(k)} - H_k g^{(k)})^T}{\alpha_k ((z^{(k)})^T g^{(k)})^2}. \qquad (13.26)$$

Note that Eq. (13.25) is equivalent to

$$\alpha_k z^{(k)} (z^{(k)})^T g^{(k)} = p^{(k)} - H_k g^{(k)}.$$

Premultiplying both sides of this equation by $(g^{(k)})^T$, we obtain

$$\alpha_k ((z^{(k)})^T g^{(k)})^2 = (g^{(k)})^T (p^{(k)} - H_k g^{(k)}).$$

Substituting this expression in the denominator of Eq. (13.26) we get

$$\alpha_k z^{(k)} (z^{(k)})^T = \frac{(p^{(k)} - H_k g^{(k)})(p^{(k)} - H_k g^{(k)})^T}{(g^{(k)})^T (p^{(k)} - H_k g^{(k)})}.$$

Thus, from (13.24), we obtain the following rank-one correction formula:

$$H_{k+1} = H_k + \frac{(p^{(k)} - H_k g^{(k)})(p^{(k)} - H_k g^{(k)})^T}{(g^{(k)})^T (p^{(k)} - H_k g^{(k)})}. \qquad (13.27)$$

One of the drawbacks of this formula is that given a positive definite H_k, the resulting matrix H_{k+1} is not guaranteed to be positive definite. Some other quasi-Newton methods, such as those described next, do guarantee such a property.

13.7.2 Other correction formulas

One of the most popular classes of quasi-Newton methods uses the following correction formula:

$$H_{k+1} = H_k + \frac{p^{(k)} (p^{(k)})^T}{(p^{(k)})^T g^{(k)}} - \frac{H_k g^{(k)} (g^{(k)})^T H_k}{(g^{(k)})^T H_k g^{(k)}} + \xi_k (g^{(k)})^T H_k g^{(k)} h^{(k)} (h^{(k)})^T,$$

where

$$h^{(k)} = \frac{p^{(k)}}{(p^{(k)})^T g^{(k)}} - \frac{H_k g^{(k)}}{(g^{(k)})^T H_k g^{(k)}};$$

$$p^{(k)} = x^{(k+1)} - x^{(k)}, g^{(k)} = \nabla_{k+1} - \nabla_k;$$

the parameters ξ_k satisfy

$$0 \leq \xi_k \leq 1$$

for all k, and H_0 is an arbitrary positive definite matrix. The scalars ξ_k parameterize the method (different values of ξ_k yield different algorithms). If $\xi_k = 0$ for all k, we obtain the *Davidon-Fletcher-Powell* (DFP) method (historically the first quasi-Newton method). If $\xi_k = 1$ for all k, we obtain the *Broyden-Fletcher-Goldfarb-Shanno* (BFGS) method (considered one of the best general-purpose quasi-Newton methods).

13.8   Inexact Line Search

In all methods we discussed so far, we had an iteration in the form

$$x^{(k+1)} = x^{(k)} + \alpha_k d^{(k)},$$

where

$$\alpha_k = \arg\min_{\alpha \geq 0}(f(x^{(k)} + \alpha d^{(k)})).$$

Computing α_k exactly may be expensive, therefore an *inexact line search* is used in practice, which provides adequate reductions in the value of f. In particular, one could try several values of α, and accept one that satisfies certain conditions, such as the Wolfe conditions given next.

Let $\phi_k(\alpha) = f(x^{(k)} + \alpha d^{(k)})$. To ensure a *sufficient decrease*, we require that

$$\phi_k(\alpha) \leq \phi_k(0) + c_1 \alpha \phi_k'(0)$$

for some small $c_1 \in (0, 1)$. This is the same as

$$f(x^{(k)} + \alpha d^{(k)}) \leq f(x^{(k)}) + c_1 \alpha \nabla f(x^{(k)})^T d^{(k)}.$$

This inequality is often called the *Armijo condition*.

The sufficient decrease condition may result in very small α_k. To avoid this, we introduce the *curvature condition*, which requires α to satisfy the inequality

$$\phi_k'(\alpha) \geq c_2 \phi_k'(0)$$

for some constant $c_2 \in (c_1, 1)$. Expressing this condition in terms of f, we obtain

$$\nabla f(x^{(k)} + \alpha d^{(k)})^T d^{(k)} \geq c_2 \nabla f(x^{(k)})^T d^{(k)}.$$

The sufficient decrease and the curvature condition together comprise the *Wolfe conditions*:

$$f(x^{(k)} + \alpha d^{(k)}) \leq f(x^{(k)}) + c_1 \alpha \nabla f(x^{(k)})^T d^{(k)}$$
$$\nabla f(x^{(k)} + \alpha d^{(k)})^T d^{(k)} \geq c_2 \nabla f(x^{(k)})^T d^{(k)}.$$

The *strong Wolfe conditions* make sure that the derivative $\phi_k'(\alpha)$ is not "too positive" by requiring α to satisfy the following inequalities:

$$f(x^{(k)} + \alpha d^{(k)}) \leq f(x^{(k)}) + c_1 \alpha \nabla f(x^{(k)})^T d^{(k)}$$
$$|\nabla f(x^{(k)} + \alpha d^{(k)})^T d^{(k)}| \leq c_2 |\nabla f(x^{(k)})^T d^{(k)}|.$$

To determine a value of α which would satisfy the desired conditions, a *backtracking line search* could be used, which proceeds as follows (for Armijo condition):

1: Choose $\bar{\alpha} > 0, \rho, c \in (0,1)$
2: $\alpha \leftarrow \bar{\alpha}$
3: **repeat**
4: $\qquad \alpha \leftarrow \rho\alpha$
5: **until** $f(x^{(k)} + \alpha d^{(k)}) \leq f(x^{(k)}) + c\alpha \nabla f(x^{(k)})^T d^{(k)}$
6: **return** $\alpha_k = \alpha$

Exercises

13.1. A company manufactures two similar products. The manufacturing cost is \$40 for a unit of product 1 and \$42 for a unit of product 2. Assume that the company can sell $q_1 = 150 - 2p_1 + p_2$ units of product 1 and $q_2 = 120 + p_1 - 3p_2$ units of product 2, where p_1 and p_2 are prices charged for product 1 and product 2, respectively. The company's goal is to maximize the total profit. What price should be charged for each product? How many units of each product should be produced? What is the optimal profit?

13.2. Solve the problem $\min\limits_{x \in \mathbb{R}^n} f(x)$ for the following functions:

(a) $f(x) = \frac{1}{-x^2 + 3x - 7}$, $n = 1$;

(b) $f(x) = (x_1 - x_2 - 1)^2 + (x_1 - x_2 + 1)^4$, $n = 2$;

(c) $f(x) = x_1^4 + x_2^4 - 4x_1 x_2$, $n = 2$;

(d) $f(x) = 2x_1^2 - x_1 x_2 + x_2^2 - 7x_2$, $n = 2$;

(e) $f(x) = \frac{1}{(x_1 - x_2 - 2)^2 + (x_1 - x_2 + 1)^4}$, $n = 2$.

13.3. Let f be continuously differentiable on an open set containing a compact convex set $X \subset \mathbb{R}^n$, and let x^* be an optimal solution of the problem

$$\min_{x \in X} f(x).$$

Prove that x^* is also optimal for the problem

$$\min_{x \in X} x^T \nabla f(x^*).$$

13.4. Consider a class $\mathcal{C}$ of continuously differentiable functions defined on $\mathbb{R}^n$ satisfying the following properties:

(i) For any $f \in \mathcal{C}$ and $x^* \in \mathbb{R}^n$, if $\nabla f(x^*) = 0$, then x^* is a global minimizer of $f(x)$.

(ii) For any $f_1, f_2 \in \mathcal{C}$ and $\alpha, \beta \geq 0$, we have $\alpha f_1 + \beta f_2 \in \mathcal{C}$.

(iii) Any linear function $f(x) = a^T x + b, x \in \mathbb{R}^n$ belongs to $\mathcal{C}$.

Show that a continuously differentiable function $f : \mathbb{R}^n \to \mathbb{R}$ belongs to $\mathcal{C}$ if and only if it is convex.

13.5. Solve the problem of minimizing a quadratic function $f(x) = \frac{1}{2}x^T Q x + c^T x$, where

(a) $Q = \begin{bmatrix} 2 & 1 \\ 1 & 3 \end{bmatrix}$, $c = \begin{bmatrix} 1 \\ -1 \end{bmatrix}$;

(b) $Q = \begin{bmatrix} 4 & 1 & 1 \\ 1 & 5 & 0 \\ 1 & 0 & 2 \end{bmatrix}$, $c = \begin{bmatrix} -1 \\ 1 \\ 2 \end{bmatrix}$;

(c) $Q = \begin{bmatrix} 3 & 1 & 1 \\ 1 & 2 & 0 \\ 1 & 0 & 5 \end{bmatrix}$, $c = \begin{bmatrix} 1 \\ 0 \\ 2 \end{bmatrix}$;

(d) $Q = \begin{bmatrix} 2 & -1 & 0 \\ -1 & 3 & 1 \\ 0 & 1 & 4 \end{bmatrix}$, $c = \begin{bmatrix} 1 \\ 2 \\ 1 \end{bmatrix}$;

(e) $Q = \begin{bmatrix} 3 & 0 & 1 \\ 0 & 3 & 1 \\ 1 & 1 & 2 \end{bmatrix}$, $c = \begin{bmatrix} 3 \\ 1 \\ 2 \end{bmatrix}$.

Clearly explain your solution.

13.6. Given the points $[x_1, y_1]^T, \ldots, [x_n, y_n]^T \in \mathbb{R}^2$, use the optimality conditions to prove that the solution of the problem

$$\min_{[a,b]^T \in \mathbb{R}^2} f(a,b) = \sum_{i=1}^n (ax_i + b - y_i)^2$$

is given by the solution of the system $A^T A z = A^T y$, where

$$z = \begin{bmatrix} a \\ b \end{bmatrix}, \ y = [y_1, \ldots, y_n]^T,$$

and $A^T = \begin{bmatrix} x_1 & \cdots & x_n \\ 1 & \cdots & 1 \end{bmatrix}$. The line $l(x) = ax + b$ is called the linear regression line for the points $[x_1, y_1]^T, \ldots, [x_n, y_n]^T$ and is often used in statistics.

13.7. In the final iteration of the Fibonacci search method,

$$\rho_n = 1/2,$$

therefore $a_n = b_n$, i.e., instead of two points a_n and b_n we obtain only one, which is the mid-point of the interval $[a_{n-1}, b_{n-1}]$. But we need two evaluation points in order to determine the final interval of uncertainty. To overcome this problem, we can add a new evaluation point $a'_n = b_n - \epsilon(b_{n-1} - a_{n-1})$, where ϵ is a small number. Show that with this modification, the reduction factor in the uncertainty range for the Fibonacci method is no worse than

$$\frac{1 + 2\epsilon}{F_{n+1}}$$

(therefore this drawback of Fibonacci search is of no significant practical consequence). *Hint:* Note that $b_{n-1} - a_{n-1} = \frac{2}{F_{n+1}}(b_0 - a_0)$.

13.8. Let $f(x) = \frac{1}{2}x^T Q x + c^T x$ be a quadratic function of n variables, where Q is a positive definite $n \times n$ matrix.

(a) Show that $f(x)$ has a unique global minimizer x^*.

(b) Show that if the initial point $x^{(0)}$ is such that $x^{(0)} - x^*$ is an eigenvector of Q, then the steepest descent sequence $x^{(k)}$ with initial point $x^{(0)}$ reaches x^* in one step, i.e., $x^{(1)} = x^*$.

13.9. Apply three steps of the steepest descent method to the problems in Exercise 13.5. Use the zero vector of appropriate dimension as your initial guess.

13.10. Consider the problem

$$\min f(x_1, x_2) = cx_1^2 + x_2^2.$$

(a) What is the global minimum of f?

(b) Apply three steps of the method of steepest descent to this problem with $x^{(0)} = [1, c]^T$, where $c > 0$ is some constant.

(c) What is the rate of convergence of the steepest descent sequence $\{x^{(k)} : k \geq 0\}$?

13.11. For the function of a single variable $f(x) = x^{4/3}$, show that

(a) $f(x)$ has a unique global minimizer $x^* = 0$;

(b) for any starting guess $x^{(0)} \neq 0$, Newton's method applied to $f(x)$ diverges.

13.12. The total annual cost C of operating a certain electric motor can be expressed as a function of its horsepower, x, as follows

$$C(x) = \$120 + \$1.5x + \frac{\$0.2}{x}(1,000).$$

Use Newton's method to find the motor horsepower that minimizes the total annual cost. Select an appropriate starting point and apply three iterations of the method.

13.13. Consider the problem

$$\min_{x \in \mathbb{R}^2} f(x) = x_1^2 \exp(x_1) + x_2^2 \exp(x_2) + 1.$$

(a) Apply Newton's method twice starting with $[1,1]^T$.

(b) What is the global minimum of $f(x)$? Explain your answer.

13.14. For a real symmetric positive definite $n \times n$ matrix Q and an arbitrary set of linearly independent vectors $p^{(0)}, \ldots, p^{(n-1)} \in \mathbb{R}^n$, the *Gram-Schmidt* procedure generates the set of vectors $d^{(0)}, \ldots, d^{(n-1)} \in \mathbb{R}^n$ as follows:

$$d^{(0)} = p^{(0)};$$

$$d^{(k+1)} = p^{(k+1)} - \sum_{i=0}^{k} \frac{p^{(k+1)^T} Q d^{(i)}}{d^{(i)^T} Q d^{(i)}} d^{(i)}.$$

Show that the vectors $d^{(0)}, \ldots, d^{(n-1)}$ are Q-conjugate.

13.15. Solve the problem

$$\text{minimize } \frac{1}{2} x^T Q x + c^T x,$$

where

(a) $Q = \begin{bmatrix} 2 & 0 \\ 0 & 1 \end{bmatrix}, c = \begin{bmatrix} 0 \\ 0 \end{bmatrix}, x^{(0)} = [1,1]^T, H_0 = I_2$ and

(b) $Q = \begin{bmatrix} 9 & 3 & 1 \\ 3 & 7 & 2 \\ 1 & 2 & 5 \end{bmatrix}, c = \begin{bmatrix} -8 \\ 0 \\ -9 \end{bmatrix}, x^{(0)} = [0,0,0]^T, H_0 = I_3$

using

(i) the conjugate gradient method,

(ii) the rank-one quasi-Newton method, and

(iii) the BFGS quasi-Newton method.

Use exact line search.

13.16. Illustrate the Wolfe conditions geometrically.

Chapter 14

Constrained Optimization

In discussing constrained optimization problems, we will follow the same sequence as we did for unconstrained problems in Chapter 13. Namely, we will present optimality conditions, followed by a brief outline of ideas behind algorithms for constrained optimization.

14.1   Optimality Conditions

14.1.1   First-order necessary conditions

We start by considering problems involving equality constraints only, and then move on to discussing the more general case of problems with both equality and inequality constraints.

14.1.1.1   Problems with equality constraints

We consider a problem with equality constraints in the form

$$
\begin{aligned}
\text{minimize} \quad & f(x) \\
\text{subject to} \quad & h(x) = 0,
\end{aligned}
$$

where $f(x) : \mathbb{R}^n \to \mathbb{R}$, $h(x) = [h_1(x), h_2(x), \ldots, h_m(x)]^T : \mathbb{R}^n \to \mathbb{R}^m$. We assume that $m < n$. Denote by

$$
J_h(x) = \begin{bmatrix}
\frac{\partial h_1(x)}{\partial x_1} & \frac{\partial h_1(x)}{\partial x_2} & \cdots & \frac{\partial h_1(x)}{\partial x_n} \\
\frac{\partial h_2(x)}{\partial x_1} & \frac{\partial h_2(x)}{\partial x_2} & \cdots & \frac{\partial h_2(x)}{\partial x_n} \\
\vdots & \vdots & \ddots & \vdots \\
\frac{\partial h_m(x)}{\partial x_1} & \frac{\partial h_m(x)}{\partial x_2} & \cdots & \frac{\partial h_m(x)}{\partial x_n}
\end{bmatrix} = \begin{bmatrix}
\nabla h_1(x)^T \\
\nabla h_2(x)^T \\
\vdots \\
\nabla h_m(x)^T
\end{bmatrix}
$$

the Jacobian of h at x. Let $X = \{x \in \mathbb{R}^n : h(x) = 0\}$ denote the feasible set of the considered optimization problem. We will call $x^* \in \mathbb{R}^n$ a regular point if $J_h(x^*)$ has the full rank, that is, $\mathrm{rank}(J_h(x^*)) = m$, or equivalently, $\nabla h_1(x^*), \nabla h_2(x^*), \ldots, \nabla h_m(x^*)$ are linearly independent.

We first discuss the FONC for the case with two variables and a single constraint: $n = 2$ and $m = 1$. We can write such a problem as follows:

$$\text{minimize} \quad f(x_1, x_2)$$
$$\text{subject to} \quad h_1(x_1, x_2) = 0.$$

Then the feasible set $X = \{x \in \mathbb{R}^2 : h_1(x_1, x_2) = 0\}$ is the level set of h_1 at the level 0.

Let $x^* = [x_1^*, x_2^*]^T$ be a regular point and a local minimizer of this problem. Then $h_1(x^*) = 0$, and $\nabla h_1(x^*)$ is orthogonal to the tangent line to any curve passing through x^* in the level set X. Consider an arbitrary curve $\gamma = \{y(t) = [x_1(t), x_2(t)]^T, \ t \in [\alpha, \beta]\} \subset X$ passing through x^* in X, that is, $y(t^*) = x^*$ for some $t^* \in (\alpha, \beta)$. The direction of the tangent line to γ at x^* is given by $y'(t)$, hence we have

$$\nabla h_1(x^*)^T y'(t^*) = 0.$$

Since x^* is a local minimizer of our problem, it will remain a local minimizer if we restrict the feasible region to the points of the curve γ. Therefore, t^* is a local minimizer of the problem $\min\limits_{t \in [\alpha, \beta]} f(y(t))$, and since t^* is an interior point of $[\alpha, \beta]$, t^* is a local minimizer of the single-variable unconstrained problem

$$\min_{t \in \mathbb{R}} f(y(t)).$$

Using the FONC for this unconstrained problem, we have

$$\frac{df(y(t^*))}{dt} = 0.$$

On the other hand, from the chain rule,

$$\frac{df(y(t^*))}{dt} = \nabla f(y(t^*))^T y'(t^*),$$

so

$$\nabla f(y(t^*))^T y'(t^*) = 0.$$

Thus, $\nabla f(y(t^*))$ is orthogonal to $y'(t^*)$. So, we have shown that if $\nabla f(x^*) \neq 0$, then $\nabla f(x^*)$ and $\nabla h_1(x^*)$ are both orthogonal to the same vector $y'(x^*)$. For 2-dimensional vectors, this means that $\nabla f(x^*)$ and $\nabla h_1(x^*)$ are parallel, implying that there exists a scalar λ such that

$$\nabla f(x^*) + \lambda \nabla h_1(x^*) = 0.$$

A similar property holds for the general case and is formulated in the following theorem.

Theorem 14.1 (Lagrange Theorem) *If x^* is a regular point and a local minimizer (maximizer) of the problem*

$$\begin{aligned} \text{minimize} \quad & f(x) \\ \text{subject to} \quad & h(x) = 0, \end{aligned} \qquad (14.1)$$

where $f(x) : \mathbb{R}^n \to \mathbb{R}$, $h(x) = [h_1(x), \ldots, h_m(x)]^T : \mathbb{R}^n \to \mathbb{R}^m$, then there exists $\lambda = [\lambda_1, \ldots, \lambda_m]^T \in \mathbb{R}^m$ such that

$$\nabla f(x^*) + \sum_{i=1}^{m} \lambda_i \nabla h_i(x^*) = 0. \qquad (14.2)$$

Here, $\lambda_i, i = 1, \ldots, m$, are called the *Lagrange multipliers* and the function

$$L(x, \lambda) = f(x) + \sum_{i=1}^{m} \lambda_i h_i(x)$$

is called the *Lagrangian* of the considered problem. Note that $L(x, \lambda)$ is a function of $n + m$ variables. If we apply the unconstrained FONC to this function we obtain the system

$$\begin{aligned} \nabla f(x^*) + \sum_{i=1}^{m} \lambda_i \nabla h_i(x^*) &= 0 \\ h(x^*) &= 0, \end{aligned}$$

which coincides with the FONC stated in the Lagrange theorem (the second equation just guarantees the feasibility). This system has $n + m$ variables and $n + m$ equations. Its solutions are candidate points for a local minimizer (maximizer). The system is not easy to solve in general. Moreover, like in the unconstrained case, even if we solve it, a solution may not be a local minimizer–it can be a saddle point or a local maximizer. Figure 14.1 illustrates the FONC.

Example 14.1 *Apply the FONC (Lagrange theorem) to the problem*

$$\begin{aligned} \text{minimize} \quad & x_1^2 + x_2^2 \\ \text{subject to} \quad & 4x_1^2 + x_2^2 - 1 = 0. \end{aligned}$$

Note that all feasible points are regular for this problem, so any local minimizer has to satisfy the Lagrange conditions. We have

$$L(x, \lambda) = x_1^2 + x_2^2 + \lambda(4x_1^2 + x_2^2 - 1),$$

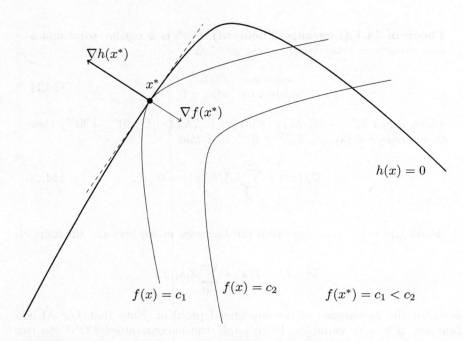

FIGURE 14.1: An illustration of the FONC for equality-constrained problems. Here x^* satisfies the FONC and is a local maximizer, which is not global.

and the Lagrange conditions give the system

$$2x_1(1 + 4\lambda) = 0$$
$$2x_2(1 + \lambda) = 0$$
$$4x_1^2 + x_2^2 = 1.$$

This system has four solutions:

(1) $\lambda_1 = -1, x^{(1)} = [0, 1]^T$;
(2) $\lambda_2 = -1, x^{(2)} = [0, -1]^T$;
(3) $\lambda_3 = -1/4, x^{(3)} = [1/2, 0]^T$;
(4) $\lambda_4 = -1/4, x^{(4)} = [-1/2, 0]^T$.

From a geometric illustration (Figure 14.2), it is easy to see that $x^{(1)}$ and $x^{(2)}$ are global maximizers, whereas $x^{(3)}$ and $x^{(4)}$ are global minimizers.

Example 14.2 *Consider the problem of optimizing $f(x) = x^T Q x$ subject to a single equality constraint $x^T P x = 1$, where P is a positive definite matrix. Apply the FONC to this problem.*

The Lagrangian is

$$L(x, \lambda) = x^T Q x + \lambda(1 - x^T P x).$$

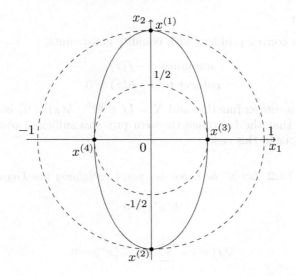

FIGURE 14.2: Illustration of Example 14.1.

From the FONC, we have

$$Qx - \lambda Px = 0.$$

Premultiplying this equation by P^{-1}, we obtain

$$(P^{-1}Q - \lambda I_n)x = 0.$$

A solution (λ^*, x^*) of this equation is an eigenpair of the matrix $P^{-1}Q$. Moreover, if we premultiply this equation for $x = x^*$ by $x^{*T}P$, we get

$$x^{*T}Qx^* - \lambda^* x^{*T}Px^* = 0,$$

and since $x^{*T}Px^* = 1$, this yields

$$x^{*T}Qx^* = \lambda^*.$$

Therefore, an eigenvector corresponding to the smallest (largest) eigenvalue of $P^{-1}Q$ is a global minimizer (maximizer) of this problem and

$$\min_{x^TPx=1} x^TQx = \lambda_{min}(P^{-1}Q),$$

$$\max_{x^TPx=1} x^TQx = \lambda_{max}(P^{-1}Q).$$

Convex case

Consider a convex problem with equality constraints,

$$\text{minimize} \quad f(x)$$
$$\text{subject to} \quad h(x) = 0,$$

where $f(x)$ is a convex function and $X = \{x \in \mathbb{R}^n : h(x) = 0\}$ is a convex set. We will show that the Lagrange theorem provides sufficient conditions for a global minimizer in this case.

> **Theorem 14.2** *Let x^* be a regular point satisfying the Lagrange theorem,*
>
> $$h(x^*) = 0 \qquad\qquad (14.3)$$
>
> *and*
>
> $$\nabla f(x^*) + \sum_{i=1}^{m} \lambda_i \nabla h_i(x^*) = 0. \qquad\qquad (14.4)$$
>
> *Then x^* is a global minimizer.*

Proof. From the first-order characterization of a convex function, we have

$$f(x) - f(x^*) \geq \nabla f(x^*)^T (x - x^*), \quad \forall x \in X. \qquad (14.5)$$

From the FONC (14.4),

$$\nabla f(x^*) = -\sum_{i=1}^{m} \lambda_i \nabla h_i(x^*).$$

So, from (14.5)

$$f(x) - f(x^*) \geq -\sum_{i=1}^{m} \lambda_i \nabla h_i(x^*)^T (x - x^*). \qquad (14.6)$$

Note that for any i, $\nabla h_i(x^*)^T (x - x^*)$ is the directional derivative of h_i at x^* in the direction $x - x^*$. Hence, using the definition of the directional derivative we obtain

$$\begin{aligned}
\nabla h_i(x^*)^T (x - x^*) &= \lim_{\alpha \to 0+} \frac{h_i(x^* + \alpha(x - x^*)) - h_i(x^*)}{\alpha} \\
&= \lim_{\alpha \to 0+} \frac{h_i(\alpha x + (1 - \alpha)x^*) - h_i(x^*)}{\alpha} \\
&= 0,
\end{aligned}$$

since $\alpha x + (1 - \alpha)x^* \in X$ due to the convexity of X and $h(y) = 0$ for any $y \in X$. Substituting this result into (14.6) we obtain

$$f(x) - f(x^*) \geq 0.$$

Since x is an arbitrary point in X, x^* is a global minimizer of the considered problem. $\qquad\square$

Example 14.3 *Let $f(x) = \frac{1}{2}x^T Q x$, where Q is a positive definite matrix, and let $h(x) = Ax - b$, where A is an $m \times n$ matrix of rank m. Consider the problem*

$$\min_{x \in X} f(x),$$

where $X = \{x : h(x) = 0\} = \{x : Ax = b\}$. This is a convex problem, since $f(x)$ is a convex function (its Hessian Q is positive definite) and X is a convex set (defined by a system of linear equations).

The Lagrangian of this problem is given by

$$L(x, \lambda) = \frac{1}{2}x^T Q x + \lambda^T (Ax - b),$$

where $\lambda \in \mathbb{R}^m$. The FONC can be expressed by the system

$$Qx + A^T\lambda = 0$$
$$Ax - b = 0.$$

From the first equation,

$$x = -Q^{-1}A^T\lambda. \qquad (14.7)$$

Premultiplying both sides of this equation by A, we obtain

$$Ax = -AQ^{-1}A^T\lambda,$$

so

$$b = -AQ^{-1}A^T\lambda$$

and

$$\lambda = -(AQ^{-1}A^T)^{-1}b.$$

Substituting this value of λ into (14.7), we obtain the global minimizer of the considered problem:

$$x^* = Q^{-1}A^T(AQ^{-1}A^T)^{-1}b.$$

In an important special case, when $Q = I_n$, the discussed problem becomes

$$\min_{Ax=b} \frac{1}{2}\|x\|^2.$$

Its solution,

$$x^* = A^T(AA^T)^{-1}b,$$

gives the solution of the system $Ax = b$ with minimum norm.

14.1.1.2　Problems with inequality constraints

We consider the following problem:

$$\begin{aligned} \text{minimize} \quad & f(x) \\ \text{subject to} \quad & h(x) = 0 \\ & g(x) \leq 0, \end{aligned}$$

where $f(x) : \mathbb{R}^n \to \mathbb{R}$, $h(x) : \mathbb{R}^n \to \mathbb{R}^m$ $(m < n)$, and $g(x) : \mathbb{R}^n \to \mathbb{R}^p$.

This problem involves two types of constraints, equality and inequality constraints. Recall that an inequality constraint $g_j(x) \leq 0$ is called *active* at x^* if $g_j(x^*) = 0$. We denote by $I(x^*) = \{j : g_j(x^*) = 0\}$ the set of indices corresponding to the active constraints for x^*. A point x^* is called a *regular point* for the considered problem if $\nabla h_i(x^*), i = 1, \ldots, m$ and $\nabla g_j(x^*), j \in I(x^*)$ form a set of linearly independent vectors. The Lagrangian of this problem is defined as

$$L(x, \lambda, \mu) = f(x) + \lambda^T h(x) + \mu^T g(x),$$

where $\lambda = [\lambda_1, \ldots, \lambda_m]^T \in \mathbb{R}^m$ and $\mu = [\mu_1, \ldots, \mu_p]^T \in \mathbb{R}^p, \mu \geq 0$. As before, the multipliers $\lambda_i, i = 1, \ldots, m$ corresponding to the equality constraints are called the Lagrange multipliers. The multipliers $\mu_j, j = 1, \ldots, p$ corresponding to the inequality constraints are called the Karush-Kuhn-Tucker (KKT) multipliers.

The first-order necessary conditions for the problems with inequality constraints are referred to as Karush-Kuhn-Tucker (KKT) conditions.

Theorem 14.3 (Karush-Kuhn-Tucker (KKT) conditions)
If x^* is a regular point and a local minimizer of the problem

$$\begin{aligned} \text{minimize} \quad & f(x) \\ \text{subject to} \quad & h(x) = 0 \\ & g(x) \leq 0, \end{aligned}$$

where all functions are continuously differentiable, then there exist $\lambda \in \mathbb{R}^m$ and a nonnegative $\mu \in \mathbb{R}^p$ such that

$$\nabla f(x^*) + \sum_{i=1}^{m} \lambda_i \nabla h_i(x^*) + \sum_{j=1}^{p} \mu_j \nabla g_j(x^*) = 0$$

and

$$\mu_j g_j(x^*) = 0, \; j = 1, \ldots, p \; \text{(Complementary slackness)}.$$

In summary, the KKT conditions can be expressed in terms of the following system of equations and inequalities:

1. $\lambda \in \mathbb{R}^m$, $\mu \in \mathbb{R}^p$, $\mu \geq 0$;

2. $\nabla f(x^*) + \sum\limits_{i=1}^{m} \lambda_i \nabla h_i(x^*) + \sum\limits_{j=1}^{p} \mu_j \nabla g_j(x^*) = 0$;

3. $\mu_j g_j(x^*) = 0$, $j = 1, \ldots, p$;

4. $h(x^*) = 0$;

5. $g(x^*) \leq 0$.

Example 14.4 *Consider the problem*

$$
\begin{aligned}
minimize \quad & x_2 - x_1 \\
subject \ to \quad & x_1^2 + x_2^2 \leq 4 \\
& (x_1 + 1)^2 + x_2^2 \leq 4.
\end{aligned}
$$

The Lagrangian is

$$L(x, \mu) = x_2 - x_1 + \mu_1(x_1^2 + x_2^2 - 4) + \mu_2((x_1 + 1)^2 + x_2^2 - 4).$$

Using the KKT conditions we have the system

$$
\begin{aligned}
-1 + 2\mu_1 x_1 + 2\mu_2(x_1 + 1) &= 0 \\
1 + 2\mu_1 x_2 + 2\mu_2 x_2 &= 0 \\
\mu_1(x_1^2 + x_2^2 - 4) &= 0 \\
\mu_2\left((x_1 + 1)^2 + x_2^2 - 4\right) &= 0 \\
x_1^2 + x_2^2 &\leq 4 \\
(x_1 + 1)^2 + x_2^2 &\leq 4 \\
\mu_1, \mu_2 &\geq 0.
\end{aligned}
$$

1. *If $\mu_1 = 0$, then the system becomes*

$$
\begin{aligned}
2\mu_2(x_1 + 1) &= 1 \\
2\mu_2 x_2 &= -1 \\
\mu_2\left((x_1 + 1)^2 + x_2^2 - 4\right) &= 0 \\
x_1^2 + x_2^2 &\leq 4 \\
(x_1 + 1)^2 + x_2^2 &\leq 4 \\
\mu_2 &\geq 0.
\end{aligned}
$$

Note that $\mu_2 \neq 0$ (from the first equation), so $\mu_2 > 0$. Adding the first two equations we get

$$x_1 = -x_2 - 1,$$

which, using the third equation, gives

$$x_2 = \pm\sqrt{2}.$$

Since $\mu_2 > 0$, from the second equation $x_2 = -\sqrt{2}$ and $\mu_2 = 1/(2\sqrt{2})$, so $x_1 = \sqrt{2}-1$. These values of x_1 and x_2 satisfy the inequality constraints, thus $x^* = [\sqrt{2}-1, -\sqrt{2}]^T$ satisfies the KKT conditions.

2. *If $\mu_2 = 0$, then the system becomes*

$$
\begin{aligned}
2\mu_1 x_1 &= 1 \\
2\mu_1 x_2 &= -1 \\
\mu_1(x_1^2 + x_2^2 - 4) &= 0 \\
x_1^2 + x_2^2 &\leq 4 \\
(x_1 + 1)^2 + x_2^2 &\leq 4 \\
\mu_1 &\geq 0.
\end{aligned}
$$

Note that $\mu_1 \neq 0$ (from the first equation), so $\mu_1 > 0$. Adding the first two equations we get

$$x_1 = -x_2,$$

which, using the third equation, gives

$$x_2 = \pm\sqrt{2}.$$

Since $\mu_1 > 0$, from the second equation $x_2 = -\sqrt{2}$ and $\mu_1 = 1/(2\sqrt{2})$, so $x_1 = \sqrt{2}$. However, these values of x_1 and x_2 do not satisfy the last inequality constraint, hence the point is infeasible and the KKT conditions are not satisfied.

3. *If $\mu_1 \neq 0, \mu_2 \neq 0$, then the system becomes*

$$
\begin{aligned}
-1 + 2\mu_1 x_1 + 2\mu_2(x_1 + 1) &= 0 \\
1 + 2\mu_1 x_2 + 2\mu_2 x_2 &= 0 \\
x_1^2 + x_2^2 - 4 &= 0 \\
(x_1 + 1)^2 + x_2^2 - 4 &= 0 \\
\mu_1, \mu_2 &\geq 0.
\end{aligned}
$$

From the last two equalities we obtain $x_1 = -1/2$, $x_2 = \pm\sqrt{15}/2$. Solving the first two equations with $x_1 = -1/2$, for $x_2 = \sqrt{15}/2$ we get $\mu_1 = -\frac{1}{2}(1+1/\sqrt{15}), \mu_2 = \frac{1}{2}(1-1/\sqrt{15})$, whereas for $x_2 = -\sqrt{15}/2$ we obtain $\mu_1 = -\frac{1}{2}(1-1/\sqrt{15}), \mu_2 = \frac{1}{2}(1+1/\sqrt{15})$. In both cases, one of the KKT multipliers is negative, so these points do not satisfy the KKT conditions.

From Figure 14.3 it is clear that the KKT point x^* is the global minimizer. The level set of the objective corresponding to the optimal value $(1 - 2\sqrt{2})$ is shown by the dashed line (a tangent to the feasible region).

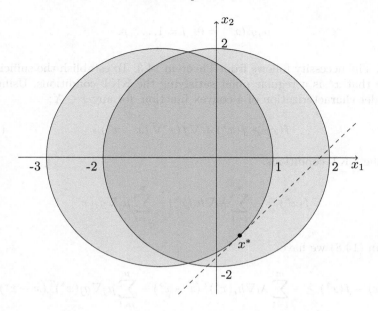

FIGURE 14.3: An illustration of Example 14.4.

Convex case

Assume that the feasible set $X = \{x : h(x) = 0, g(x) \leq 0\}$ is a convex set and $f(x)$ is a convex function over X. We will show that in this case the KKT conditions are sufficient conditions for a global minimizer.

Theorem 14.4 *Consider a convex problem*

$$\begin{array}{ll} \text{minimize} & f(x) \\ \text{subject to} & h(x) = 0 \\ & g(x) \leq 0, \end{array}$$

where all functions are continuously differentiable, the feasible set $X = \{x \in \mathbb{R}^n : h(x) = 0, g(x) \leq 0\}$ is convex, and $f(x)$ is convex on X. A regular point x^* is a global minimizer of this problem if and only if it satisfies the KKT conditions, that is, there exist $\lambda \in \mathbb{R}^m$ and a nonnegative $\mu \in \mathbb{R}^p$ such that

$$\nabla f(x^*) + \sum_{i=1}^{m} \lambda_i \nabla h_i(x^*) + \sum_{j=1}^{p} \mu_j \nabla g_j(x^*) = 0$$

and

$$\mu_j g_j(x^*) = 0, \; j = 1, \ldots, p.$$

Proof. The necessity follows from Theorem 14.3. To establish the sufficiency, assume that x^* is a regular point satisfying the KKT conditions. Using the first-order characterization of a convex function, for any $x \in X$:

$$f(x) \geq f(x^*) + \nabla f(x^*)^T (x - x^*). \tag{14.8}$$

From the KKT conditions,

$$\nabla f(x^*) = -\sum_{i=1}^m \lambda_i \nabla h_i(x^*) - \sum_{j=1}^p \mu_j \nabla g_j(x^*).$$

So, from (14.8) we have

$$f(x) - f(x^*) \geq -\sum_{i=1}^m \lambda_i \nabla h_i(x^*)^T (x - x^*) - \sum_{j=1}^p \mu_j \nabla g_j(x^*)^T (x - x^*).$$

When we considered convex problems with equality constraints, we proved that

$$\nabla h_i(x^*)^T (x - x^*) = 0, \; i = 1, \ldots, m.$$

Next we show that

$$\mu_j \nabla g_j(x^*)^T (x - x^*) \leq 0.$$

Indeed, $\nabla g_j(x^*)^T (x - x^*)$ is the directional derivative of $g_j(x)$ at x^* in the direction $x - x^*$, hence

$$
\begin{aligned}
\mu_j \nabla g_j(x^*)^T (x - x^*) &= \mu_j \lim_{\alpha \to 0+} \frac{g_j(x^* + \alpha(x - x^*)) - g_j(x^*)}{\alpha} \\
&= \lim_{\alpha \to 0+} \frac{\mu_j g_j(x^* + \alpha(x - x^*)) - \mu_j g_j(x^*)}{\alpha} \\
&= \lim_{\alpha \to 0+} \frac{\mu_j g_j(\alpha x + (1 - \alpha)x^*)}{\alpha} \\
&\leq 0.
\end{aligned}
$$

Here we used the complementary slackness $(\mu_j g_j(x^*) = 0)$.

So, $f(x) - f(x^*) \geq 0$ for any $x \in X$, thus x^* is a global minimizer of our problem. $\square$

Example 14.5 *The problem in Example 14.4 is convex, hence, we can conclude that the only KKT point x^* is the global minimizer for this problem.*

14.1.2   Second-order conditions

14.1.2.1   Problems with equality constraints

For the surface $X = \{x \in \mathbb{R}^N : h(x) = 0\}$, the *tangent space* $T(x^*)$ at a point x^* is defined as the null space of the Jacobian of $h(x)$ at x^*:

$$T(x^*) = \{y \in \mathbb{R}^n : J_h(x^*)y = 0\},$$

and the corresponding *tangent plane* at x^* is given by

$$TP(x^*) = \{y \in \mathbb{R}^n : J_h(x^*)(y - x^*) = 0\}.$$

Here we assume that $f(x)$ and $h(x)$ are twice continuously differentiable. Denote by $\nabla^2_{xx}L(x, \lambda)$ the Hessian of $L(x, \lambda)$ as a function of x:

$$\nabla^2_{xx}L(x, \lambda) = \nabla^2 f(x) + \sum_{i=1}^{m} \lambda_i \nabla^2 h_i(x).$$

The second-order optimality conditions in the case of equality-constrained problems are formulated similarly to the unconstrained case. However, in the equality-constrained case, they are restricted to the tangent space at x^* only.

Theorem 14.5 (SONC) *If a regular point x^* is a local minimizer of the problem $\min\limits_{h(x)=0} f(x)$, then there exists $\lambda = [\lambda_1, \ldots, \lambda_m]^T \in \mathbb{R}^m$ such that*

(1) $\nabla f(x^*) + \sum\limits_{i=1}^{m} \lambda_i \nabla h_i(x^*) = 0;$

(2) for any $y \in T(x^*)$:

$$y^T \nabla^2_{xx}L(x^*, \lambda)y \geq 0. \qquad (14.9)$$

Theorem 14.6 (SOSC) *If a regular point x^* satisfies the SONC stated in Theorem 14.5 with $\lambda = \lambda^*$ and for any $y \in T(x^*), y \neq 0$:*

$$y^T \nabla^2_{xx}L(x^*, \lambda^*)y > 0, \qquad (14.10)$$

then x^* is a strict local minimizer of the problem $\min\limits_{h(x)=0} f(x)$.

Since the problem $\max\limits_{h(x)=0} f(x)$ is equivalent to the problem $\min\limits_{h(x)=0} (-f(x))$, it is easy to check that by reversing the sign of the inequalities in (14.9) and (14.10) we will obtain the corresponding second-order optimality conditions for the maximization problem $\max\limits_{h(x)=0} f(x)$.

Example 14.6 *Consider the problem from Example 14.1 (page 353): $f(x) = x_1^2 + x_2^2$; $h(x) = 4x_1^2 + x_2^2 - 1$. There are four candidates for local optimizers:*
(1) $\lambda_1 = -1, x^{(1)} = [0, 1]^T$;
(2) $\lambda_2 = -1, x^{(2)} = [0, -1]^T$;
(3) $\lambda_3 = -1/4, x^{(3)} = [1/2, 0]^T$;
(4) $\lambda_4 = -1/4, x^{(4)} = [-1/2, 0]^T$.
We apply the second-order conditions to these points. We have

$$L(x, \lambda) = x_1^2 + x_2^2 + \lambda(4x_1^2 + x_2^2 - 1),$$

$$\nabla_{xx}^2 L(x, \lambda) = \begin{bmatrix} 2 + 8\lambda & 0 \\ 0 & 2 + 2\lambda \end{bmatrix}.$$

Next we need to find the tangent space for each of the candidate points.

$$\nabla h(x) = [8x_1, 2x_2]^T,$$

so $\nabla h(x^{(1)}) = [0, 2]^T$ and

$$
\begin{aligned}
T(x^{(1)}) &= \{y = [y_1, y_2]^T : \nabla h(x^{(1)})^T y = 0\} \\
&= \{y : 0y_1 + 2y_2 = 0\} \\
&= \{y : y = [c, 0]^T, c \in \mathbb{R}\}.
\end{aligned}
$$

Thus,

$$y^T \nabla_{xx}^2 L(x^{(1)}, \lambda_1)y = -6c^2 < 0, \ \forall c \neq 0,$$

and $x^{(1)}$ is a local maximizer. It is easy to check that $T(x^{(2)}) = T(x^{(1)})$ and $\nabla_{xx}^2 L(x^{(2)}, \lambda_2) = \nabla_{xx}^2 L(x^{(1)}, \lambda_1)$, so $x^{(2)}$ is also a local maximizer.
Similarly, $T(x^{(3)}) = T(x^{(4)}) = \{y : y = [0, c]^T, c \in \mathbb{R}\}$ and

$$y^T \nabla_{xx}^2 L(x^{(3)}, \lambda_3)y = y^T \nabla_{xx}^2 L(x^{(4)}, \lambda_4)y = \frac{3}{2}c^2 > 0, \ \forall c \neq 0.$$

Thus, $x^{(3)}$ and $x^{(4)}$ are local minimizers.

14.1.2.2   Problems with inequality constraints

As before, we consider a problem with inequality constraints in the form

$$
\begin{aligned}
\text{minimize} \quad & f(x) \\
\text{subject to} \quad & h(x) = 0 \\
& g(x) \leq 0,
\end{aligned}
\tag{P}
$$

where $f : \mathbb{R}^n \to \mathbb{R}, h : \mathbb{R}^n \to \mathbb{R}^m, g : \mathbb{R}^n \to \mathbb{R}^p$. For $x^* \in \mathbb{R}^n$, $I(x^*)$ denotes the set of indices corresponding to active constraints at x^*.

Given the Lagrangian of the problem under consideration,

$$L(x, \lambda, \mu) = f(x) + \sum_{i=1}^{m} \lambda_i h_i(x) + \sum_{j=1}^{p} \mu_j g_j(x),$$

we denote by

$$\nabla_{xx}^2 L(x, \lambda, \mu) = \nabla^2 f(x) + \sum_{i=1}^{m} \lambda_i \nabla^2 h_i(x) + \sum_{j=1}^{p} \mu_j \nabla^2 g_j(x).$$

Let

$$T(x^*) = \left\{ \begin{array}{ll} y \in \mathbb{R}^n : & \nabla h_i(x^*)^T y = 0, \quad i = 1, \ldots, m \\ & \nabla g_j(x^*)^T y = 0, \qquad j \in I(x^*) \end{array} \right\}.$$

Theorem 14.7 (SONC) *If a regular point x^* is a local minimizer of* (P), *then there exist $\lambda \in \mathbb{R}^m$ and nonnegative $\mu \in \mathbb{R}^p$ such that*

(I) the KKT conditions are satisfied for x^*, λ, and μ, and

(II) for any $y \in T(x^*)$: $y^T \nabla_{xx}^2 L(x, \lambda, \mu) y \geq 0$.

Before we state the second-order sufficient conditions, we need to introduce another notation. Given $x^* \in \mathbb{R}^n$ and $\mu \in \mathbb{R}^p$, we denote by

$$T(x^*, \mu) = \left\{ \begin{array}{ll} y \in \mathbb{R}^n : & \nabla h_i(x^*)^T y = 0, \qquad\qquad i = 1, \ldots, m \\ & \nabla g_j(x^*)^T y = 0, \quad j \in I(x^*) \text{ and } \mu_j > 0 \end{array} \right\}.$$

Note that $T(x^*) \subseteq T(x^*, \mu)$.

Theorem 14.8 (SOSC) *If x^* is a regular point satisfying the KKT conditions and the SONC in Theorem 14.7 with some $\lambda \in \mathbb{R}^m$ and nonnegative $\mu \in \mathbb{R}^p$, and for any nonzero $y \in T(x^*, \mu)$:*

$$y^T \nabla_{xx}^2 L(x^*, \lambda, \mu) y > 0,$$

then x^* is a strict local minimizer of (P).

Example 14.7 *Consider the problem*

$$\begin{array}{ll} minimize & x_2 - x_1 \\ subject\ to & x_1^2 + x_2^2 = 4 \\ & (x_1 + 1)^2 + x_2^2 \leq 4. \end{array}$$

The Lagrangian is

$$L(x, \lambda, \mu) = x_2 - x_1 + \lambda(x_1^2 + x_2^2 - 4) + \mu((x_1 + 1)^2 + x_2^2 - 4).$$

Using the KKT conditions we have the system

$$
\begin{aligned}
-1 + 2\lambda x_1 + 2\mu(x_1 + 1) &= 0 \\
1 + 2\lambda x_2 + 2\mu x_2 &= 0 \\
\mu\left((x_1 + 1)^2 + x_2^2 - 4\right) &= 0 \\
x_1^2 + x_2^2 &= 4 \\
(x_1 + 1)^2 + x_2^2 &\leq 4 \\
\mu &\geq 0.
\end{aligned}
$$

To solve this system, we consider two cases: $\mu = 0$ and $(x_1 + 1)^2 + x_2^2 - 4 = 0$.

1. *In the first case, the KKT system reduces to*

$$
\begin{aligned}
-1 + 2\lambda x_1 &= 0 \\
1 + 2\lambda x_2 &= 0 \\
x_1^2 + x_2^2 &= 4 \\
(x_1 + 1)^2 + x_2^2 &\leq 4.
\end{aligned}
$$

From the first two equations, noting that λ cannot be zero, we obtain $x_1 = -x_2$, and considering the third equation we have $x_1 = -x_2 = \pm\sqrt{2}$. Taking into account the inequality constraint, we obtain a unique solution to the above system,

$$x_1^{(1)} = -\sqrt{2}, x_2^{(1)} = \sqrt{2}, \lambda^{(1)} = -1/(2\sqrt{2}), \mu^{(1)} = 0.$$

2. *In the second case, $(x_1 + 1)^2 + x_2^2 - 4 = 0$, the KKT system becomes*

$$
\begin{aligned}
-1 + 2\lambda x_1 + 2\mu(x_1 + 1) &= 0 \\
1 + 2\lambda x_2 + 2\mu x_2 &= 0 \\
(x_1 + 1)^2 + x_2^2 &= 4 \\
x_1^2 + x_2^2 &= 4 \\
\mu &\geq 0.
\end{aligned}
$$

From the last two equalities we obtain $x_1 = -1/2$, $x_2 = \pm\sqrt{15}/2$, and using the first two equations we obtain the following two solutions:

$$x_1^{(2)} = -1/2, x_2^{(2)} = \sqrt{15}/2, \lambda^{(2)} = -\frac{1}{2}(1 + 1/\sqrt{15}), \mu^{(2)} = \frac{1}{2}(1 - 1/\sqrt{15});$$

$$x_1^{(3)} = -1/2, x_2^{(3)} = -\sqrt{15}/2, \lambda^{(3)} = -\frac{1}{2}(1 - 1/\sqrt{15}), \mu^{(3)} = \frac{1}{2}(1 + 1/\sqrt{15}).$$

Thus, there are three points satisfying the KKT conditions, $x^{(1)}, x^{(2)}$, and $x^{(3)}$. Next, we apply the second-order optimality conditions to each of these points. Let $h(x) = x_1^2 + x_2^2 - 4$, $g(x) = (x_1+1)^2 + x_2^2 - 4$. The Hessian of the Lagrangian as the function of x is

$$\nabla_{xx}^2 L(x, \lambda, \mu) = \begin{bmatrix} 2(\lambda + \mu) & 0 \\ 0 & 2(\lambda + \mu) \end{bmatrix}.$$

For the first KKT point,

$$x_1^{(1)} = -\sqrt{2}, x_2^{(1)} = \sqrt{2}, \lambda^{(1)} = -1/(2\sqrt{2}), \mu^{(1)} = 0,$$

since the inequality constraint is inactive at $x^{(1)}$, it can be ignored, and we have

$$
\begin{aligned}
T(x^{(1)}) = T(x^{(1)}, \mu^{(1)}) &= \{y \in \mathbb{R}^2 : \nabla h(x^{(1)})^T y = 0\} \\
&= \{y \in \mathbb{R}^2 : 2x_1^{(1)} y_1 + 2x_2^{(1)} y_2 = 0\} \\
&= \{y \in \mathbb{R}^2 : -2\sqrt{2} y_1 + 2\sqrt{2} y_2 = 0\} \\
&= \{y \in \mathbb{R}^2 : y_1 = y_2\} \\
&= \{y = [c, c]^T : c \in \mathbb{R}\}.
\end{aligned}
$$

In this case,

$$\nabla_{xx}^2 L(x^{(1)}, \lambda^{(1)}, \mu^{(1)}) = \begin{bmatrix} -1/\sqrt{2} & 0 \\ 0 & -1/\sqrt{2} \end{bmatrix},$$

and for any $y \in T(x^{(1)})$, we have $y^T \nabla_{xx}^2 L(x^{(1)}, \lambda^{(1)}, \mu^{(1)}) y = -\sqrt{2} c^2 < 0$ (if $c \neq 0$). Therefore, $x^{(1)}$ is a strict local maximizer. It should be noted that we would arrive at the same conclusion for any tangent space $T(x^{(1)})$ since $\nabla_{xx}^2 L(x^{(1)}, \lambda^{(1)}, \mu^{(1)})$ is clearly negative definite and the inequality constraint is inactive at $x^{(1)}$.

For the second KKT point,

$$x_1^{(2)} = -1/2, x_2^{(2)} = \sqrt{15}/2, \lambda^{(2)} = -\frac{1}{2}(1 + 1/\sqrt{15}), \mu^{(2)} = \frac{1}{2}(1 - 1/\sqrt{15}),$$

we have

$$
\begin{aligned}
T(x^{(2)}) = T(x^{(2)}, \mu^{(2)}) &= \left\{ y \in \mathbb{R}^2 : \begin{array}{c} \nabla h(x^{(2)})^T y = 0 \\ \nabla g(x^{(2)})^T y = 0 \end{array} \right\} \\
&= \left\{ y \in \mathbb{R}^2 : \begin{array}{c} 2x_1^{(2)} y_1 + 2x_2^{(2)} y_2 = 0 \\ 2(x_1^{(2)} + 1) y_1 + 2x_2^{(2)} y_2 = 0 \end{array} \right\} \\
&= \left\{ y \in \mathbb{R}^2 : \begin{array}{c} -y_1 + \sqrt{15} y_2 = 0 \\ y_1 + \sqrt{15} y_2 = 0 \end{array} \right\} \\
&= \{[0, 0]^T\}.
\end{aligned}
$$

Hence, the tangent space has no nonzero elements, the SONC and SOSC are automatically satisfied, and $x^{(2)}$ is a strict local minimizer.

Finally, for the third KKT point,

$$x_1^{(3)} = -1/2, x_2^{(3)} = -\sqrt{15}/2, \lambda^{(3)} = -\frac{1}{2}(1 - 1/\sqrt{15}), \mu^{(3)} = \frac{1}{2}(1 + 1/\sqrt{15}),$$

we also have $T(x^{(3)}) = T(x^{(3)}, \mu^{(3)}) = \{[0, 0]^T\}$, *implying that* $x^{(3)}$ *is a strict local minimizer.*

Note that the feasible region of this problem is a compact set, therefore, a global minimizer exists. Since there are two local minimizers, one of them must be global. Comparing the objective function $f(x) = x_2 - x_1$ at the points of local minimum, $f(x^{(2)}) = (\sqrt{15} + 1)/2$, $f(x^{(3)}) = (-\sqrt{15} + 1)/2$, we conclude that the global minimum is achieved at $x^{(3)}$.

14.2   Duality

Consider the functions $f : X \to \mathbb{R}$, $g : Y \to \mathbb{R}$, and $F : X \times Y \to \mathbb{R}$, where $X \subseteq \mathbb{R}^m$ and $Y \subseteq \mathbb{R}^n$. We assume that the global minima and maxima do exist in all cases discussed below in this section. Suppose that $f(x) \leq g(y)$ for all $(x, y) \in X \times Y$. Then it is clear that

$$\max_{x \in X} f(x) \leq \min_{y \in Y} g(y).$$

Under certain conditions, the above inequality can be satisfied as an equality

$$\max_{x \in X} f(x) = \min_{y \in Y} g(y).$$

A result of this kind is called a *duality theorem*. It is easy to prove that the following inequality holds:

$$\max_{y \in Y} \min_{x \in X} F(x, y) \leq \min_{x \in X} \max_{y \in Y} F(x, y).$$

Under certain conditions, we can prove that

$$\max_{y \in Y} \min_{x \in X} F(x, y) = \min_{x \in X} \max_{y \in Y} F(x, y).$$

A result of this type is called a *minimax theorem*.

The point $(x^*, y^*) \in X \times Y$ is a *saddle point* of F (with respect to maximizing in Y and minimizing in X) if

$$F(x^*, y) \leq F(x^*, y^*) \leq F(x, y^*) \text{ for all } (x, y) \in X \times Y.$$

Theorem 14.9 *The point* $(x^*, y^*) \in X \times Y$ *is a saddle point of* F *if and only if*

$$F(x^*, y^*) = \max_{y \in Y} \min_{x \in X} F(x, y) = \min_{x \in X} \max_{y \in Y} F(x, y). \qquad (14.11)$$

Proof. First assume that (x^*, y^*) is a saddle point of F. Then

$$F(x^*, y) \leq F(x^*, y^*) \leq F(x, y^*) \text{ for all } (x, y) \in X \times Y,$$

hence

$$\begin{aligned}
F(x^*, y^*) &\leq \min_{x \in X} F(x, y^*) \leq \max_{y \in Y} \min_{x \in X} F(x, y) \leq \min_{x \in X} \max_{y \in Y} F(x, y) \\
&\leq \max_{y \in Y} F(x^*, y) \leq F(x^*, y^*),
\end{aligned}$$

implying that all the above inequalities must hold with equality.

Next assume that

$$F(x^*, y^*) = \max_{y \in Y} \min_{x \in X} F(x, y) = \min_{x \in X} \max_{y \in Y} F(x, y).$$

Then we have

$$\min_{x \in X} F(x, y^*) = F(x^*, y^*) = \max_{y \in Y} F(x^*, y),$$

which implies that (x^*, y^*) is a saddle point. $\qquad \square$

Consider now the *primal* optimization problem

$$\begin{aligned}
\text{minimize} \quad & f(x) \\
\text{subject to} \quad & g(x) \leq 0 \qquad\qquad\qquad\qquad \text{(P)} \\
& x \in X,
\end{aligned}$$

where $g : \mathbb{R}^n \to \mathbb{R}^p$ and $X \subseteq \mathbb{R}^n$.

The Lagrangian of (P) is

$$L(x, \mu) = f(x) + \mu^T g(x), \mu \in \mathbb{R}^p, \mu \geq 0, x \in X.$$

Note that

$$\sup_{\mu \geq 0} L(x, \mu) = \sup_{\mu \geq 0} \{ f(x) + \mu^T g(x) \} = \begin{cases} f(x), & \text{if } g(x) \leq 0 \\ +\infty, & \text{otherwise,} \end{cases}$$

and the problem (P) can be restated in the form

$$\min_{x \in X} \max_{\mu \geq 0} L(x, \mu).$$

For $\mu \geq 0$, define the *dual function*

$$d(\mu) = \min_{x \in X} L(x, \mu). \qquad (14.12)$$

It can be shown (Exercise 14.15) that $d(\mu)$ is a concave function, independently of whether the problem (P) is convex. Then the *dual* problem of (P) is defined to be the following optimization problem:

$$\max_{\mu \geq 0} d(\mu) = \max_{\mu \geq 0} \min_{x \in X} L(x, \mu). \tag{D}$$

The problem (D) is called the *dual problem*.

Example 14.8 *Consider the quadratic programming (QP) problem*

$$\begin{array}{ll} \text{minimize} & \frac{1}{2}x^T Q x + c^T x \\ \text{subject to} & Ax \leq b, \end{array}$$

where Q is an $n \times n$ symmetric positive definite matrix, $c \in \mathbb{R}^n$, $A \in \mathbb{R}^{m \times n}$, $b \in \mathbb{R}^m$. The corresponding Lagrangian function is

$$\begin{aligned} L(x, \mu) &= \frac{1}{2}x^T Q x + c^T x + \mu^T (Ax - b) \\ &= \frac{1}{2}x^T Q x + (A^T \mu + c)^T x - b^T \mu. \end{aligned}$$

The minimum of $L(x, \mu)$ with respect to x occurs at the point x^* where $\nabla L(x^*, \mu) = 0$, that is,

$$x^* = -Q^{-1}(c + A^T \mu).$$

Hence, the dual function

$$d(\mu) = \min_{x \in \mathbb{R}^n} L(x, \mu) = -\frac{1}{2}\mu^T \left(A Q^{-1} A^T \right) \mu - (A Q^{-1} c + b)^T \mu - \frac{1}{2}c^T Q^{-1} c,$$

and the dual problem is given by

$$\max_{\mu \geq 0} d(\mu) = -\frac{1}{2}\mu^T M \mu + d^T \mu,$$

where $M = A Q^{-1} A^T$ and $d = -(A Q^{-1} c + b)$. Hence, the dual of a convex QP problem is a concave QP problem.

Theorem 14.10 (Weak Duality Theorem) *Let x^* be a global minimizer of the primal problem, and let μ^* be a global maximizer of the dual problem. Then for any $\mu \geq 0$,*

$$d(\mu) \leq d(\mu^*) \leq f(x^*).$$

Proof. By definition of the dual, $d(\mu) = \min_{x \in X} L(x, \mu)$, hence for any $\mu \geq 0$ and $x^* \in X$, we have $d(\mu) \leq f(x^*) + \mu^T g(x^*)$. This implies that $d(\mu) \leq$

$\max\limits_{\mu \geq 0} d(\mu) = d(\mu^*)$. Since $\mu \geq 0$ and $g(x^*) \leq 0$, it follows that $\mu^T g(x^*) \leq 0$ for any μ, hence, $d(\mu^*) = f(x^*) + \mu^{*T} g(x^*) \leq f(x^*)$. $\qquad\qquad\qquad\square$

The difference $f(x^*) - d(\mu^*)$ is called the *duality gap*.

The following result follows directly from Theorem 14.10.

Theorem 14.11 *A point* $(x^*, \mu^*) \in X \times \mathbb{R}^p_+$ *is a saddle point of the Lagrangian* $L(x, \mu)$ *if and only if* x^* *is a global minimum point of the primal problem* (P), μ^* *is a global maximum point of the dual* (D), *and the optimal values* $f(x^*)$ *of* (P) *and* $d(\mu^*)$ *of* (D) *coincide.*

14.3 Projected Gradient Methods

Consider a set-constrained problem

$$\min_{x \in X} f(x).$$

In methods for unconstrained problems, we used an iteration in the form

$$x^{(k+1)} = x^{(k)} + \alpha_k d^{(k)}, \ k \geq 0,$$

where $d^{(k)}$ is a direction and α_k is a step size. If such an iteration is used for the set-constrained problem with $x^{(k)} \in X$, we may obtain $x^{(k+1)} \notin X$. Therefore we take the projection of this point onto X as our $x^{(k+1)}$:

$$x^{(k+1)} = \Pi_X(x^{(k)} + \alpha_k d^{(k)}),$$

where $\Pi_X(y)$ denotes the projection of y onto set X. The projection of y onto X can be defined as

$$\Pi_X(y) = \arg \min_{z \in X} \|z - y\|.$$

This definition may not be valid since such a minimizer may not exist or may not be unique in general. Even if it does exist, it may be as difficult to find as to solve the original optimization problem. However, in some cases the projection can be easily computed.

Example 14.9 *For* $X = \{x : a_i \leq x_i \leq b_i, \ i = 1, \ldots, n\}$ *and* $y \in \mathbb{R}^n$, *the* i^{th} *component of the projection of* y *onto* X *is given by*

$$[\Pi_X(y)]_i = \begin{cases} y_i, & \text{if } a_i \leq y_i \leq b_i, \\ a_i, & \text{if } y_i < a_i, \\ b_i, & \text{if } y_i > b_i. \end{cases}$$

Example 14.10 *Given an $m \times n$ matrix A of rank $m(m < n)$, consider $X = \{x : Ax = b\}$. Given $y \in \mathbb{R}^n, y \notin X$, the projection of y onto X is given by a solution to the problem*

$$\min_{Az=b} \|z - y\|.$$

Changing the variable, $\xi = z - y$, we obtain an equivalent problem

$$\min_{A\xi=b-Ay} \|\xi\|.$$

This is the problem of finding a solution of the system $A\xi = b - Ay$ with minimum norm. The solution to this problem is (see Example 14.3 at page 357)

$$\xi = A^T(AA^T)^{-1}(b - Ay).$$

Thus,

$$z = y + \xi = y + A^T(AA^T)^{-1}(b - Ay) = (I_n - A^T(AA^T)^{-1}A)y + A^T(AA^T)^{-1}b.$$

Consider an iteration for the problem $\min\limits_{x \in X} f(x)$:

$$y = x^{(k)} + \alpha_k d^{(k)}, \ x^{(k)} \in X.$$

The projection of y onto X is

$$
\begin{aligned}
\Pi_X(y) &= (I_n - A^T(AA^T)^{-1}A)(x^{(k)} + \alpha_k d^{(k)}) + A^T(AA^T)^{-1}b \\
&= x^{(k)} - A^T(AA^T)^{-1}Ax^{(k)} + \alpha_k Pd^{(k)} + A^T(AA^T)^{-1}b \\
&= x^{(k)} + \alpha_k Pd^{(k)},
\end{aligned}
$$

where $P = I_n - A^T(AA^T)^{-1}A$ is the *orthogonal projector onto the null space* $\{x : Ax = 0\}$ of A. Note that for any $x \in \mathbb{R}^n$ we have $A(Px) = (A - AA^T(AA^T)^{-1}A)x = 0$.

Thus, we can write our iteration for the constrained problem in the form

$$x^{(k+1)} = x^{(k)} + \alpha_k Pd^{(k)}.$$

In other words, instead of direction $d^{(k)}$ used for the unconstrained problem, we will use the direction $Pd^{(k)}$ in the constrained problem. This direction is the projection of $d^{(k)}$ onto the null space of A.

Recall that in the gradient methods for unconstrained problems we used a step

$$x^{(k+1)} = x^{(k)} + \alpha_k d^{(k)},$$

where

$$d^{(k)} = -\nabla_k.$$

Consider a nonlinear programming problem

$$\min_{Ax=b} f(x).$$

In the projected gradient methods we use

$$d^{(k)} = P(-\nabla_k),$$

where $P = I_n - A^T(AA^T)^{-1}A$, as defined before. If we define

$$\alpha_k = \arg\min_{\alpha \geq 0} f(x^{(k)} - \alpha P\nabla_k),$$

we obtain an iteration of the *projected steepest descent*:

$$
\begin{aligned}
x^{(k+1)} &= x^{(k)} - \alpha_k P\nabla_k, \ k \geq 0, \\
P &= I_n - A^T(AA^T)^{-1}A, \\
\alpha_k &= \arg\min_{\alpha \geq 0} f(x^{(k)} - \alpha P\nabla_k).
\end{aligned}
$$

Theorem 14.12 *Given* $x^* \in \mathbb{R}^n$, $P\nabla f(x^*) = 0$ *if and only if* x^* *satisfies the Lagrange conditions,* $\nabla f(x^*) + A^T\lambda = 0$ *for some* $\lambda \in \mathbb{R}^m$.

Proof. Assume that $P\nabla f(x^*) = 0$, then we have

$$(I_n - A^T(AA^T)^{-1}A)\nabla f(x^*) = 0,$$

that is

$$\nabla f(x^*) - A^T((AA^T)^{-1}A\nabla f(x^*)) = 0,$$

so, the Lagrange conditions are satisfied with $\lambda = (AA^T)^{-1}A\nabla f(x^*)$.

On the other hand, assuming that there exists $\lambda \in \mathbb{R}^m$ such that

$$\nabla f(x^*) + A^T\lambda = 0,$$

we obtain

$$
\begin{aligned}
P\nabla f(x^*) &= -PA^T\lambda = -(I_n - A^T(AA^T)^{-1}A)A^T\lambda \\
&= -(A^T - A^T(AA^T)^{-1}AA^T)\lambda = 0.
\end{aligned}
$$

$\square$

Descent property

Consider the kth step of the projected steepest descent method and denote by

$$\phi_k(\alpha) = f(x^{(k)} - \alpha P\nabla_k).$$

The derivative of this function is

$$\phi_k'(\alpha) = -\nabla f(x^{(k)} - \alpha P\nabla_k)^T P\nabla_k,$$

so,

$$\phi_k'(0) = -\nabla_k^T P \nabla_k.$$

Using the following properties of the projector P,

$$P^T = P, \quad P^2 = P,$$

we obtain

$$\phi_k'(0) = -\nabla_k^T P \nabla_k = -\nabla_k^T P^T P \nabla_k = -\|P\nabla_k\|^2.$$

Thus, if $P\nabla_k \neq 0$, then $\phi_k'(0) < 0$, so there exists $\bar{\alpha} > 0$ such that for any $\alpha \in (0, \bar{\alpha})$: $\phi_k(\alpha) < \phi_k(0)$. Hence, $f(x^{(k+1)}) < f(x^{(k)})$ and we have the descent property.

Next we show that if $x^{(k)} \to x^*$, where $\{x^{(k)} : k \geq 0\}$ is the sequence generated by the projected steepest descent method, then $P\nabla f(x^*) = 0$. Consider $\phi_k(\alpha) = f(x^{(k)} - \alpha P\nabla_k)$. Since in the steepest descent method α_k minimizes $\phi_k(\alpha)$, by the FONC,

$$\phi_k'(\alpha_k) = 0.$$

On the other hand, using the chain rule:

$$\phi_k'(\alpha_k) = \left. \frac{df(x^{(k)} - \alpha P\nabla_k)}{d\alpha} \right|_{\alpha = \alpha_k} = \nabla_{k+1}^T P \nabla_k = (P\nabla_{k+1})^T (P\nabla_k).$$

So, $(P\nabla_{k+1})^T (P\nabla_k) = 0$, $k \geq 0$ and

$$\|P\nabla f(x^*)\|^2 = \lim_{k \to \infty} (P\nabla_{k+1})^T (P\nabla_k) = 0.$$

Note that if $f(x)$ is a convex function, then the problem $\min_{Ax=b} f(x)$ is a convex problem. In this case, if the projected steepest descent converges, then it converges to a global minimizer.

14.3.1   Affine scaling method for LP

To illustrate the projected gradient method, we discuss its variation applied to linear programming (LP). The resulting *affine scaling method* belongs to the class of *interior point* methods, in which the points generated by the algorithms always lie in the interior of the feasible region. It should be noted that, unlike some other variations of interior point methods for LP, the method we discuss here is not efficient in general.

Recall that a linear program in the standard form can be written as

$$\begin{array}{ll} \text{minimize} & c^T x \\ \text{subject to} & Ax = b \\ & x \geq 0, \end{array} \qquad (14.13)$$

where $c \in \mathbb{R}^n, b \in \mathbb{R}^m, A \in \mathbb{R}^{m \times n}$. We also assume that $\text{rank}(A) = m$.

We assume that all points generated by the method are interior, $x^{(k)} > 0, k \geq 0$. In the affine scaling method, the original LP is transformed to an equivalent LP, so that the current point is "better" positioned for the projected steepest descent method. It is based on observation that if the current point is close to the "center" of the feasible region, then there is more space for move in a descent direction; therefore a larger step toward the minimum can be made. An appropriate choice for the "center" is the point $e = [1, 1, \ldots, 1]^T \in \mathbb{R}^n$, which has equal distance to all bounds given by $x_i = 0, i = 1, \ldots, n$.

Denote by

$$D_k = \operatorname{diag}(x^{(k)}) = \begin{bmatrix} x_1^{(k)} & 0 & \cdots & 0 \\ 0 & x_2^{(k)} & \cdots & 0 \\ \vdots & \vdots & \ddots & \vdots \\ 0 & 0 & \cdots & x_n^{(k)} \end{bmatrix}.$$

Then, introducing a new variable y such that

$$y = D_k^{-1} x \quad \Leftrightarrow \quad x = D_k y,$$

results in the *affine scaling* that transforms x to y and, in particular, $x^{(k)}$ to $y^{(k)} = e$. Matrix D_k is called the *scaling matrix*.

Expressing problem (14.13) in terms of y we obtain the equivalent problem

$$\begin{aligned} \text{minimize} \quad & c_k^T y \\ \text{subject to} \quad & A_k y = b \\ & y \geq 0, \end{aligned} \tag{14.14}$$

where

$$c_k = D_k c \quad \text{and} \quad A_k = A D_k.$$

We have $y^{(k)} = D_k^{-1} x^{(k)} = e$. Next, we make an iteration in the direction of the projected steepest descent for problem (14.14). Since $\nabla(c_k^T y) = c_k$, we obtain

$$y^{(k+1)} = y^{(k)} - \alpha_k d^{(k)}, \tag{14.15}$$

where $y^{(k+1)}$ denotes the result of the iteration, and

$$d^{(k)} = P_k c_k,$$

$$P_k = I_n - A_k^T (A_k A_k^T)^{-1} A_k.$$

We will select $\alpha_k > 0$ such that the step is as large as possible, but still results in an interior point $y^{(k+1)}$. In other words, when moving in the direction of projected steepest descent, we need to stop just short of the boundary (the method is not defined on the boundary since for $y^{(k+1)}$ on the boundary we would have $y_i^{(k+1)} = 0$ for some i, and, as a result, D_{k+1}^{-1} would not exist).

Consider the ith component of (14.15):

$$y_i^{(k+1)} = y_i^{(k)} - \alpha_k d_i^{(k)} = 1 - \alpha_k d_i^{(k)}. \tag{14.16}$$

Observe that if $d_i^{(k)} \leq 0$, then $y_i^{(k+1)} > 0$ for any $\alpha_k > 0$. If, on the other hand, $d_i^{(k)} > 0$, then

$$y_i^{(k+1)} > 0 \quad \Leftrightarrow \quad 1 - \alpha_k d_i^{(k)} > 0 \quad \Leftrightarrow \quad \alpha_k < 1/d_i^{(k)}.$$

It is easy to see that the above inequality is satisfied for α_k given by

$$\alpha_k = \alpha \min_{i:d_i^{(k)}>0} \left\{ 1/d_i^{(k)} \right\},$$

where $0 < \alpha < 1$ (typically $\alpha = 0.9$ or 0.99 is chosen). Note that if $d^{(k)} \neq 0$ and $d^{(k)} \leq 0$, then the problem is unbounded (there is no minimizer). Finally, having computed $y_i^{(k+1)}$, we need to find the corresponding feasible point of the original problem (14.13) by applying the inverse scaling operation,

$$x^{(k+1)} = D_k y^{(k+1)} = D_k(e - \alpha_k d^{(k)}).$$

The affine scaling method is summarized in Algorithm 14.1.

Algorithm 14.1 The affine scaling algorithm for solving LP (14.13).

1: **Input:** $A, b, c, x^{(0)} > 0$ such that $Ax^{(0)} = b$, α
2: **Output:** an approximate solution x^* of the LP
3: $k = 0$
4: **repeat**
5:     $D_k = \text{diag}(x^{(k)})$
6:     $c_k = D_k c$
7:     $A_k = A D_k$
8:     $P_k = I_n - A_k^T (A_k A_k^T)^{-1} A_k$
9:     $d^{(k)} = P_k c_k$
10:     $\alpha_k = \alpha \min_{i:d_i^{(k)}>0} \left\{ 1/d_i^{(k)} \right\}$
11:     $x^{(k+1)} = D_k(e - \alpha_k d^{(k)})$
12:     $k = k + 1$
13: **until** a stopping criterion is satisfied
14: **return** $x^{(k)}$

Example 14.11 *We apply the affine scaling algorithm to the following LP starting from the point $[1, 2, 1]^T$:*

$$\begin{aligned} minimize \quad & 3x_1 + 2x_2 + 4x_3 \\ subject\ to \quad & x_1 + 2x_2 + 3x_3 = 8 \\ & x_1 - x_2 + x_3 = 0 \\ & x_1, x_2, x_3 \geq 0. \end{aligned}$$

We use the step-length coefficient $\alpha = 0.9$. *We have:*

$$A = \begin{bmatrix} 1 & 2 & 3 \\ 1 & -1 & 1 \end{bmatrix}, b = [8,0]^T, c = [3,2,4]^T, x^{(0)} = [1,2,1]^T,$$

hence,

$$D_0 = \begin{bmatrix} 1 & 0 & 0 \\ 0 & 2 & 0 \\ 0 & 0 & 1 \end{bmatrix}.$$

Using the affine scaling we obtain

$$A_0 = AD_0 = \begin{bmatrix} 1 & 4 & 3 \\ 1 & -2 & 1 \end{bmatrix};$$

$$P_0 = I_3 - A_0^T(A_0 A_0^T)^{-1} A_0 = \begin{bmatrix} 5/7 & 1/7 & -3/7 \\ 1/7 & 1/35 & -3/35 \\ -3/7 & -3/35 & 9/35 \end{bmatrix};$$

$$d^{(0)} = P_0 D_0 c = [1, 1/5, -3/5]^T.$$

Next we need to determine the step length. We have

$$\min_{i:d_i^{(0)}>0} \left\{ \frac{1}{d_i^{(0)}} \right\} = \min\{1,5\} = 1,$$

so $\alpha_0 = \alpha \cdot 1 = 0.9$. *Finally,*

$$x^{(1)} = D_0(e - 0.9d^{(0)}) = [0.1, 1.64, 1.54]^T.$$

Table 14.1 shows the approximate solution $x^{(k)}$, *the corresponding objective function value* $c^T x^{(k)}$, *and the error* $\|x^{(k)} - x^*\|$ *for 8 steps of the affine scaling method applied to the considered LP. Here* $x^* = [0, 1.6, 1.6]^T$ *is the exact optimal solution of the problem.*

14.4   Sequential Unconstrained Minimization

Consider the problem

$$\begin{aligned} \text{minimize} \quad & f(x) \\ \text{subject to} \quad & x \in X, \end{aligned}$$

where $X = \{x \in \mathbb{R}^n : g_j(x) \le 0, j = 1, \ldots, m\}$ and $f, g_j : \mathbb{R}^n \to \mathbb{R}$, $j = 1, \ldots, m$ are continuously differentiable functions.

We will briefly review two types of schemes of *sequential unconstrained minimization*, in which a solution to this problem is approximated by a sequence of solutions to some auxiliary unconstrained minimization problems; namely, the penalty function methods and the barrier methods.

TABLE 14.1: Affine scaling iterations for the LP in Example 14.11. All the values are rounded to 7 decimal places.

k	$x^{(k)T}$	$c^T x^{(k)}$	$\|x^{(k)} - x^*\|$
1	[0.1000000, 1.6400000, 1.5400000]	9.7400000	0.1232883
2	[0.0100000, 1.6040000, 1.5940000]	9.6140000	0.0123288
3	[0.0010000, 1.6004000, 1.5994000]	9.6014000	0.0012329
4	[0.0001000, 1.6000400, 1.5999400]	9.6001400	0.0001233
5	[0.0000100, 1.6000040, 1.5999940]	9.6000140	0.0000123
6	[0.0000010, 1.6000004, 1.5999994]	9.6000014	0.0000012
7	[0.0000001, 1.6000000, 1.5999999]	9.6000001	0.0000001
8	[0.0000000, 1.6000000, 1.6000000]	9.6000000	0.0000000

14.4.1   Penalty function methods

A continuous function $\Phi(x)$ is called a *penalty function* (or simply, a *penalty*) for a closed set X if $\Phi(x) = 0$ for any $x \in X$ and $\Phi(x) > 0$ for any $x \notin X$.

Example 14.12 *Denote by* $(a)_+ = \max\{a, 0\}$. *Then for*

$$X = \{x \in \mathbb{R}^n : g_j(x) \leq 0, j = 1, \ldots, m\}$$

we have the following quadratic penalty:

$$\Phi(x) = \sum_{j=1}^{m} ((g_j(x))_+)^2.$$

It is easy to see that the penalty functions have the following property: If $\Phi_1(x)$ is a penalty for X_1 and $\Phi_2(x)$ is a penalty for X_2, then $\Phi_1(x) + \Phi_2(x)$ is a penalty for the intersection $X_1 \cap X_2$.

The general scheme of penalty function methods is described in Algorithm 14.2. The following theorem establishes its convergence.

Theorem 14.13 *Let x^* be a global optimal solution to the considered constrained problem. If there exists a value $\bar{t} > 0$ such that the set*

$$S = \{x \in \mathbb{R}^n : f(x) + \bar{t}\Phi(x) \leq f(x^*)\}$$

is bounded, then

$$\lim_{k \to \infty} f(x^{(k)}) = f(x^*), \quad \lim_{k \to \infty} \Phi(x^{(k)}) = 0.$$

Algorithm 14.2 A general scheme of the penalty function method for solving $\min\limits_{x\in X} f(x)$.

1: choose a penalty $\Phi(x)$ of X
2: choose $x_0 \in \mathbb{R}^n$
3: choose a sequence of penalty coefficients $0 < t_k < t_{k+1}$ such that $t_k \to \infty$
4: **for** $k = 1, 2, \ldots$ **do**
5:     find a solution $x^{(k)}$ to $\min\limits_{x\in\mathbb{R}^n} \{f(x) + t_{k-1}\Phi(x)\}$ starting with x_{k-1}
6:     **if** a stopping criterion is satisfied **then**
7:         **return** $x^{(k)}$
8:     **end if**
9: **end for**

Proof: Denote by $\Psi_k(x) = f(x) + t_k\Phi(x)$, $\Psi_k^* = \min\limits_{x\in\mathbb{R}^n} \Psi_k(x)$. We have $\Psi_k^* \le \Psi_k(x^*) = f(x^*)$, and for any $x \in \mathbb{R}^n$, $\Psi_{k+1}(x) \ge \Psi_k(x) \Rightarrow \Psi_{k+1}^* \ge \Psi_k^*$. Hence, there exists a limit $\lim\limits_{k\to\infty} \Psi_k^* = \Psi^* \le f(x^*)$. If $t_k > \bar{t}$, then $f(x^{(k+1)}) + \bar{t}\Phi(x^{(k+1)}) \le f(x^{(k+1)}) + t_k\Phi(x^{(k+1)}) = \Psi_k^* \le f(x^*)$, implying that $x^{(k+1)} \in S$. By Bolzano-Weierstrass theorem (which states that every bounded, infinite set of real numbers has a limit point), since S is bounded, the sequence $\{x^{(k)} : k \ge 1\}$ has a limit point x_*. Since $\lim\limits_{k\to\infty} t_k = +\infty$ and $\lim\limits_{k\to\infty} \Psi_k^* \le f(x^*)$, we have $\Phi(x_*) = 0$ and $f(x_*) \le f(x^*)$. The fact that $\Phi(x_*) = 0$ implies that $x_* \in X$, so x_* is a global minimizer of the problem. $\square$

To apply a penalty function method in practice, we need to choose an appropriate penalty function and a sequence of penalty coefficients $\{t_k\}$. In addition, a method for solving the unconstrained penalty problems needs to be selected. These important decisions are not easy to address in general.

14.4.2 Barrier methods

Let X be a closed set with a nonempty interior. A continuous function $F(x) : X \to \mathbb{R}$ is called a *barrier function* (or simply, a barrier) for X if $F(x) \to \infty$ when x approaches the boundary of X.

Example 14.13 *Let $X = \{x \in \mathbb{R}^n : g_j(x) \le 0, j = 1,\ldots,m\}$ satisfy the Slater condition: there exists $\bar{x}$ such that $g_j(\bar{x}) < 0, j = 1,\ldots,m$. Then the following functions are barriers for X.*

- *Power-function barrier:* $F(x) = \sum\limits_{j=1}^m \frac{1}{(-g_j(x))^p}, p \ge 1$.

- *Logarithmic barrier:* $F(x) = -\sum\limits_{j=1}^m \ln(-g_j(x))$.

- *Exponential barrier:* $F(x) = \sum\limits_{j=1}^{m} \exp\left(\frac{1}{-g_j(x)}\right)$.

Barriers have the following property. If $F_1(x)$ is a barrier for X_1 and $F_2(x)$ is a barrier for X_2, then $F_1(x) + F_2(x)$ is a barrier for the intersection $X_1 \cap X_2$. The general scheme of barrier methods is outlined in Algorithm 14.3.

Algorithm 14.3 A general scheme of the barrier function method for solving $\min\limits_{x \in X} f(x)$.

1: choose a barrier $F(x)$ of X
2: choose x_0 in the interior of X
3: choose a sequence of coefficients $t_k > t_{k+1} > 0$ such that $t_k \to 0$, $k \to \infty$
4: **for** $k = 1, 2, \ldots$ **do**
5:    find a solution $x^{(k)}$ to the barrier problem $\min\limits_{x \in X}\{f(x) + t_{k-1}F(x)\}$
6:    **if** a stopping criterion is satisfied **then**
7:       **return** $x^{(k)}$
8:    **end if**
9: **end for**

Denote by $\Psi_k(x) = f(x) + t_k F(x)$, $\Psi_k^* = \min\limits_{x \in X} \Psi_k(x)$.

Theorem 14.14 *Let the barrier $F(x)$ be bounded from below on X. Then $\lim\limits_{k \to \infty} \Psi_k^* = f^*$, where f^* is the optimal objective value of the considered problem.*

Proof: Let $F(x) \geq F^*$ for all $x \in X$. For any $\bar{x}$ in the interior of X we have

$$\limsup_{k \to \infty} \Psi_k^* \leq \lim_{k \to \infty} [f(\bar{x}) + t_k F(\bar{x})] = f(\bar{x}),$$

thus $\limsup\limits_{k \to \infty} \Psi_k^* \leq f^*$ (since the opposite would yield $\limsup\limits_{k \to \infty} \Psi_k^* > f(\bar{x})$ for some $\bar{x}$ in the interior of X).
On the other hand,

$$\Psi_k^* = \min_{x \in X}\{f(x) + t_k F(x)\} \geq \min_{x \in X}\{f(x) + t_k F^*\} = f^* + t_k F^*,$$

so $\liminf\limits_{k \to \infty} \Psi_k^* \geq f^*$. Thus, $\lim\limits_{k \to \infty} \Psi_k^* = f^*$. $\square$

Again, several important issues need to be addressed in order to use this method, such as finding a starting point x_0, choosing the barrier function, updating the penalty coefficients, and solving the barrier problems. These issues can be effectively addressed for *convex* optimization problems, leading to a powerful framework of *interior point methods*.

14.4.3 Interior point methods

Interior point methods take advantage of fast convergence of Newton's method for approximately solving the barrier problems. They are generally preferred to other available techniques for convex optimization due to their superior theoretical convergence properties and excellent practical performance. Consider the problem

$$\begin{aligned} \text{minimize} \quad & f(x) \\ \text{subject to} \quad & h(x) = 0 \\ & g(x) \leq 0, \end{aligned}$$

where $f(x) : \mathbb{R}^n \to \mathbb{R}$, $h(x) : \mathbb{R}^n \to \mathbb{R}^m$ $(m < n)$, and $g(x) : \mathbb{R}^n \to \mathbb{R}^p$. Introducing p slack variables given by a p-dimensional vector s, we obtain the following equivalent problem:

$$\begin{aligned} \text{minimize} \quad & f(x) \\ \text{subject to} \quad & h(x) = 0 \\ & g(x) + s = 0 \\ & s \geq 0. \end{aligned}$$

Using the logarithmic barrier for the nonnegativity constraints, we write the corresponding barrier problem

$$\begin{aligned} \text{minimize} \quad & f(x) - t \sum_{i=1}^{p} \log s_i \\ \text{subject to} \quad & h(x) = 0 \\ & g(x) + s = 0. \end{aligned}$$

To apply the barrier method, we need to choose the sequence of positive coefficients $\{t_k : k \geq 0\}$ that converges to zero and solve the above barrier problem. The Lagrangian for the barrier problem is

$$L(x, s, \lambda, \mu) = f(x) - t \sum_{i=1}^{p} \log s_i - \lambda h(x) - \mu(g(x) + s).$$

KKT conditions are given by

$$\begin{aligned} \nabla f(x) - J_h^T(x)\lambda - J_g^T(x)\mu &= 0 \\ -tS^{-1}e - \mu &= 0 \\ h(x) &= 0 \\ g(x) + s &= 0, \end{aligned}$$

where $S = \text{diag}(s)$ is the diagonal matrix with the diagonal given by s, and e is the vector of all ones. The last system can be rewritten as

$$\begin{aligned} \nabla f(x) - J_h^T(x)\lambda - J_g^T(x)\mu &= 0 \\ S\mu + te &= 0 \\ h(x) &= 0 \\ g(x) + s &= 0, \end{aligned}$$

which can be written as $F_t(y) = 0$, where $y = [x, s, \lambda, \mu]^T$ and

$$F_t(y) = \begin{bmatrix} \nabla f(x) - J_h^T(x)\lambda - J_g^T(x)\mu \\ S\mu + te \\ h(x) \\ g(x) + s \end{bmatrix}.$$

Recall that to derive the $(k + 1)^{\text{st}}$ iteration of Newton's method for solving this system, we consider a linear approximation

$$F_t(y) \approx F_t(y^{(k)}) + J_{F_t}(y^{(k)})(y - y^{(k)}),$$

where $y^{(k)}$ is the solution after k steps. Solve the system $J_{F_t}(y^{(k)})z = -F_t(y^{(k)})$ for z, and set $y^{(k+1)} = y^{(k)} + z$. To ensure some desired convergence and numerical stability properties, the *primal-dual* interior point methods use step-length coefficients that may be different for different components of Newton's direction z. More specifically, the system $J_{F_t}(y^{(k)})z = -F_t(y^{(k)})$ is given by

$$\begin{bmatrix} \nabla_{xx}^2 L(y) & 0 & -J_h^T(x) & -J_g^T(x) \\ 0 & M & 0 & S \\ J_h(x) & 0 & 0 & 0 \\ J_g(x) & I_p & 0 & 0 \end{bmatrix} \begin{bmatrix} z_x \\ z_s \\ z_\lambda \\ z_\mu \end{bmatrix} = - \begin{bmatrix} \nabla_x L(y) \\ S\mu + te \\ h(x) \\ g(x) + s \end{bmatrix},$$

where $y = [x, s, \lambda, \mu] = y^{(k)} = [x^{(k)}, s^{(k)}, \lambda^{(k)}, \mu^{(k)}]^T$, $M = \text{diag}(\mu^{(k)})$, $S = \text{diag}(s^{(k)})$, and $L(y) = f(x) - t \sum_{i=1}^{p} \log s_i - \lambda h(x) - \mu(g(x) + s)$. Then after solving this system for $z = [z_x, z_s, z_\lambda, z_\mu]^T$, the $(k + 1)^{\text{st}}$ iterate $y^{(k+1)} = [x^{(k+1)}, s^{(k+1)}, \lambda^{(k+1)}, \mu^{(k+1)}]^T$ is given by

$$\begin{bmatrix} x^{(k+1)} \\ s^{(k+1)} \end{bmatrix} = \begin{bmatrix} x^{(k)} \\ s^{(k)} \end{bmatrix} + \alpha_s \begin{bmatrix} z_x \\ z_s \end{bmatrix}, \quad \begin{bmatrix} \lambda^{(k+1)} \\ \mu^{(k+1)} \end{bmatrix} = \begin{bmatrix} \lambda^{(k)} \\ \mu^{(k)} \end{bmatrix} + \alpha_\lambda \begin{bmatrix} z_\lambda \\ z_\mu \end{bmatrix}$$

with appropriately selected coefficients α_s and α_λ.

Exercises

14.1. For a twice continuously differentiable function $f : \mathbb{R}^n \to \mathbb{R}$, let $x^{(k)} \in \mathbb{R}^n$ be such that $\nabla f(x^{(k)}) \neq 0$ and $\nabla^2 f(x^{(k)})$ is positive definite. Consider the following constrained problem:

$$\begin{array}{ll} \text{minimize} & f(d) = \nabla f(x^{(k)})^T d \\ \text{subject to} & d^T \nabla^2 f(x^{(k)}) d \leq 1. \end{array}$$

Show that the solution d^* of this problem is the Newton's direction for $f(x)$.

14.2. Given the equality constraints $h(x) = 0, x \in \mathbb{R}^n$, where

    (a) $h(x) = 2x_1^2 + x_2^2 - 1 = 0$ $(n = 2)$;

    (b) $h(x) = Ax - b$, where A is an $m \times n$ matrix with $\mathrm{rank}(A) = m$, $m < n$;

    (c) $h(x) = 1 - x^T P x$, where P is a symmetric positive definite matrix,

check that all feasible points are regular in each case.

14.3. The figure below illustrates a problem of minimizing a concave function f over the set given by $\{x : g(x) \leq 0\}$. Which of the following statements are true/false?

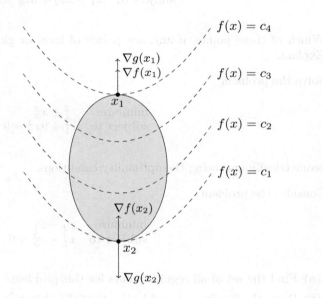

    (a) $c_1 \geq c_2 \geq c_3 \geq c_4$;

    (b) x_1 is a local minimizer;

    (c) x_2 is a local minimizer;

    (d) x_1 is a global minimizer;

    (e) x_2 is a global minimizer.

14.4. Let x^* be an optimal solution to the problem

$$\begin{aligned} \text{minimize} \quad & f(x) \\ \text{subject to} \quad & h(x) = 0, \end{aligned}$$

where $f, h : \mathbb{R}^3 \to \mathbb{R}$, $\nabla f(x) = [10x_1 + 2x_2, 2x_1 + 8x_2 - 5, -1]^T$, and $\nabla h(x^*) = [-12, -5, 1]^T$. Find $\nabla f(x^*)$.

14.5. Consider the problem

$$\begin{aligned} \text{minimize} \quad & 5x_1^2 + 2x_1x_2 + 4x_2^2 - 5x_2 - x_3 \\ \text{subject to} \quad & 12x_1 + 5x_2 - x_3 = 16. \end{aligned}$$

(a) Apply the Lagrange theorem to find all stationary points.

(b) Is there a global minimizer? Explain.

14.6. Find all points satisfying FONC for the following problem:

$$\begin{aligned} \text{minimize} \quad & 2x_1^2 + x_2^2 + x_3^2 \\ \text{subject to} \quad & x_1 + 2x_2 + 3x_3 = 4. \end{aligned}$$

Which of these points, if any, are points of local or global minimum? Explain.

14.7. Solve the problem

$$\begin{aligned} \text{minimize} \quad & x_1^2 + x_2^2 \\ \text{subject to} \quad & x_1^2 + 9x_2^2 = 9 \end{aligned}$$

geometrically and using the optimality conditions.

14.8. Consider the problem

$$\begin{aligned} \text{minimize} \quad & x_2 - x_1 \\ \text{subject to} \quad & x_1^2 - x_2^3 = 0. \end{aligned}$$

(a) Find the set of all regular points for this problem.

(b) Draw the feasible set and level sets of the objective function corresponding to local minima.

(c) Does $x^* = [0, 0]^T$ satisfy the FONC? Is x^* a point of local minimum? Explain.

14.9. Formulate the FONC (KKT conditions), SONC and SOSC for the maximization problem with inequality constraints:

$$\begin{aligned} \text{maximize} \quad & f(x) \\ \text{subject to} \quad & h(x) = 0 \\ & g(x) \leq 0, \end{aligned}$$

where $f : \mathbb{R}^n \to \mathbb{R}, h : \mathbb{R}^n \to \mathbb{R}^m, g : \mathbb{R}^n \to \mathbb{R}^p$ are twice continuously differentiable functions.

14.10. Use KKT conditions to find all stationary points for the following problems:

    (a)   minimize $2x - x^2$
         subject to $0 \le x \le 3$

    (b)   minimize $-(x_1^2 + x_2^2)$
         subject to $x_1 \le 1$

    (c)   minimize $x_1 - (x_2 - 2)^3 + 3$
         subject to $x_1 \ge 1$.

Which of these points, if any, are points of local or global minimum? Explain.

14.11. Check that the following problem is convex and use the KKT conditions to find all its solutions.

$$\begin{aligned} \text{minimize} \quad & \exp\{-(x_1 + x_2)\} \\ \text{subject to} \quad & \exp(x_1) + \exp(x_2) \le 10 \\ & x_2 \ge 0. \end{aligned}$$

14.12. Use the SOSC to find all local minimizers of the problem

$$\begin{aligned} \text{minimize} \quad & x_1^2 + x_2^2 \\ \text{subject to} \quad & x_1^2 - x_2 \le 4 \\ & x_2 - x_1 \le 2. \end{aligned}$$

14.13. Consider the following problem:

$$\begin{aligned} \text{minimize} \quad & x_2 \\ \text{subject to} \quad & x_1^2 + (x_2 - 4)^2 \le 16 \\ & (x_1 - 3)^2 + (x_2 - 3)^2 = 18. \end{aligned}$$

    (a) Find all points satisfying the KKT conditions.

    (b) Apply the second-order conditions to determine the type of each KKT point found.

    (c) Solve the problem graphically.

14.14. Find all points satisfying the KKT conditions for the following quadratic programming problem

$$\begin{aligned} \text{minimize} \quad & \tfrac{1}{2}x^T Q x \\ \text{subject to} \quad & Ax \le b, \end{aligned}$$

where Q is a symmetric positive definite matrix, A is a $m \times n$ matrix, and $b \geq 0$.

14.15. Show that the dual function $d(\mu)$ defined in Eq. (14.12) at page 369 is concave.

14.16. Apply two iterations of the affine scaling method to the following problem:

$$
\begin{aligned}
\text{minimize} \quad & 2x_1 + x_2 + 3x_3 \\
\text{subject to} \quad & x_1 + 2x_2 - x_3 = 1 \\
& x_1 + x_2 + x_3 = 1 \\
& x_1, x_2, x_3 \geq 0.
\end{aligned}
$$

Use $x^{(0)} = [1/4, 1/2, 1/4]^T$ as the initial point. At each iteration, use the step-length coefficient $\alpha = 0.9$.

Notes and References

There are many excellent books for further reading on the topics so concisely introduced in this text. Here we provide several pointers based on our personal preferences, which, we hope, will help the reader to avoid getting overwhelmed by the variety of available choices.

The following texts are good choices for learning more about numerical methods for problems discussed in Part II of this text, as well as their computer implementations: [9,14,18,24]. The classical book by Cormen et al. [13] provides a great, in-depth introduction to algorithms. See [16, 28] for more information on computational complexity.

The discussion in Sections 9.5 and 9.6 is based on [20] and [25], respectively. The simplex-based approach to establishing the fundamental theorem of LP presented in Chapter 10 is inspired by Chvátal [11]. Textbooks [19,31] contain an extensive collection of LP exercises and case studies. Other recommended texts on various aspects of linear and nonlinear optimization include [1–6,10, 15,17,23,26,27,29].

Bibliography

[1] M. Avriel. *Nonlinear Programming: Analysis and Methods*. Dover Publications, 2003.

[2] M. S. Bazaraa, J. J. Jarvis, and H. D. Sherali. *Linear Programming and Network Flows*. John Wiley & Sons, 4th edition, 2010.

[3] M. S. Bazaraa, H. D. Sherali, and C. M. Shetty. *Nonlinear Programming: Theory and Algorithms*. Wiley-Interscience, 2006.

[4] D. Bertsekas. *Nonlinear Programming*. Athena Scientific, Belmont, MA, 2nd edition, 1999.

[5] D. Bertsimas and J. N. Tsitsiklis. *Introduction to Linear Optimization*. Athena Scientific, Belmont, MA, 1997.

[6] S. Boyd and L. Vandenberghe. *Convex Optimization*. Cambridge University Press, 2004.

[7] S. Brin and L. Page. The anatomy of a large-scale hypertextual web search engine. *Computer Networks*, 30:107–117, 1998.

[8] K. Bryan and T. Leise. The $25,000,000,000 eigenvector: The linear algebra behind Google. *SIAM Review*, 48:569–581, 2006.

[9] R. Butt. *Introduction to Numerical Analysis Using MATLAB*. Infinity Science Press, 2007.

[10] E. K. P. Chong and S. H. Żak. *An Introduction to Optimization*. Wiley-Interscience, 3rd edition, 2008.

[11] V. Chvátal. *Linear Programming*. W. H. Freeman, New York, 1980.

[12] S. Cook. The complexity of theorem proving procedures. In Proceedings of the Third Annual ACM Symposium on Theory of Computing, pages 151-158, 1971.

[13] T. H. Cormen, C. E. Leiserson, R. L. Rivest, and C. Stein. *Introduction to Algorithms*. MIT Press and McGraw-Hill, 3rd edition, 2009.

[14] G. Dahlquist and Å. Björck. *Numerical Methods*. Dover Publications, 2003.

[15] D.-Z. Du, P. M. Pardalos, and W. Wu. *Mathematical Theory of Optimization*. Kluwer Academic Publishers, 2001.

[16] M.R. Garey and D.S. Johnson. *Computers and Intractability: A Guide to the Theory of NP-Completeness*. W.H. Freeman and Company, New York, 1979.

[17] I. Griva, S. Nash, and A. Sofer. *Linear and Nonliner Optimization*. SIAM, Philadelphia, 2nd edition, 2009.

[18] R. W. Hamming. *Numerical Methods for Scientists and Engineers*. Dover Publications, 1987.

[19] F. S. Hillier and G. J. Lieberman. *Introduction to Operations Research*. McGraw-Hill, 9th edition, 2010.

[20] R. Horst, P. Pardalos, and N. Thoai. *Introduction to Global Optimization*. Kluwer Academic Publishers, 2nd edition, 2000.

[21] W. W. Leontief. Input-output economics. *Scientific American*, 185:15–21, October 1951.

[22] D. G. Luenberger. *Investment Science*. Oxford University Press, 1997.

[23] O. L. Mangasarian. *Nonlinear Programming*. SIAM, Philadelphia, 1987.

[24] J. Mathews and K. Fink. *Numerical Methods Using MATLAB*. Prentice Hall, 4th edition, 2004.

[25] Y. Nesterov. *Introductory Lectures on Convex Optimization: A Basic Course*. Kluwer Academic Publishers, 2003.

[26] Y. Nesterov and A. Nemirovski. *Interior-Point Polynomial Algorithms in Convex Programming*. SIAM, Philadelphia, 1994.

[27] J. Nocedal and S. J. Wright. *Numerical Optimization*. Springer, 2nd edition, 2006.

[28] C. H. Papadimitriou. *Computational Complexity*. Addison Wesley, 1994.

[29] A. L. Peressini, F. E. Sullivan, and J. J. Uhl. *The Mathematics of Nonlinear Programming*. Springer, 1993.

[30] T. L. Saaty. *The Analytic Hierarchy Process*. McGraw Hill Company, New York, 1980.

[31] W. L. Winston and M. Venkataramanan. *Introduction to Mathematical Programming: Applications and Algorithms*. Brooks/Cole, 4th edition, 2002.

Index

$C(\cdot)$, 28
$C^{(k)}(\cdot)$, 28
$O(\cdot)$, 30
$\mathcal{NP}$, 193
$\mathcal{NP}$-complete, 195
$\mathcal{NP}$-hard, 198
$\mathcal{P}$, 191
$\in$, 3
inf, 25, 167
max, 25
min, 25
$\propto$, 194
$\mathbb{R}^n$, 8
$\subset$, 3
$\subseteq$, 3
sup, 25, 167
$o(\cdot)$, 30
$\mathcal{P}$ vs. $\mathcal{NP}$, 193
3-SAT, 196
3-satisfiability, 196

accumulation point of a sequence, 26
active constraint, 165, 358
additivity assumption, 222
adjacency list, 187
adjacency matrix, 187
adjacent basic feasible solutions, 252
adjacent vertices, 187
admissible region, 164
affine scaling method, 374
algorithm, 185
Analytic Hierarchy Process, 81
annuity-due equation, 87, 91
approximate inverse matrix, 70
Armijo condition, 346
artificial variable, 255
augmented matrix, 59

auxiliary problem, 257
average running time, 189

back substitution, 58
backtracking line search, 346
barrier function, 379
barrier methods, 379
base, 37, 45
basic feasible solution, 239
basic solution, 238
basic variable, 238
basis, 10
BFGS method, 345
bfs, 239
big M, 257
big O notation, 30
big-M method, 255
bijection, 5
binary point, 38
binary relation, 5
binary system, 38
binding constraint, 165
bisection method, 93
Bland's rule, 248
boundary point, 165
bounded from above, 25
bounded from below, 25
bounded set, 26
bracketing methods, 92

Cantor's diagonal method, 5
certainty assumption, 221
chain rule, 29
characteristic polynomial, 19
Chebyshev nodes, 122
Chebyshev polynomials, 121
chopping, 47

391

For Product Safety Concerns and Information please contact our
EU representative GPSR@taylorandfrancis.com Taylor & Francis
Verlag GmbH, Kaufingerstraße 24, 80331 München, Germany